Recent Advances in Lactic Acid Bacteria and Metabolite for Food Quality

Recent Advances in Lactic Acid Bacteria and Metabolite for Food Quality

Guest Editors

Luis M. Medina
Fernando Pérez-Rodríguez

MDPI

Basel • Beijing • Wuhan • Barcelona • Belgrade • Novi Sad • Cluj • Manchester

Guest Editors
Luis M. Medina
Department of Food Science
and Technology
Universidad de Córdoba
Córdoba
Spain

Fernando Pérez-Rodríguez
Department of Food Science
and Technology
University of Cordoba
Córdoba
Spain

Editorial Office
MDPI AG
Grosspeteranlage 5
4052 Basel, Switzerland

This is a reprint of the Special Issue, published open access by the journal *Foods* (ISSN 2304-8158), freely accessible at: https://www.mdpi.com/journal/foods/special_issues/833K640U21.

For citation purposes, cite each article independently as indicated on the article page online and as indicated below:

Lastname, A.A.; Lastname, B.B. Article Title. *Journal Name* **Year**, *Volume Number*, Page Range.

ISBN 978-3-7258-7520-7 (Hbk)
ISBN 978-3-7258-7521-4 (PDF)
https://doi.org/10.3390/books978-3-7258-7521-4

Contents

About the Editors

Luis M. Medina

Luis M. Medina earned a degree in Veterinary Medicine (1989), specializing in Food Science, Health, and Food Technology, and he holds a Ph.D. in Veterinary Medicine (1993), having also received an Extraordinary Doctorate Award from the University of Córdoba, Spain. He holds a Postgraduate Diploma from the University of Jaén and is recognized as an Expert in University Policy and Management (2021) by the International University of Andalusia (UNIA). Throughout his career, he has been a member of various research groups and entities, such as the International Food Protection Association, the Spanish Society of Microbiology, and the Andalusian Association of Food Science and Technology Professionals (ACTA-A). He is currently the Director of the Department of Food Science and Technology at the University of Córdoba, the Head of the "Food Microbiology" Research Group at the University of Córdoba and the Regional Government of Andalusia, an associate member of the Competitive Research Unit on Zoonoses and Emerging Diseases of the University of Córdoba (ENZOEM), and a member of the International Agrifood Campus of Excellence (ceiA3). He has also served as an evaluator for various ANECA programs and other national and international entities, as well as for EAEVE. His research spans multiple areas, mainly in the microbiology of meat and dairy products, focusing on the control of foodborne pathogens, on one hand, and on functional and probiotic products on the other.

Fernando Pérez-Rodríguez

Fernando Pérez-Rodríguez undertook a degree in Biological Science and in Food Science and Technology from the University of Cordoba in 1999 and 2002, respectively. He completed his PhD at the University of Córdoba (2007), which dealt with quantitative microbiological risk assessment and cross-contamination in foods. He has published more than 100 articles in some of the most prestigious international food microbiology journals and several books concerning predictive microbiology in foods. Due to his expertise, he has been involved as a scientific advisor in several expert panels at both a national and international level, for example, for the Spanish Food Safety Agency, and as a Risk Assessment Expert for the European Food Safety Authority (EFSA) and Food and Agriculture Organization (FAO), providing scientific advice and reports. Currently, Fernando is a member of the EFSA Biohazard Panel (2024–2029), providing the European Commission with scientific opinions in the field of microbiological food safety. He is also a member of the Joint FAO/WHO Expert Meetings on Microbiological Risk Assessment (JEMRA) Roster of Experts from 2018, having renewed his membership for the period 2023–2027. Furthermore, he serves as the chair at the International Committee of Predictive Modelling in Food. Currently, he is a full professor of Food Microbiology and Hygiene at UCO, and he serves as a vice-chair of the competitive research unit for Zoonosis and Emergence Disease (ENZOEM) and AGR-170, HIBRO: Food Safety and Quality. In addition, he participates as an invited teacher in various national and international master's programs and courses related to predictive microbiology and food safety.

Preface

The reprint belongs to the Special Issue 'Recent Advances in Lactic Acid Bacteria and Metabolite for Food Quality'.

As previously described, lactic acid bacteria (LAB) have been of great interest and have generally been at the center of many of the most interesting challenges in food innovation, functional food, and food culture in many populations.

This publication compiles some of the research on lactic acid bacteria (LAB) that expands the frontiers regarding their knowledge, uses, and applications.

Luis M. Medina and Fernando Pérez-Rodríguez
Guest Editors

foods

MDPI

Editorial

Lactic Acid Bacteria: An Inexhaustible Source of Scientific Knowledge and Food Innovation

Luis M. Medina * and Fernando Pérez-Rodríguez

Food Science and Technology Department, ENZOEM, Competitive Research Unit on Zoonoses and Emerging Diseases of the University of Cordoba, International Agrifood Campus of Excellence (ceiA3), Charles Darwin Annex Building, Campus de Rabanales, E-14071 Cordoba, Spain; b42perof@uco.es

* Correspondence: luismedina@uco.es

For thousands of years, lactic acid bacteria (LAB) have been involved in food processes that have given rise to a wide range of traditional foods, mainly fermented foods. Their empirical formulation in food production has led to important innovations in the sector, particularly from the 1980s and 1990s onwards.

In the 1990s, conferences and congresses dedicated specifically to LAB, their role, and potential applications in the food industry became frequent, often with the dairy industry at the forefront.

Today, LAB are still the subject of continuous research and innovative applications, which, in parallel with the development of omics, have enabled the investigation of viable, but nonculturable bacteria (VBNC), and low prevalent microbial groups and their consequences in the configuration of microbial ecosystems. These advances have bridged the gap between taxonomies, taking into account the genomic diversity of the genus *Lactobacillus*, enriching and updating the perspectives from which they are studied.

The leading role of LAB in the world of probiotics, their ability to produce bacteriocins and other bioactive substances, combined with a human microbiome-based knowledge, bring about a wide range of applications of prime interest for these bacteria, which continues to be one of the most fertile fields of knowledge in food science.

This Special Issue looks at some of these future perspectives.

Overview of contents of the Special Issue

This Special Issue comprises ten contributions, seven of them research articles and three reviews. These works have been carried out in diverse universities and research centers of six different countries.

The review articles provide updates on very relevant aspects related to LAB, and in particular, delineate the future directions for their respective topics.

Without a doubt, LAB have played a role in fermentation processes for millennia, and Huang et al. [1], highlight the limitations and challenges of applying microbial fermentation in lactic acid production. Not without their own challenges, the authors report useful solutions for the industrial production of lactic acid.

Furthermore, LAB can also present biotechnological features for the production of other compounds, such as mannitol, a sugar alcohol, which is of interest due to its potential to reduce the use of sugars, given its low calories content. This is the case of *Leuconostoc citreum*, a promising microorganism for this purpose, which has been revised by Rosca et al. [2], especially for its use in sourdough-based baked products, extending into other benefits from its application.

A Special Issue devoted to LAB and future trends regarding them must always include new perspectives on probiotics. In this Special Issue, a review on the use of probiotics is presented, underscoring the need to demonstrate their safety when they are intended for use by pregnant women or during lactation. This topic has been revised by Fernández et al. [3].

Regarding the possibilities of applying LAB to control pathogens in foods, this Special Issue demonstrates the effects of selected autochthonous LAB (two *Lacticaseibacillus casei* strains) against *Listeria monocytogenes*. This work was carried out by Martín-Miguelez et al. [4] in two elaborations of soft-ripened cheese. The use of these strains, alone or combined, led to reductions of 2 to 4 log CFU/g in *L. monocytogenes* over 60 days of ripening, which could be a suitable strategy for the biocontrol of pathogen contamination during cheese processing. Furthermore, in another article [5], this research team reported that these selected LAB strains improved the sensory profile of the soft-ripened cheeses, decreasing the values of texture parameters such as hardness, gumminess, and chewiness for a softer texture, and increasing the umami taste and floral and lactic odor attributes. A sensory analysis revealed that consumers perceived differences between inoculated and non-inoculated cheeses, although the overall acceptability was not affected.

Sometimes, innovation can arise from the valorization of resources that have been forgotten or unused for years. Alba et al. [6] isolated and characterized a strain of *Ligilactobacillus salivarius* (*L. salivarius* SP36) from a cheese from a Pyrenean village sealed in 1936 and produced via household production. Overall, the phenotypic and genetic features of this strain support its high potential as an adjunct culture in cheesemaking. Other work in this Special Issue addresses the aforementioned aspects. Arias et al. [7] report that *L. salivarius* SP36 increased the formation of 25 volatile compounds, receiving higher scores for aroma intensity and quality than control cheeses.

When considering cheeses as food matrices, among the key attributes that make artisanal cheeses of particular value are their diverse aromatic profiles that impart distinctive characteristics to each product. In their article, Ruiz et al. [8] report the relationship between the activity of the autochthonous lactic bacteria (*Lacticaseibacillus paracasei*, *Lactiplantibacillus plantarum*, *Leuconostoc mesenteroides*, and *Lactococcus lactis* subsp. *hordniae*) involved in raw sheep milk cheeses, with the volatile compounds determined by GC-IMS, a technique with which these authors have been working for the last decade. The occurrence of *Lactococcus lactis* subsp. *hordniae* from cheese was reported for the first time.

Some of the most novel contributions to bacteriocin studies are represented in this Special Issue by the work of Feito et al. [9], a group led by Prof. Hernández-Cruza, from the Complutense University in Madrid (Spain). The work describes the application of *Lactococcus lactis* strains producing Garvicin A and/or Garvicin Q, either alone or together with Nisin A or Nisin Z, and high antimicrobial activity against *Lactococcus garvieae*, a main ichthyopathogen in rainbow trout (*Oncorhynchus mykiss*, Walbaum) farming.

As for the use of novel LAB found in natural sources such as fermented foods, Nazareth et al. [10] describe the isolation, identification, characterization, and quantification of LAB with antifungal activity, in this case in dry-cured sausages, for application in the field of meat products. The authors found promising biological candidates corresponding to *Lactiplantibacillus plantarum* and *Pediococcus pentosaceus*. *P. pentosaceus* C15 was distinguished from other bacteria thanks to its production of antifungal compounds such as nonanoic acid and phenyl ethyl alcohol, as well as its higher production of lactic and acetic acid.

We therefore present ten relevant examples of the inexhaustible source of scientific knowledge and innovative applications of lactic acid bacteria in product development, health, and food safety.

Author Contributions: Conceptualization, writing—original draft preparation, writing—review and editing, visualization, and supervision, L.M.M. and F.P.-R. All authors have read and agreed to the published version of the manuscript.

Acknowledgments: We would like to express our sincere gratitude to all authors and co-authors who submitted manuscripts to this Special Issue. Also, we want to thank the support and assistance of *Foods*.

Conflicts of Interest: The authors declare no conflicts of interest.

References

1. Huang, Y.; Wang, Y.; Shang, N.; Li, P. Microbial Fermentation Processes of Lactic Acid: Challenges, Solutions, and Future Prospects. *Foods* **2023**, *12*, 2311. [CrossRef] [PubMed]
2. Rosca, M.-F.; Paucean, A.; Man, S.M.; Chis, M.S.; Pop, C.R.; Pop, A.; Farcas, A.C. *Leuconostoc citreum*: A Promising Sourdough Fermenting Starter for Low-Sugar-Content Baked Goods. *Foods* **2024**, *13*, 96. [CrossRef]
3. Fernández, L.; Orgaz, B.; Rodríguez, J.M. The Safety of Probiotics Intended for Use in Pregnant and Lactating Women: From a Desirable to a Required Task. *Foods* **2024**, *13*, 4024. [CrossRef] [PubMed]
4. Martín-Miguélez, J.M.; Robledo, J.; Martín, I.; Castaño, C.; Delgado, J.; Córdoba, J.J. Biocontrol of *L. monocytogenes* with Selected Autochthonous Lactic Acid Bacteria in Raw Milk Soft-Ripened Cheese under Different Water Activity Conditions. *Foods* **2024**, *13*, 172. [CrossRef] [PubMed]
5. Martín Miguélez, J.M.; Martín, I.; Robledo, J.; Ventanas, S.; Córdoba, J.J. Effect of Artisanal Processing on Volatile Compounds and Sensory Characteristics of Traditional Soft-Ripened Cheeses Matured with Selected Lactic Acid Bacteria. *Foods* **2025**, *14*, 231. [CrossRef] [PubMed]
6. Alba, C.; Arroyo, R.; Fernández, L.; Narbad, A.; Rodríguez, J.M. Characterization of a *Ligilactobacillus salivarius* Strain Isolated from a Cheese Seal Which Was Last Used in 1936. *Foods* **2024**, *13*, 2005. [CrossRef] [PubMed]
7. Arias, R.; Alba, C.; Calzada, J.; Jiménez, L.; Fernández, L.; Ávila, M.; Roman, M.; Roman, J.; Rodríguez, J.M.; Garde, S. Application of *Ligilactobacillus salivarius* SP36, a Strain Isolated from an Old Cheese Seal, as an Adjunct Culture in Cheesemaking. *Foods* **2024**, *13*, 2296. [CrossRef] [PubMed]
8. Ruiz, M.J.; Salatti-Dorado, J.A.; Cardador, M.J.; Frizzo, L.; Jordano, R.; Arce, L.; Medina, L.M. Relationship between Volatile Organic Compounds and Microorganisms Isolated from Raw Sheep Milk Cheeses Determined by Sanger Sequencing and GC–IMS. *Foods* **2023**, *12*, 372. [CrossRef] [PubMed]
9. Feito, J.; Araújo, C.; Arbulu, S.; Contente, D.; Gómez-Sala, B.; Díaz-Formoso, L.; Muñoz-Atienza, E.; Borrero, J.; Cintas, L.M.; Hernández, P.E. Design of *Lactococcus lactis* Strains Producing Garvicin A and/or Garvicin Q, Either Alone or Together with Nisin A or Nisin Z and High Antimicrobial Activity against *Lactococcus garvieae*. *Foods* **2023**, *12*, 1063. [CrossRef] [PubMed]
10. De Melo Nazareth, T.; Calpe, J.; Luz, C.; Mañes, J.; Meca, G. Manufacture of a Potential Antifungal Ingredient Using Lactic Acid Bacteria from Dry-Cured Sausages. *Foods* **2023**, *12*, 1427. [CrossRef] [PubMed]

foods

MDPI

Article

Effect of Artisanal Processing on Volatile Compounds and Sensory Characteristics of Traditional Soft-Ripened Cheeses Matured with Selected Lactic Acid Bacteria

José M. Martín Miguélez [1], Irene Martín [1], Jurgen Robledo [1], Sonia Ventanas [2] and Juan J. Córdoba [1,*]

[1] Higiene y Seguridad Alimentaria, Instituto Universitario de Investigación de Carne y Productos Cárnicos (IProCar), Facultad de Veterinaria, Avda. de la Universidad s/n., 10003 Cáceres, Spain; jmmm@unex.es (J.M.M.M.); iremartint@unex.es (I.M.); jrobledo@unex.es (J.R.)

[2] Tecnología y Calidad de Alimentos, Instituto Universitario de Investigación de Carne y Productos Cárnicos (IProCar), Facultad de Veterinaria, Avda. de la Universidad s/n., 10003 Cáceres, Spain; sanvenca@unex.es

* Correspondence: jcordoba@unex.es

Abstract: The present research evaluated the effect of selected *Lacticaseibacillus casei* strains with anti-*Listeria monocytogenes* properties on the characteristics of traditional soft-ripened cheeses produced in two different seasons. Physicochemical, microbiological, texture, volatile compound, and sensory evaluations were performed on the cheeses after 60 days of ripening. The inoculation with protective cultures of selected LAB did not negatively affect the physicochemical parameters of the cheeses. Thirty-two volatile compounds were identified, including acids, alcohols, ketones, aldehydes, and esters, with differences between productions and inoculated batches. The selected LAB strains improved the sensory profile of the soft-ripened cheeses, decreasing values of texture parameters such as hardness, gumminess, and chewiness related to a softer texture and increasing umami taste and floral and lactic odor attributes. Sensory analysis revealed that consumers perceived differences between inoculated and non-inoculated cheeses, although the overall acceptability was not affected. This study provides valuable information for the artisanal cheese industry, demonstrating that it is possible to use selected protective lactic acid bacteria to assure food safety without compromising traditional flavor and even improving sensorial attributes.

Keywords: *Lacticaseibacillus casei*; artisanal cheeses; volatile compounds; sensorial analysis

1. Introduction

The use of safe microorganisms as a biocontrol strategy against pathogenic bacteria is currently common due to its effectiveness and adaptation to consumer trends as a "clean label" [1]. This approach aligns with the growing demand for natural food preservation methods and minimal processing, reflecting modern consumer preferences for healthier and less artificially preserved foods [2,3]. Lactic acid bacteria (LAB) are the most common microbial group utilized as protective cultures of ripened products because of their antimicrobial activity and their adaptation to pH and salt stress conditions [4]. The concept of "protective cultures" refers to the deliberate use of specific microorganisms to inhibit pathogenic and spoilage bacteria, thereby enhancing food safety and extending shelf life naturally [1]. In addition, the metabolism of LAB could favor the development of sensory characteristics on the ripened products [5]. Thus, specific LAB strains have been selected, not only for achieving the intended biocontrol strategy but also for improving the quality of the ripened foods [6,7].

However, some Protected Designations of Origin (P.D.O.) regulations do not allow the addition of selected LAB to artisanal ripened products such as soft-ripened cheeses, due to the possible negative repercussion of added bacteria to the sensorial characteristic of cheeses [8]. The term "artisanal" in this context refers to cheese production methods that adhere to traditional practices, often involving small-scale production and manual techniques [9]. This fact delimitates the use of a biocontrol strategy in products whose physicochemical characteristics may allow the contamination of pathogenic microorganisms, such as *Listeria monocytogenes*, which poses a risk for consumers [10,11].

In addition, most of the above soft-ripened cheeses are carried out in artisanal conditions within a wide range of relative humidity conditions (from 95 to 85%), leading to water activity (a_w) variations throughout the ripening process, that could impact the growth and the effect of added LAB on the sensorial characteristics of the cheeses [12]. Thus, it is necessary to evaluate the effect of selected LAB strains on the sensorial characteristic of traditional soft-ripened cheeses in different conditions of a_w, before being proposed as protective cultures, to avoid negative effects on this product.

The present study aims to identify the sensory profile modifications related to the utilization of different selected anti-*L. monocytogenes* LAB strains, to introduce their use in a traditional industry of soft-ripened cheese [6]. Two elaborations in different seasons of the year have been performed to evaluate the variability related to the artisanal processing of the cheese. Sensory analyses with consumers and volatile compound identification have been carried out to evaluate the produced changes within the batches and elaborations. This approach seeks to balance the preservation of traditional cheese-making practices with the implementation of innovative biocontrol strategies, addressing both food safety concerns and consumer demands for minimally processed, naturally preserved foods.

2. Materials and Methods

2.1. Microbial Cultures

Lacticaseibacillus casei strains 31 and 116 isolated from "Torta del Casar" soft-ripened cheese and selected based on their anti-*L. monocytogenes* activity were used [6]. The authors assessed the inhibitory effect of *L. casei* 31 and 116 against *L. monocytogenes* S7-2 (serotype 4b) using both agar spot testing and cheese model systems. They were sub-cultured twice into 10 mL of Man Rogosa Sharpe (MRS; Conda Pronadisa, Madrid, Spain), from a frozen stock culture (−80 °C) of the same medium containing 20% (*w*/*v*) of glycerol (Thermo Fisher Scientific, Waltham, MA, USA). Then, the cultures were centrifugated at 10,621 g for 10 min, supernatants were discarded, and cells were dissolved in 10 mL of Phosphate-Buffered Saline (PBS; 0.32 g of sodium dihydrogen phosphate (Scharlau Chemie S.A., Barcelona, Spain), 1.09 g of disodium hydrogen phosphate (Scharlau Chemie S.A.), and 9 g of NaCl (Thermo Fisher Scientific) in 1 L of distilled water). A concentration of ≈11 log CFU/mL was achieved to reach ≈6 log CFU/mL in the milk to be processed.

2.2. Preparation of Soft-Ripened Cheeses

Soft-ripened cheeses evaluated in this study were elaborated in the facilities of one "Torta del Casar" P.D.O. industry from Casar de Cáceres (Extremadura region, Spain). Two elaborations performed at different seasons (first elaboration in winter, second elaboration in spring) of the year were carried out to evaluate the action of the selected LAB in an artisanal production facility. The industry's materials were utilized to elaborate the studied cheeses to replicate the actual processing.

A scheme of the cheese-making process is displayed in Figure 1. Four batches were performed in each elaboration: A (uninoculated batch), B (inoculated with LAB 116), C (LAB 31), and D (LAB 116 + 31). In each batch, 50 L of raw ewe milk collected from several farms was added to 10 mL of PBS with a LAB concentration of 11 log CFU/mL to reach approximately 6 log CFU/mL [12]. Previous counts in MRS agar were performed to confirm the concentration added. Cheese curling was carried out for 1 h at 28 °C with a vegetable coagulant from the flowers of *Cynara cardunculus* L. Thirty fresh pieces of cheese of approximately 500 g were obtained by batch, which were pressed for 1 h and salted in brine 18% NaCl (expressed as a weight-to-weight basis) for 2 h. Hygienic measures were taken to avoid contamination between batches, following the industry-established protocol and using their disinfection materials too. Cheeses were ripened for 60 days following the ripening conditions of P.D.O. regulation of "Torta del Casar" soft-ripened cheese (6 °C during the first 35 days and then 8 °C for 10 days, 9 °C for 10 days, and 5 days at 10 °C) [12]. The conditions of relative humidity were slightly different between elaboration 1 (90% RH during the first 35 days and the remaining days at 70% RH) and maturation 2 (90% RH during the first 35 days and the rest of the days at 80% RH).

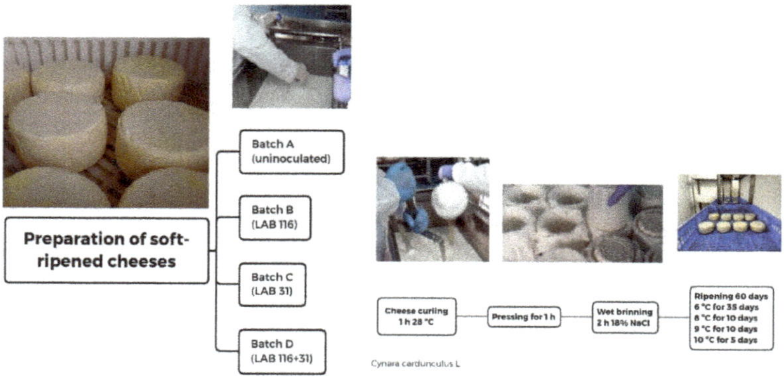

Figure 1. The flow chart of the cheese-making preparation. LAB—lactic acid bacteria.

In each maturation, 3 cheeses of each batch were taken to perform physicochemical and microbiological analyses at the end of ripening process (Figure 2).

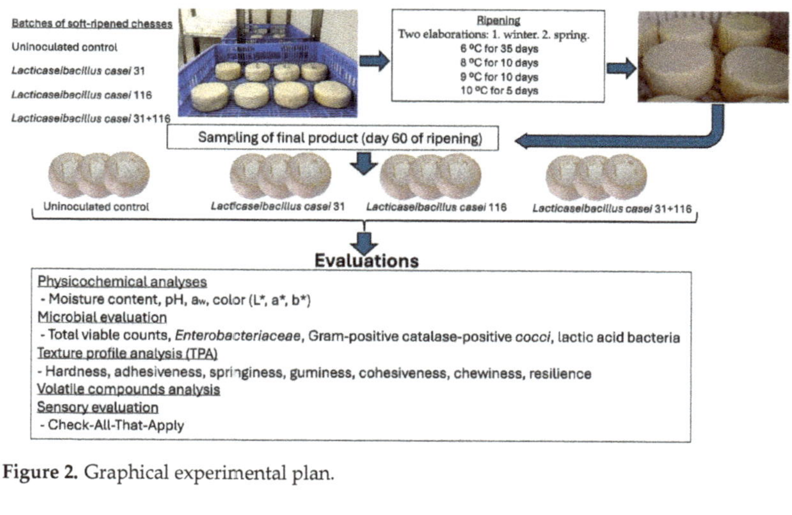

Figure 2. Graphical experimental plan.

2.3. Physicochemical Analyses

Physicochemical analyses were executed following the guidelines of the Association of Official Analytical Chemists [13]. These analyses were conducted by triplicate for each production on the final days of ripening for each batch, utilizing the homogenized cheese employed in microbial analyses. The samples were homogenized in ultrapure water and measured using a pH meter (Hach Lange Spain S.L.U., Barcelona, Spain), previously calibrated with standard solutions from Scharlau Chemie S.A. The a_w was assessed using a Novasina Lab Master Water Activity Meter model SPRINT-TH 300 (Novasina AG, Lachen, Switzerland). The moisture content was determined through the dehydration of samples at 105 °C, supplemented with washed sea sand (Scharlau Chemie S.A.).

Instrumental color assessment was conducted using a Minolta CR-300 colorimeter (Konica Minolta, Osaka, Japan) equipped with illuminant D65, a 0° standard observer, and a 2.5 cm port/viewing area. The measurements were executed on the transversal section of the samples, following the principles outlined by the Commission Internationale de L'Éclairage [14]. Each sample underwent five measurements. The color coordinates measured included lightness (L*), redness (a*), and yellowness (b*).

2.4. Microbial Analyses

Microbiological analyses were conducted by triplicate through the extraction of the central cube of each cheese encompassing 16 cm^2 (excluding the outer layers) and homogenizing the cheese using Stomacher homogenizer bags (Interscience, Saint-Germain-en-Laye, France). Subsequently, 25 g of cheese from each batch was employed to verify the absence of *L. monocytogenes* following the International Organization for Standardization (ISO) 11290-1 standard [15]. Furthermore, 10 g of each homogenized cheese from every batch underwent analysis with 90 mL of 1% (w/v) peptone water (Conda Pronadisa). Decimal dilutions of peptone water were then prepared, and agar plates from various media, corresponding to the sampled microbial group, were inoculated. Plate Count Agar (PCA; Conda Pronadisa), Mannitol Salt Agar (MSA; Conda Pronadisa), and MRS were utilized for sampling total viable count (TVC), Gram-positive catalase-positive cocci (GC+), and LAB, respectively. These plates were incubated at 30 °C for 24–48 h, except for MRS, which was incubated for 48 h under microaerophilic conditions, employing a plastic bag to create an oxygen-deprived environment. Malt Extract Agar (MEA; 20 g of malt extract (Conda Pronadisa), 20 g of glucose (Labbox Labware S.L., Barcelona, Spain), 1 g of bacteriological peptone (Conda Pronadisa), and 20 g of bacteriological agar (Conda Pronadisa) in 1 L of distilled water) was employed to assess yeasts, with an incubation period of 25 °C for 48 h. Violet Red Bile Glucose Agar (VRBG; Conda Pronadisa) was incubated at 37 °C for 24 h to determine Enterobacteriaceae (EB).

2.5. Texture Profile Analysis

Texture Profile Analysis (TPA) was conducted to quantify variations in texture exhibited by the samples, employing a texturometer, Texture Analyser TA XT Plus (Stable Micro Systems Ltd., Godalming, UK). Five measurements of 1 × 1 × 2 cm were taken from each sample at 20.5 °C, using a 5 cm diameter cylindrical probe applying a vertical compression of 40% of the original height. Parameters calculated included hardness (g), adhesiveness (g × s), springiness (g), guminess (g), cohesiveness (dimensionless), chewiness (g), and resilience ($J \times m^{-3}$), following methodologies described in studies by other authors [16,17].

2.6. Volatile Compound Analysis

The volatile compounds in soft cheeses were extracted using solid-phase microextraction (SPME) after being heated to 37 °C for 30 min. A divinylbenzene–carboxen–polydimethylsiloxane

(DVB/CAR/PDMS) 50/30 µm fiber (Merck; Darmstadt, Germany) was employed for the extraction. The extracted compounds were then analyzed using gas chromatography–mass spectrometry (GC-MS) on a Gas Chromatograph 6890 GC (Agilent Technologies, Santa Clara, CA, USA), equipped with an HP-5 column (5% phenyl—95% dimethylpolysiloxane) and coupled to a mass spectrometer (MS) detector, 5975C (Agilent Technologies).

The oven temperature started at 40 °C for 5 min and was then increased to 280 °C at a rate of 7 °C/min. Desorption was carried out for 30 min at 250 °C, and the transfer line temperature was set at 280 °C. Helium (Air Liquide, Madrid, Spain) served as the carrier gas with a flow rate of 1.2 mL/min. Mass spectrometry detection was performed in full-scan mode (50–350 amu). For data treatment, automated peak search and spectral deconvolution were employed. The identification of volatile compounds was accomplished by comparing their mass spectra with the NIST/EPA/NIH library (Institute of Standards and Technology, Gaithersburg, MD, USA).

2.7. Sensory Evaluation

In this study, a triangular test in each elaboration was carried out to evaluate the differences between the uninoculated batch and batches B and C individually. Later, Check-All-That-Apply (CATA) was employed in all the batches in both elaborations, due to its cost-effectiveness and high efficacy in gathering results from a consumer group [18]. Thirty-six semi-trained panelists were enlisted from the University of Extremadura research facilities. Samples were presented individually, and unsalted crackers along with a glass of water were provided between samples to clean the palate.

A questionnaire featuring 19 randomly presented terms related to the sensory attributes of soft-ripened cheese was used (Table 1). These terms were selected from prior research on cheese products [19–21]. Additionally, a seven-point hedonic test was conducted to gauge the overall liking of the samples (1 = dislike very much; 7 = like very much). All procedures performed in this study involving human participants were approved based on the ethical standards of the Delegation of the Bioethics and Biosafety Commission of the University of Extremadura (Delegación de Comisión de Bioética y Bioseguridad de la Universidad de Extremadura) and the 1964 Helsinki declaration and its later amendments or comparable ethical standards with the approval number 140/2024.

Table 1. Attributes and descriptions used on RATA and CATA questionnaires.

Attribute	Description
Brightness [a]	Reflection of light
Intense color [a]	Strength perceived of a color
Vegetal odor [o]	Odor perception characteristic of vegetables
Lactic odor [o]	Odor perception characteristic of milk
Floral odor [o]	Odor perception characteristic of flowers
Spicy [t]	Spicy mouth sensation
Smoothness [te]	Lack of gritty or grainy particles perceived in the mass while chewing
Adherence [te]	Adherence to the palate when pressed against the roof of the mouth by the tongue
Mouth coating [te]	Impression of lubricated food during chewing
Acid taste [ta]	Primary acid taste
Saltiness [ta]	Primary salty taste
Bitter taste [ta]	Primary bitter taste
Sweetness [ta]	Primary sweet taste
Umami taste [ta]	Primary umami taste
Intense flavor [f]	Intensity of flavor as a whole
Persistent flavor [f]	Continued flavor perception after swallowing the sample
Ripened flavor [f]	Flavor perception characteristic of ripened products
Rancid flavor [f]	Flavor perception characteristic of oxidized products
Pungent flavor [f]	Flavor perception characteristic of traditional spices

([a]) appearance; ([o]) odor; ([t]) trigeminal sensation; ([te]) texture; ([f]) flavor; ([ta]) taste.

2.8. Statistical Analysis

The results obtained underwent analyses using the Shapiro–Wilk test for data normal distribution and the Levene test for equality of variances. In the cases where more than two variables were evaluated, Tukey's honest significance test (HSD) was employed for normal and homogeneously distributed data, T3 Dunnett's for normal but not homogeneously distributed data, and Kruskall–Wallis' test for data not following normal distribution. When only two variables were evaluated, T-student's and Mann–Whitney U tests were employed for normal and not normal distributed data, respectively. IBM SPSS Statistics software v.22 (IBM Co., New York, NY, USA) was used to perform these analyses. Volatile compounds' heat map was generated with MetaboAnalyst https://www.metaboanalyst.ca/ (accessed on 1 January 2025) without filtering data and through an autoscaling data mean-centered and divided by the standard deviation of each variable.

Triangular tests were analyzed through binomial distribution 1/3. CATA-obtained sensory results were evaluated through Cochran's Q test, McNemar pairwise multiple comparisons, Correspondence Analysis (CA), and Penalty Analysis (PA) using XLSTAT 2023.2.0 software (Addinsoft, Paris, France). Statistical differences in all analyses were considered significant at the 5% level ($p \leq 0.05$).

3. Results

3.1. Physicochemical Characteristics

The moisture content of the first elaboration showed levels around 41%, and the ones from the second elaboration displayed 44–47% (Table 2). There were no significant differences between batches in moisture content within each elaboration. However, the moisture content was significantly higher ($p \leq 0.05$) in all batches of the second elaboration than in the first one. The pH of the first and second elaborations reached values from 5.51 to 5.74 and from 5.16 to 5.49, respectively, showing values significantly higher ($p \leq 0.05$) in the first elaboration than in the second one (Table 2). In the second elaboration, batch D showed values of pH significantly higher than those of batches B and C. The first elaboration obtained a_w values between 0.935 and 0.948 and the second one between 0.948 and 0.952 (Table 2).

Table 2. Moisture content, pH, a_w, and instrumental color at day 60 of ripening of cheeses matured in two different traditional elaborations.

Batches	A	B	C	D
		First elaboration		
Moisture content (%)	40.87 ± 0.58 *	40.89 ± 0.46 *	40.07 ± 0.47 *	41.06 ± 0.20 *
pH	5.70 ± 0.08 *	5.74 ± 0.04 *	5.62 ± 0.06 *	5.51 ± 0.14
a_w	0.939 ± 0.005	0.935 ± 0.001 *	0.947 ± 0.004	0.948 ± 0.001
L*	78.11 ± 0.59 *	79.51 ± 3.90	77.97 ± 0.97 *	82.92 ± 2.19 *
a*	−1.46 ± 0.07 ^{b*}	−1.54 ± 0.07 ^{b*}	−1.83 ± 0.02 ^{a*}	−1.87 ± 0.02 ^{a*}
b*	19.85 ± 0.24 *	20.39 ± 0.62 *	21.37 ± 0.63 *	20.59 ± 0.63 *
		Second elaboration		
Moisture content (%)	47.24 ± 1.72 *	47.13 ± 1.44 *	45.61 ± 0.00 *	44.86 ± 1.08 *
pH	5.26 ± 0.03 ^{ab*}	5.25 ± 0.09 ^{a*}	5.16 ± 0.02 ^{a*}	5.49 ± 0.01 ^{b}
a_w	0.948 ± 0.002	0.948 ± 0.003 *	0.952 ± 0.002	0.951 ± 0.001
L*	89.53 ± 0.70 ^{b*}	83.85 ± 0.28 ^{a}	90.37 ± 1.15 ^{b*}	89.58 ± 0.45 ^{b*}
a*	−0.70 ± 0.05 ^{a*}	−0.24 ± 0.03 ^{b*}	−0.79 ± 0.04 ^{a*}	−0.712 ± 0.07 ^{a*}
b*	16.83 ± 0.20 ^{a*}	17.84 ± 0.25 ^{ab*}	18.00 ± 0.21 ^{ab*}	18.36 ± 0.33 ^{b*}

A (uninoculated control), B (inoculated with LAB 116), C (inoculated with LAB 31), and D (inoculated with LAB 116 and 31). L* (lightness), a* (redness), and b* (yellowness). Values are expressed as the mean ± standard error of the mean. Significant differences ($p \leq 0.05$) among batches from the same elaboration are shown with letters as superscripts ($^{a\text{-}b}$) in the same row. Significant differences ($p \leq 0.05$) between elaborations from the same batch are shown with asterisks (*) in the same column.

Regarding color parameters, similar studies showed similar soft-ripened cheese lightness (L*), redness (a*), and yellowness (b*) values around 100, −1.5, and 5.5, respectively [22]. The lightness and redness showed significantly lower values on the first elaboration than in the second one, while yellowness showed higher values in the first elaboration (Table 2). The color results were different between elaborations ($p \leq 0.05$).

3.2. Microbial Population

The absence of *L. monocytogenes* was confirmed in all the sampled batches. Table 3 shows the microbial counts of the evaluated groups. The TVC of both elaborations showed levels from 8.62 to 9.27 log CFU/g. EB levels went from 4.83 to 6.72 log CFU/g in any batch from both elaborations. The GC+ concentrations from the first elaboration showed levels from 5.96 to 6.28 log CFU/g, while the second one had concentrations from 4.68 to 5.36 log CFU/g (Table 3). LAB showed levels from 8.32 to 9.20 log CFU/g. Only small differences ($p \leq 0.05$) between batches were observed in some of the microbial groups. The most relevant differences are the higher level of LAB in batch C with respect to batch B in the first elaboration and with respect to batch D in the second elaboration, and the lower counts of EB in batch C with respect to batch A in the first elaboration (Table 3). However, no significant differences were found between batch A and batches B and C for any of the evaluated microbial groups in both elaborations. When differences are analyzed between elaborations, in batch B, higher levels ($p \leq 0.05$) of TVC and LAB and lower ($p \leq 0.05$) counts of EB and GC+ were observed in the second elaboration with respect to the first elaboration. In the remaining batches, the higher counts of LAB and GC+ were only observed in batch D in the first elaboration with respect to the second one.

Table 3. Microbial counts at day 60 of ripening of cheeses matured in two different traditional elaborations.

Batches	A	B	C	D
		First elaboration (log CFU/g)		
TVC	9.09 ± 0.14	8.74 ± 0.07 *	9.25 ± 0.03	9.10 ± 0.14
EB	5.45 ± 0.46 ^{ab}	6.72 ± 0.13 ^{b*}	4.83 ± 0.29 ^{a}	5.06 ± 0.18 ^{a}
GC+	5.96 ± 0.34	6.24 ± 0.09 *	6.28 ± 0.49	6.15 ± 0.27 *
LAB	9.06 ± 0.20 ^{ab}	8.67 ± 0.06 ^{a*}	9.20 ± 0.03 ^{b}	9.08 ± 0.12 ^{ab*}
		Second elaboration (log CFU/g)		
TVC	9.06 ± 0.14 ^{ab}	9.27 ± 0.05 ^{b*}	9.25 ± 0.08 ^{b}	8.62 ± 0.15 ^{a}
EB	5.46 ± 0.29	5.02 ± 0.28 *	4.99 ± 0.16	5.26 ± 0.19
GC+	5.36 ± 0.20	4.79 ± 0.11 *	5.01 ± 0.30	4.68 ± 0.16 *
LAB	9.10 ± 0.13 ^{b}	9.10 ± 0.05 ^{b*}	9.09 ± 0.10 ^{b}	8.32 ± 0.10 ^{a*}

A (uninoculated control), B (inoculated with LAB 116), C (inoculated with LAB 31), and D (inoculated with LAB 116 and 31). TVC (total viable count), EB (Enterobacteriaceae), GC+ (Gram-positive catalase-positive cocci), and LAB (lactic acid bacteria). Values are expressed as the mean ± standard error of the mean. Significant differences ($p \leq 0.05$) among batches from the same elaboration are shown with letters as superscripts ($^{a-b}$) in the same row. Significant differences ($p \leq 0.05$) between elaborations from the same batch are shown with asterisks (*) in the same column.

3.3. Texture Profile

The texture profile is shown in Table 4. In general, no differences were found between the control batch and the ones inoculated with the LAB individually. Batch D showed lower ($p \leq 0.05$) gumminess, 293.42 g, and chewiness, 182.71 g, than the control batch in the first elaboration, which showed 397.85 g and 276.94 g of gumminess and chewiness, respectively. Batch D showed lower ($p \leq 0.05$) cohesiveness, 0.51, in the second elaboration in comparison to the uninoculated batch, which showed 0.72.

When the two elaborations are compared, the LAB-inoculated cheeses of the first elaborations showed higher ($p \leq 0.05$) levels of hardness, cohesiveness, and gumminess than those of the second elaboration (Table 4).

Table 4. Texture profile results at day 60 of ripening of cheeses matured in two different traditional elaborations.

Batches	A	B	C	D
		First elaboration		
Hardness (g)	828.23 ± 120.71 *	1083.47 ± 157.91 *	1029.99 ± 135.91 *	675.90 ± 105.12 *
Adhesiveness (g × s)	−15.74 ± 7.21	−4.07 ± 1.69	−2.68 ± 1.20	−7.57 ± 3.57
Springiness (g)	0.69 ± 0.02	0.67 ± 0.04	0.60 ± 0.04	0.62 ± 0.03
Cohesiveness	0.51 ± 0.05 *	0.43 ± 0.02	0.38 ± 0.03 *	0.45 ± 0.04
Gumminess (g)	397.85 ± 20.80 ^{b*}	451.84 ± 44.30 ^{b*}	382.50 ± 30.60 ^{ab*}	293.42 ± 18.90 ^{a*}
Chewiness (g)	276.94 ± 15.14 ^{b}	302.00 ± 30.21 ^{ab*}	228.15 ± 18.39 ^{ab*}	182.71 ± 7.31 ^{a*}
Resilience (J x m^{-3})	0.14 ± 0.02	0.11 ± 0.01	0.11 ± 0.01 *	0.12 ± 0.01
		Second elaboration		
Hardness (g)	225.13 ± 102.57 *	290.84 ± 66.89 *	260.24 ± 92.00 *	194.53 ± 79.04 *
Adhesiveness (g × s)	−28.60 ± 14.02	−9.81 ± 5.42	−6.76 ± 3.17	−7.85 ± 2.94
Springiness (g)	0.79 ± 0.10	0.65 ± 0.03	0.67 ± 0.03	0.62 ± 0.06
Cohesiveness	0.72 ± 0.04 ^{b*}	0.47 ± 0.04 ^{a}	0.53 ± 0.01 ^{a*}	0.51 ± 0.02 ^{a}
Gumminess (g)	162.20 ± 70.86 *	125.63 ± 21.77 *	138.12 ± 47.63 *	96.42 ± 38.72 *
Chewiness (g)	142.64 ± 65.68	81.83 ± 15.69 *	95.63 ± 35.43 *	64.64 ± 27.84 *
Resilience (J x m^{-3})	0.19 ± 0.04	0.12 ± 0.01	0.15 ± 0.01 *	0.13 ± 0.01

A (uninoculated control), B (inoculated with LAB 116), C (inoculated with LAB 31), and D (inoculated with LAB 116 and 31). Values are expressed as the mean ± standard error of the mean. Significant differences ($p \leq 0.05$) among batches from the same elaboration are shown with letters as superscripts ($^{a-b}$) in the same row. Significant differences ($p \leq 0.05$) between elaborations from the same batch are shown with asterisks (*) in the same column.

3.4. Volatile Compound Identification

A total of 31 volatile compounds were identified in the analyzed cheeses, including acids, alcohols, ketones, aldehydes, and esters, among other molecules (Table 5; Figure 3). The first elaboration showed a lower variety of volatile compounds independent of the analyzed batches. Some compounds such as acetic acid, propanoic acid, 2-heptanol, or 2-propanone were not even identified in the first elaboration. Moreover, some compounds such as dimethyl disulfide showed higher concentrations in the second elaboration.

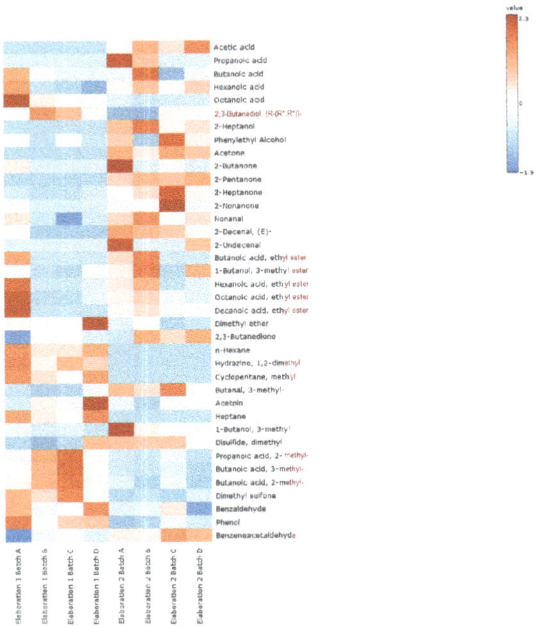

Figure 3. Volatile compounds' heat map of cheeses matured in two different traditional elaborations. Red color establishes a higher detection of a compound in the variable than blue color.

Table 5. Volatile compounds (AU × 10^6) of cheeses matured in two different traditional elaborations.

	First Elaboration				Second Elaboration			
Batches	**A**	**B**	**C**	**D**	**A**	**B**	**C**	**D**
Acids								
Acetic acid	n.d. *	n.d.	n.d. *	n.d. *	3.72 ± 0.77 *	6.76 ± 1.64	4.57 ± 1.12 *	8.18 ± 0.80 *
Propanoic acid	n.d.	n.d.	n.d.	n.d.	0.09 ± 0.04	0.05 ± 0.01	n.d.	n.d.
Butanoic acid	1.63 ± 0.05	1.24 ± 0.05 *	1.18 ± 0.07	1.18 ± 0.18	1.19 ± 0.25 ^{ab}	1.83 ± 0.09 ^{b*}	0.89 ± 0.24 ^{a}	1.21 ± 0.06 ^{ab}
Hexanoic acid	2.38 ± 0.09 ^{b}	1.32 ± 0.14 ^{a*}	1.22 ± 0.06 ^{a}	0.99 ± 0.26 ^{a*}	1.84 ± 0.33	2.20 ± 0.11 *	1.76 ± 0.53	2.07 ± 0.05 *
Octanoic acid	0.32 ± 0.04 ^{b*}	0.17 ± 0.01 ^{a*}	0.16 ± 0.02 ^{a}	0.14 ± 0.04 ^{a}	0.11 ± 0.05 *	0.09 ± 0.02 *	0.10 ± 0.02	0.10 ± 0.00
2-methylpropanoic acid	0.74 ± 0.11 ^{a}	1.75 ± 0.25 ^{ab*}	2.20 ± 0.04 ^{b*}	0.84 ± 0.06 ^{a*}	0.44 ± 0.13	0.34 ± 0.02 *	0.66 ± 0.09 *	0.29 ± 0.01 *
3-methylbutanoic acid	6.37 ± 0.22 ^{a}	9.79 ± 0.45 ^{b*}	12.21 ± 0.82 ^{b*}	6.39 ± 0.22 ^{a*}	3.82 ± 1.13	3.47 ± 0.32 *	4.68 ± 0.61 *	2.77 ± 0.08 *
2-methylbutanoic acid	0.58 ± 0.05 ^{ab}	0.86 ± 0.05 ^{b*}	1.09 ± 0.12 ^{b*}	0.46 ± 0.02 ^{a*}	0.33 ± 0.11	0.16 ± 0.02 *	0.40 ± 0.07 *	0.14 ± 0.00 *
Aldehydes								
3-methylbutanal	0.08 ± 0.01	0.06 ± 0.01 *	0.06 ± 0.02	0.12 ± 0.02	0.21 ± 0.03	0.17 ± 0.03 *	0.25 ± 0.09	0.15 ± 0.02
Nonanal	0.11 ± 0.01 ^{b}	0.08 ± 0.01 ^{ab}	n.d. ^{a*}	0.08 ± 0.01 ^{ab}	0.11 ± 0.02	0.15 ± 0.02	0.09 ± 0.00 *	0.11 ± 0.02
2-Decenal, (E)-	0.06 ± 0.01	n.d. *	n.d. *	n.d.	0.08 ± 0.00	0.07 ± 0.00 *	0.07 ± 0.00 *	0.07 ± 0.00
Benzaldehyde	0.12 ± 0.01 ^{ab*}	0.07 ± 0.01 ^{a}	0.07 ± 0.01 ^{ab}	0.14 ± 0.02 ^{b*}	0.06 ± 0.00 ^{b*}	0.06 ± 0.01 ^{ab}	0.09 ± 0.04 ^{ab}	n.d. ^{a*}
Benzeneacetaldehyde	n.d. ^{a*}	0.06 ± 0.01 ^{b}	0.06 ± 0.01 ^{b*}	0.05 ± 0.00 ^{b*}	0.08 ± 0.01 *	0.07 ± 0.01	0.11 ± 0.01 *	0.10 ± 0.01 *
Alcohols								
2,3-Butanediol [R-(R*,R*)]-	2.59 ± 0.14	3.51 ± 0.70	3.17 ± 0.43	2.64 ± 0.09	1.58 ± 0.52	1.55 ± 0.07	2.68 ± 0.27	2.37 ± 0.22
3-methyl-1-butanol	0.77 ± 0.06	0.72 ± 0.01 *	0.74 ± 0.04	0.68 ± 0.01 *	1.28 ± 0.22	0.89 ± 0.02 *	0.82 ± 0.05	0.79 ± 0.04 *
2-Heptanol	n.d.	n.d. *	n.d.	n.d.	0.08 ± 0.03	0.07 ± 0.02 *	0.05 ± 0.00	0.06 ± 0.01
Phenylethyl Alcohol	0.82 ± 0.06	0.54 ± 0.08	0.82 ± 0.26	0.57 ± 0.08 *	1.12 ± 0.08	0.60 ± 0.01	1.58 ± 0.43	0.96 ± 0.07 *
Ketones								
2-Propanone	n.d. *	n.d. *	n.d. *	n.d. *	0.12 ± 0.00 *	0.08 ± 0.01 *	0.12 ± 0.03 *	0.10 ± 0.01 *
2-Butanone	0.13 ± 0.06	n.d.	n.d.	n.d.	0.42 ± 0.22	n.d.	0.06 ± 0.02	n.d.
2-Pentanone	n.d.	n.d.	n.d. *	n.d. *	0.08 ± 0.03	0.10 ± 0.03	0.10 ± 0.02 *	0.11 ± 0.04 *
2-Heptanone	0.08 ± 0.01 ^{b*}	n.d. ^{a}	n.d. ^{a}	0.10 ± 0.03 ^{ab}	0.38 ± 0.08 *	0.51 ± 0.20	1.11 ± 0.74	0.41 ± 0.20
2-Nonanone	0.22 ± 0.04	n.d. *	n.d.	n.d.	0.19 ± 0.08	0.37 ± 0.11 *	1.50 ± 1.32	0.14 ± 0.05
Esters								
Butanoic acid, ethyl ester	0.09 ± 0.01 ^{b}	n.d. ^{a*}	n.d. ^{a}	n.d. ^{a}	0.06 ± 0.01 ^{ab}	0.11 ± 0.01 ^{b*}	n.d. ^{a}	n.d. ^{a}
1-Butanol, 3-methyl-. acetate	0.17 ± 0.01	0.15 ± 0.00 *	0.15 ± 0.01	0.17 ± 0.01	0.18 ± 0.00	0.21 ± 0.02 *	0.15 ± 0.01	0.20 ± 0.01
Hexanoic acid, ethyl ester	0.22 ± 0.04	0.06 ± 0.01 *	0.05 ± 0.00	0.07 ± 0.00	0.13 ± 0.01 ^{bc}	0.18 ± 0.01 ^{c*}	0.06 ± 0.00 ^{ab}	0.07 ± 0.00 ^{a}
Octanoic acid, ethyl ester	0.25 ± 0.04 ^{b*}	n.d. ^{a*}	n.d. ^{a*}	0.06 ± 0.00 ^{a}	0.11 ± 0.02 ^{bc*}	0.14 ± 0.01 ^{c*}	0.05 ± 0.00 ^{ab*}	0.06 ± 0.00 ^{a}
Decanoic acid, ethyl ester	0.33 ± 0.04 ^{b*}	n.d. ^{a*}	n.d. ^{a*}	n.d. ^{a*}	0.12 ± 0.02 ^{bc*}	0.15 ± 0.02 ^{c*}	0.07 ± 0.00 ^{a*}	0.07 ± 0.01 ^{ab*}

Table 5. *Cont.*

	First Elaboration				Second Elaboration			
Batches	**A**	**B**	**C**	**D**	**A**	**B**	**C**	**D**
Others								
Acetoin	4.72 ± 0.38	6.37 ± 0.63 *	5.50 ± 0.67	9.88 ± 1.84	5.20 ± 1.16	4.13 ± 0.29 *	5.57 ± 0.50	5.78 ± 0.06
2,3-Butanedione	1.37 ± 0.18 ^{a*}	3.13 ± 0.40 ^{b*}	3.44 ± 0.16 ^{b}	2.74 ± 0.20 ^{b*}	2.49 ± 0.34 ^{a*}	4.61 ± 0.31 ^{b*}	3.91 ± 0.34 ^{b}	4.64 ± 0.15 ^{b*}
Dimethyl disulfide	1.25 ± 0.30 ^{a*}	0.85 ± 0.34 ^{a*}	1.29 ± 0.50 ^{a}	3.62 ± 0.18 ^{b*}	3.52 ± 0.24 *	3.37 ± 0.66 *	3.43 ± 0.75	2.33 ± 0.31 *
Dimethyl sulfone	0.08 ± 0.00	0.07 ± 0.00	0.10 ± 0.04	0.06 ± 0.00 *	0.04 ± 0.00	0.05 ± 0.00	0.04 ± 0.00	0.04 ± 0.00 *

A (uninoculated control), B (inoculated with LAB 116), C (inoculated with LAB 31), and D (inoculated with LAB 116 and 31). Values are expressed as the mean ± standard error of the mean. Significant differences ($p \leq 0.05$) among batches from the same elaboration are shown with letters as superscripts ($^{a\text{-}b}$) in the same row. Significant differences ($p \leq 0.05$) between elaborations from the same batch are shown with asterisks (*) in the same column.

When differences between batches are analyzed, it can be observed that, in general, some volatile compounds such as butanoic, hexanoic and octanoic acids, 2-decenal, 2-butanone, 2-nonanone, butanoic acid ethyl ester, or decanoic acid ethyl ester were found in high levels in the control batch than in LAB-inoculated cheeses (Table 5). However, 2-methylpropanoic acid and 3- and 2-methylbutanoic acids are in higher concentrations in the LAB-inoculated than in control batches.

3.5. Sensory Evaluation

In the first elaboration, the cheeses from batches B and C were perceived as different ($p \leq 0.05$) than the ones from batch A. However, the triangular test of the second elaboration only identified differences ($p \leq 0.05$) between batches A and C.

CATA analysis was performed to analyze the qualitative characteristics of each batch and compare them within the same elaboration. Cochran's Q test evaluated the differences between the four batches in each of the elaborations (Table 6). There were statistical differences ($p \leq 0.05$) among the frequencies of citation of the four evaluated batches in floral and lactic odors in the first elaboration and brightness in the second one.

Table 6. Cochran's Q test of the CATA-evaluated attributes of cheeses matured in two different traditional elaborations.

Attributes	*p*-Values	
	First Elaboration	Second Elaboration
Adherence	0.991	0.439
Brightness	0.110	0.050 *
Ripened flavor	0.140	0.910
Intense flavor	0.782	0.555
Persistent flavor	0.603	0.218
Pungent flavor	0.359	0.252
Rancid flavor	0.083	0.543
Intense color	0.623	0.815
Floral odor	0.015 *	0.912
Lactic odor	0.048 *	0.364
Vegetal odor	0.494	0.753
Spicy	0.317	0.889
Acid taste	0.841	0.140
Bitter taste	0.914	0.576
Sweetness	0.697	0.468
Umami taste	0.456	0.699
Saltiness	0.409	0.682
Mouth coating	0.494	0.227
Smoothness	0.116	0.392

Statistical differences ($p \leq 0.05$) among some of the four evaluated batches are represented with an asterisk (*) in each of the elaborations' frequencies of perception.

McNemar pairwise multiple comparisons were performed to analyze the differences among each batch individually within the same elaboration, though no differences in the citation frequencies were found (Table 7). There were no differences in the acceptability of the samples either.

CA of the first elaboration is shown in Figure 4, in which the two first dimensions explained 85.48% of the total variance. The four analyzed batches are related to different attributes. Batch A is closer to intense flavor, adherence, and mouth coating. Batch B shows a spatial relation with ripened flavor, persistent flavor, and spicy. Batch C is related to lactic odor, brightness, pungent, or rancid flavor. Batch D shows a relationship with umami taste and smoothness.

Table 7. McNemar pairwise multiple comparisons of the CATA-evaluated attributes (%) and acceptability (over seven points) of cheeses matured in two different traditional elaborations.

	First Elaboration				Second Elaboration			
Attributes	**A**	**B**	**C**	**D**	**A**	**B**	**C**	**D**
Adherence	41.67	38.89	38.89	38.89	44.44	47.22	50.00	58.33
Brightness	69.44	55.56	72.22	50.00	50.00	41.67	61.11	66.67
Ripened flavor	47.22	44.44	30.56	38.89	44.44	44.44	41.67	38.89
Intense flavor	66.67	69.44	61.11	61.11	66.67	69.44	61.11	55.56
Persistent flavor	61.11	55.56	47.22	52.78	63.89	75.00	63.89	55.56
Pungent flavor	22.22	8.33	19.44	19.44	25.00	33.33	22.22	16.67
Rancid flavor	8.33	13.89	16.67	5.56	19.44	16.67	11.11	13.89
Intense color	16.67	19.44	11.11	13.89	13.89	11.11	11.11	16.67
Floral odor	2.78	0.00	2.8	13.89	5.56	5.56	8.33	5.56
Lactic odor	44.44	44.44	63.89	38.89	47.22	61.11	50.00	47.22
Vegetal odor	0.00	2.78	2.78	5.56	8.33	8.33	5.56	11.11
Spicy	19.44	22.22	13.89	11.11	22.22	19.44	19.44	16.67
Acid taste	47.22	44.44	50.00	41.67	50.00	63.89	47.22	38.89
Bitter taste	22.22	25.00	27.78	25.00	33.33	25.00	22.22	22.22
Sweetness	5.56	8.33	5.56	11.11	2.78	8.33	2.78	5.56
Umami taste	19.44	11.11	13.89	22.22	22.22	19.44	27.78	25.00
Saltiness	27.78	27.78	19.44	33.33	30.56	27.78	27.78	36.11
Mouth coating	69.44	58.33	55.56	61.11	55.56	47.22	63.89	61.11
Smoothness	69.44	50.00	58.33	75.00	77.78	63.89	77.78	77.78
Acceptability	4.97 ± 1.30	5.14 ± 1.33	5.05 ± 1.28	5.30 ± 1.21	4.55 ± 1.50	4.36 ± 1.39	4.69 ± 1.30	5.11 ± 1.52

A (uninoculated control), B (inoculated with LAB 116), C (inoculated with LAB 31), and D (inoculated with LAB 116 and 31).

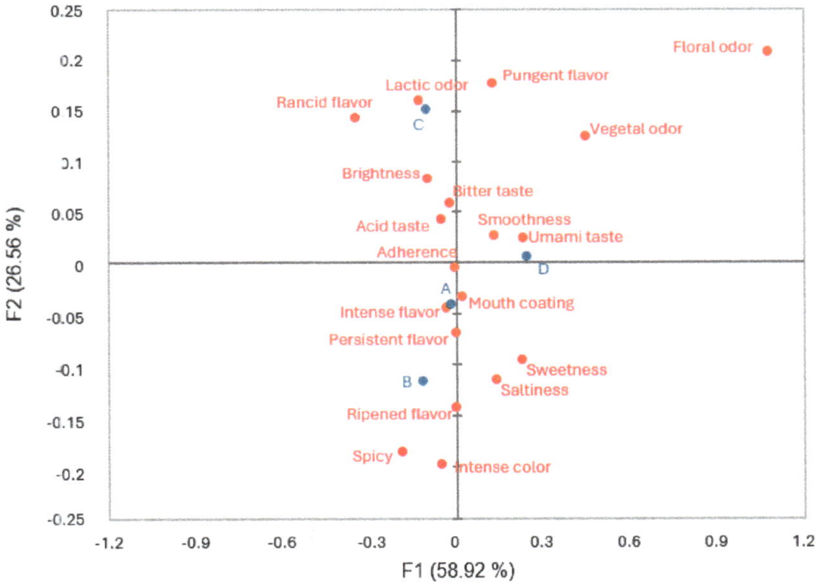

Figure 4. Correspondence Analysis (CA) of the CATA-evaluated attributes in the first elaboration of cheeses at day 60 of ripening. A (uninoculated control), B (inoculated with LAB 116), C (inoculated with LAB 31), and D (inoculated with LAB 116 and 31). Batches are represented with blue color and attributes with red color.

CA of the second elaboration is represented in Figure 5, in which the two first dimensions explained 85.39% of the total variance. Each of the batches analyzed is placed in one of the four quadrants of the plot that represented the differences among all of them. Batch A is related to spicy, bitter taste and ripened flavor. Batch B is closer to acid taste. Batch C shows a spatial relation with floral odor and smoothness. Batch D shows a relationship with intense color and umami taste.

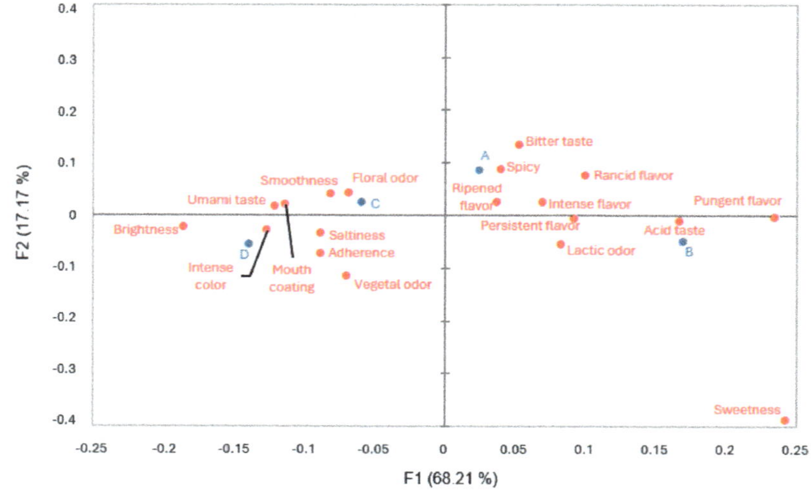

Figure 5. Correspondence Analysis (CA) of the CATA-evaluated attributes in the second elaboration of cheeses at day 60 of ripening. A (uninoculated control), B (inoculated with LAB 116), C (inoculated with LAB 31), and D (inoculated with LAB 116 and 31). Batches are represented with blue color and attributes with red color.

In the first elaboration, there were no attributes in which statistical differences were identified in the PA, though Figure 6 shows how they would positively or negatively affect the mean liking. For example, umami taste would have a significant impact close to 1 point over 7.

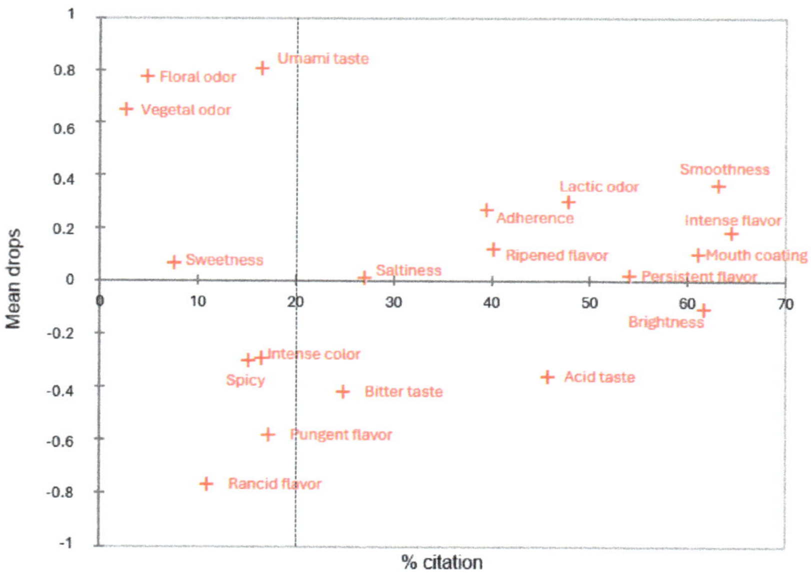

Figure 6. Penalty Analysis (PA) from CATA, in which the attributes are distributed according to the percentage of citation during the sensory evaluation of the first elaboration of cheeses at day 60 of ripening and how they affect the mean liking. The attributes whose presence affects the acceptability mean ($p \leq 0.05$) are circled in red or blue based on whether they reduce or increase it, respectively.

In the second elaboration, umami taste has a positive impact on the overall acceptability, being able to increase it even 1 point over 7 according to the PA (Figure 7). Acid taste, pungent flavor, and bitter taste were attributes that acted as drivers of disliking and thus negatively affected the acceptability of the product by −0.8, −1.0, and −1.1 points over 7, respectively.

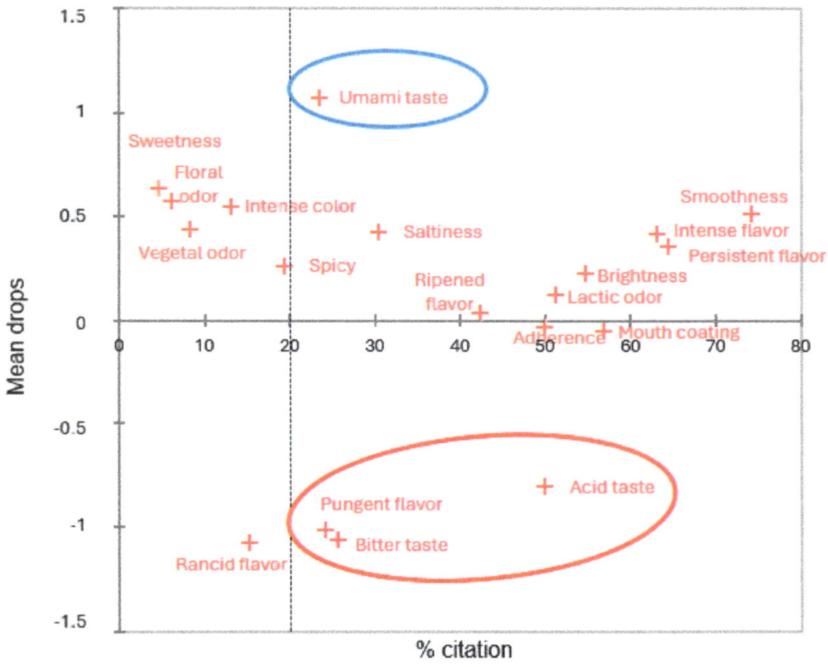

Figure 7. Penalty Analysis (PA) from CATA, in which the attributes are distributed according to the percentage of citation during the sensory evaluation of the second elaboration of cheeses at day 60 of ripening and how they affect the mean liking. The attributes whose presence affects the acceptability mean ($p \leq 0.05$) are circled in red or blue based on whether they reduce or increase it, respectively.

4. Discussion

4.1. Physicochemical Characteristics

The evaluation of physicochemical parameters did not show consistent differences between batches in both elaborations that could be justified by the effect of the inoculation of selected LAB. The moisture content close to 50% of the cheese in both elaborations are in the reported values for this kind of cheese related to the creamy effect intended in soft-ripened cheeses [23]. The pH values are also usual at the final day of ripening in the evaluate cheeses probably due to the acidification provoked by the LAB throughout the maturation and by the slight increase in pH reported in this product as consequence of the proteolysis that occurs mainly at the end of processing [24,25]. No consistent differences in the pH values between LAB-inoculated and non-inoculated batches were observed. However, differences in this parameter were detected in cheeses of the two elaborations studied, probably due to the small variations in conditions of temperature and relative humidity of both productions, as have been reported in artisanal productions [11,24]. The a_w values could be also considered normal for this kind of cheese [26], and they showed differences between elaborations due to the artisanal elaboration process, but not between batches that could be attributed to the inoculation of LAB strains.

Regarding the color, the results obtained are more similar to cheeses ripened for several months [27]. No differences in color between LAB-inoculated and non-inoculated batches were observed. The only differences were observed between elaborations and are probably related to the characteristics of artisanal production.

4.2. Microbial Population

The LAB was the dominant microbial population even in the uninoculated control, with these bacteria being derived from the microbial population of raw milk [25,28,29]. In both elaborations and batches, *L. monocytogenes* was not detected. The high levels of TVC and EB observed in both elaborations are usual in soft-ripened cheeses elaborated with raw milk, as other authors have proved [28,30]. No consistent differences between batches in the same elaboration in any of the microbial groups evaluated were observed, probably because the LAB was the dominant microbial population.

The small differences observed in B and D batches between elaborations in LAB, TVC, and GC+ may be related to the slight variations in temperature and relative humidity conditions between elaborations, especially in conditions of artisanal production, that could favor the growth of LAB and the decrease in TVC and GC+ [31].

4.3. Texture Profile

The values of the texture parameters observed are like those reported for "Torta del Casar" [22]. No differences between LAB-inoculated and non-inoculated cheeses were found. However, batch D, co-inoculated with *L. casei* 31 and 116, showed lower values of hardness, gumminess, and chewiness within each of the elaborations, proving to be a softer batch. This fact is probably due to the proteolytic activity that the inoculated LAB performed in the matrix, though the differences found between batches are often associated with the proteolytic effect on the rennet and the artisanal character of the elaboration [7, 32]. Usually, creamier cheeses with lower hardness are preferred by consumers, so the inoculation of both LAB may help in the development of more desired cheeses even though differences with the control batch have been observed [33]. In addition, when the two elaborations were compared, higher values of hardness, cohesiveness, and gumminess were found in the cheeses of the first elaboration in comparison to the cheeses of the second one, probably due to the lower moisture content found in the cheeses of the first elaboration [27].

4.4. Volatile Compounds

The analysis of the volatile compounds in the cheeses showed differences mainly due to elaboration and at a minor level due to the LAB inoculation level as it is discussed below for each family.

4.4.1. Acids

Acetic acid and propanoic acid could be related to major LAB activity due to the fermentation of lactose during the ripening of cheeses in the case of propanoic acid, and because of lactose and casein biochemical transformation in the case of acetic acid [34,35]. Though a higher concentration of butanoic, hexanoic, and octanoic acids may indicate that their production is influenced by the lipolytic activity of microorganisms, it is mainly attributed to the native milk lipases present in the raw milk, or the plant rennet utilized in the traditional elaboration of the studied cheeses [36,37]. However, the higher concentrations of these acids in the present study may indicate that the wild microbial population derived from the natural contamination of raw milk, which developed in the uninoculated batches, is showing some lipolytic activity. This fact could be a problem in the sensory evaluation since the presence of acids in the final product is related to sourness, like vinegar, in the case of acetic or propanoic acid presence, or closer to a rancid, fruity, or goaty flavor in the case of butanoic, hexanoic, or octanoic acid, respectively [27,37]. The LAB-inoculated batches showed a higher concentration of methyl-carboxylic acids such as 2-methylpropanoic, 3-methylbutanoic, and 2-methylbutanoic acids, probably due to the Strecker reactions that occur during the cheese-ripening process, which are recognized as sweaty and ripened flavors in the aroma development of cheeses during maturation [37,38].

4.4.2. Aldehydes

Nonanal, 2-decenal, and benzaldehyde have a stronger presence in the uninoculated cheeses and has been proven in other studies to have a similar green-grass aroma or an almond flavor in the case of the latter [39–42]. Nevertheless, the inoculated batches showed a higher concentration of benzeneacetaldehyde, which is related to a rose-like aroma [43]. However, aldehydes, in general, have a low presence in the volatile profile of the evaluated cheeses due to their instability, as they are reduced into alcohols or oxidized into acids [42].

4.4.3. Alcohols

2-heptanol was just identified in the cheeses from the second elaboration, and it is related to herbaceous and fruity odor descriptors [44]. Phenylethyl alcohol and 3-methyl-1-butanol are mainly found in the second elaboration, presenting an odor impression of honey and rose in the case of the former and herbaceous odor in the case of the latter [39,44–46]. 2,3-butanediol had higher concentrations in the inoculated batches, usually appearing through the reduction of acetoin and showing a positive correlation with ripened cheeses [40,47].

4.4.4. Ketones

Most of the ketones were not identified in the analyzed cheeses from the first elaboration, though they appeared in the second one. The 2-propanone, 2-butanone, 2-heptanone, and 2-nonanone were related to microbial activity in other studies, being mainly found in batch C [22,48,49]. The 2-butanone has a buttery flavor, while 2-heptanone and 2-nonanone are linked with a cheesy flavor [27,46]. The 2-pentanone and 2-propanone are related to orange peel and hay odor descriptors, respectively [44].

4.4.5. Esters

The presence of the esters identified in this study has been confirmed by other authors who have analyzed the same kind of cheese in previous studies, affirming that the microbial presence of GC+ and Enterobacteriaceae are correlated with the appearance of esters [22,47,48,50]. Even though there were no differences between the counts of these microbial groups, no remarkable differences have been found in the ester production of the analyzed samples. The ethyl esters found in the present study have been identified as sweet, fruity, and ice-cream flavored, specifically, pineapple-like in the case of hexanoic acid and ethyl ester; coconut-like for octanoic acid and ethyl ester; or banana-like in the case of 1-butanol, 3-methyl, and acetate [44,51].

4.4.6. Other Molecules

Dimethyl disulfide and dimethyl sulfone are derived from methionine and are contributors to cheese flavor, though their odor descriptor is sulfurous, and their low threshold allows to identify them as garlic, cabbage, or onion [44,52]. A higher concentration of dimethyl disulfide has been found in the cheeses from the second elaboration. Acetoin is found in this study in high amounts, especially in the batch D, though it is usually found only in the initial stages of cheeses due to its transformation into diacetyl [53]. The 2,3-butanedione (diacetyl) was found in higher concentrations in the second elaboration, and it has been observed mainly in the inoculated batches, being related to a butter-like flavor [54].

4.5. Sensory Evaluation

The inoculation of *L. casei* caused some sensory differences in the final product, despite the ones of the second elaboration not being perceived in some cheeses.

Cochran's Q test showed differences in the floral odor that may be related to the presence of benzeneacetaldehyde, an aldehyde with a rose-like odor that has not been detected in the uninoculated batch of the first elaboration. On the other hand, the lactic odor differences can relate to the 2-butanone or 2,3-butanedione identification, because these compounds with a butter odor have been identified in different proportions in the first elaboration.

McNemar pairwise multiple comparisons showed that some of the attributes of the first and second elaboration are directly related to the volatile molecules identified, such as 3-methylbutanoic acid, which can be identified with a pungent flavor and shows the highest values in batch C in the first elaboration. Other attributes, however, cannot be related to a specific volatile molecule due to the sensory interactions between the totality of aroma and taste inputs [55]. However, physical properties such as the texture profile can be related to the sensory results, as CA identified batch D as the smoother sample, similar to what the TPA showed previously in the first elaboration. The inoculation of starter cultures has shown in other studies an improvement in the creamy and soft texture of cheeses [7], with batches D and C being the ones sensory related with the attribute "smoothness" in the first and second elaborations, respectively, in the present research. This fact may indicate that the utilization of LAB could increase the smoothness of the cheeses.

In both elaborations, only batch D has shown a common relationship with an attribute, umami taste, despite the artisanal character of the evaluated cheeses. This attribute has a positive impact on the overall acceptability in both sensory analyses. Acid and bitter tastes were some of the attributes that negatively affected the mean liking and were related to batches B and A, in that order. Therefore, the inoculation of selected LAB must be evaluated in soft-ripened cheeses, because the sensory profile developed during ripening could be different depending on the bacteria inoculated.

Despite the artisanal variations in the processing that affect the physicochemical, microbiological, and texture characteristics of the product, batch D seemed sensory-stable thanks to the combined inoculation of *L. casei* 31 and 116. This batch showed the highest acceptability mean values in both elaborations, and for the softest texture according to the TPA, and it was related to umami taste, a positive attribute according to the PA. Therefore, the co-inoculations of *L. casei* 31 and 116 used in batch D are the ones in which better results have been obtained, and it is the recommended strategy to obtain more homogeneous soft-ripened cheeses by improving sensorial characteristics.

5. Conclusions

The maturation of soft-ripened cheeses with selected protective cultures of *L. casei* did not negatively affect the physicochemical, volatile compound profile, and sensorial parameters of the cheeses. On the contrary, the use of individual cultures of *L. casei* 31 and 116 or especially a mixed culture of both LAB strains could improve the sensory profile of the soft-ripened cheeses, thus decreasing the values of texture parameters such as hardness, gumminess, and chewiness related to a softer texture and increasing umami taste and floral and lactic odor attributes. Sensory analysis revealed that consumers perceived differences between inoculated and non-inoculated cheeses, although the overall acceptability was not affected. This study provides valuable information for the artisanal cheese industry, demonstrating that it is possible to use selected protective lactic acid bacteria to assure food safety without compromising traditional flavor and even improving sensorial attributes.

6. Patents

Martín, I., Rodríguez, A., and Córdoba, J.J., inventor 2024. Nueva cepa de *Lacticaseibacillus casei* 116 con actividad antagonista frente a *Listeria monocytogenes* para su uso como cultivo protector en quesos madurados de pasta blanda elaborados con leche cruda de oveja. No. ES 2942554 B2

Author Contributions: Conceptualization, I.M. and J.J.C.; methodology, S.V. and J.J.C.; software, S.V.; formal analysis, J.M.M.M.; investigation, J.M.M.M.; resources, J.R. and J.J.C.; data curation, J.M.M.M.; writing—original draft preparation, J.M.M.M.; writing—review and editing, I.M., J.R., S.V. and J.J.C.; supervision, S.V. and J.J.C.; project administration, I.M., J.R. and J.J.C.; funding acquisition, J.R. and J.J.C. All authors have read and agreed to the published version of the manuscript.

Funding: This research was funded by Junta de Extremadura and FEDER grants IDA1-21-0130-3 and GR18056. This study was supported by grant 2022/00254/001 funded by MCIN/AEI/10.13039/501100011033.

Data Availability Statement: The original contributions presented in this study are included in the article. Further inquiries can be directed to the corresponding author.

Acknowledgments: Torta del Casar Protected Designation of Origin, the Doña Francisca cheese factory (Casar de Cáceres, Spain), Laboratory Hidromante (Plasencia, Spain), and Facility of Innovation and Analysis in Animal Source Foodstuffs (SiPA) of SAIUEx of the University of Extremadura (Cáceres, Spain) for the support analytical services.

Conflicts of Interest: The authors declare no conflicts of interest.

References

1. Delgado, J.; Álvarez, M.; Cebrián, E.; Martín, I.; Roncero, E.; Rodríguez, M. Biocontrol of Pathogen Microorganisms in Ripened Foods of Animal Origin. *Microorganisms* **2023**, *11*, 1578. [CrossRef] [PubMed]
2. Szűcs, V.; Szabó, E.; Guerrero, L.; Tarcea, M.; Bánáti, D. Modelling of Avoidance of Food Additives: A Cross Country Study. *Int. J. Food Sci. Nutr.* **2019**, *70*, 1020–1032. [CrossRef] [PubMed]
3. Zhong, Y.; Wu, L.; Chen, X.; Huang, Z.; Hu, W. Effects of Food-Additive-Information on Consumers' Willingness to Accept Food with Additives. *Int. J. Environ. Res. Public. Health* **2018**, *15*, 2394. [CrossRef] [PubMed]
4. Yang, H.; He, M.; Wu, C. Cross Protection of Lactic Acid Bacteria during Environmental Stresses: Stress Responses and Underlying Mechanisms. *LWT* **2021**, *144*, 111203. [CrossRef]
5. Xia, L.; Qian, M.; Cheng, F.; Wang, Y.; Han, J.; Xu, Y.; Zhang, K.; Tian, J.; Jin, Y. The Effect of Lactic Acid Bacteria on Lipid Metabolism and Flavor of Fermented Sausages. *Food Biosci.* **2023**, *56*, 103172. [CrossRef]
6. Martín, I.; Rodríguez, A.; Alía, A.; Martínez, R.; Córdoba, J.J. Selection and Characterization of Lactic Acid Bacteria with Activity against *Listeria monocytogenes* from Traditional RTE Ripened Foods. *LWT* **2022**, *163*, 113579. [CrossRef]
7. Fusco, V.; Quero, G.M.; Poltronieri, P.; Morea, M.; Baruzzi, F. Autochthonous and Probiotic Lactic Acid Bacteria Employed for Production of "Advanced Traditional Cheeses". *Foods* **2019**, *8*, 412. [CrossRef]
8. Official Journal of the European Union. Publication of an Amendment Application Pursuant to Article 50(2)(a) of Regulation (EU) No 1151/2012 of the European Parliament and of the Council on Quality Schemes for Agricultural Products and Foodstuffs. Available online: https://eur-lex.europa.eu/legal-content/ES/TXT/?uri=uriserv:OJ.C_.2015.235.01.0005.01.SPA&toc=OJ:C:2015:235:TOC (accessed on 30 September 2023).

9. West, H.G. Crafting Innovation: Continuity and Change in the "Living Traditions" of Contemporary Artisan Cheesemakers. *Food Foodways* **2020**, *28*, 91–116. [CrossRef]
10. Costanzo, N.; Ceniti, C.; Santoro, A.; Clausi, M.T.; Casalinuovo, F. Foodborne Pathogen Assessment in Raw Milk Cheeses. *Int. J. Food Sci.* **2020**, *2020*, 3616713. [CrossRef]
11. Campagnollo, F.B.; Margalho, L.P.; Kamimura, B.A.; Feliciano, M.D.; Freire, L.; Lopes, L.S.; Alvarenga, V.O.; Cadavez, V.A.P.; Gonzales-Barron, U.; Schaffner, D.W.; et al. Selection of Indigenous Lactic Acid Bacteria Presenting Anti-Listerial Activity, and Their Role in Reducing the Maturation Period and Assuring the Safety of Traditional Brazilian Cheeses. *Food Microbiol.* **2018**, *73*, 288–297. [CrossRef]
12. Martín-Miguélez, J.M.; Robledo, J.; Martín, I.; Castaño, C.; Delgado, J.; Córdoba, J.J. Biocontrol of *L. monocytogenes* with Selected Autochthonous Lactic Acid Bacteria in Raw Milk Soft-Ripened Cheese under Different Water Activity Conditions. *Foods* **2024**, *13*, 172. [CrossRef] [PubMed]
13. AOAC Association of Official Analytical Chemists. *Official Methods of Analysis of the AOAC International*; AOAC: Washington, DC, USA, 2000.
14. Commission Internationale de L'Éclairage Colorimetry. *Color Research & Application*, 2nd ed.; John Wiley & Sons, Ltd.: Hoboken, NJ, USA, 1988; Volume 13, pp. 64–65.
15. *ISO 11290-1:2017*; Microbiology of the Food Chain—Horizontal Method for the Detection and Enumeration of Listeria monocytogenes and of *Listeria* spp.—Part 1: Detection Method. International Organization for Standardization: Geneva, Switzerland, 2017.
16. Jiménez, A.; González-Mohino, A.; Rufo, M.; Paniagua, J.M.; Antequera, T.; Perez-Palacios, T. Dry-Cured Loin Characterization by Ultrasound Physicochemical and Sensory Parameters. *Eur. Food Res. Technol.* **2022**, *248*, 2603–2613. [CrossRef]
17. González-Mohino, A.; Jiménez, A.; Rufo, M.; Paniagua, J.M.; Antequera, T.; Perez-Palacios, T. Ultrasound Parameters Used to Characterize Iberian Fresh Pork Loins of Different Feeding Systems. *J. Food Eng.* **2022**, *314*, 110795 [CrossRef]
18. Ventanas, S.; González-Mohino, A.; Estévez, M.; Carvalho, L. Innovation in Sensory Assessment of Meat and Meat Products. In *Meat Quality Analysis: Advanced Evaluation Methods, Techniques, and Technologies*; Academic Press: Cambridge, MA, USA, 2020; pp. 393–418. [CrossRef]
19. Pérez Elortondo, F.J.; Torre, P. Queso. In *Análisis Sensorial de Alimentos y Respuesta del Consumidor*; Pérez Elortondo, F.J., Salvador Moya, M.a.D., Eds.; Editorial Acribia, S.A.: Zaragoza, Spain, 2022; pp. 209–229, ISBN 978-84-200-1279-7.
20. Torres, F.R.; Esmerino, E.A.; Carr, B.T.; Ferrão, L.L.; Granato, D.; Pimentel, T.C.; Bolini, H.M.A.; Freitas, M.Q.; Cruz, A.G. Rapid Consumer-Based Sensory Characterization of Requeijão Cremoso, a Spreadable Processed Cheese: Performance of New Statistical Approaches to Evaluate Check-All-That-Apply Data. *J. Dairy Sci.* **2017**, *100*, 6100–6110. [CrossRef]
21. Meals, S.E.; Schiano, A.N.; Drake, M.A. Drivers of Liking for Cheddar Cheese Shreds. *J. Dairy Sci.* **2020**, *103*, 2167–2185. [CrossRef] [PubMed]
22. Martín, I.; Rodríguez, A.; García, C.; Córdoba, J.J. Evolution of Volatile Compounds during Ripening and Final Sensory Changes of Traditional Raw Ewe's Milk Cheese "Torta Del Casar" Maturated with Selected Protective Lactic Acid Bacteria. *Foods* **2022**, *11*, 2658. [CrossRef]
23. Fu, W.; Nakamura, T. Moisture Content Impact Creaming Effect and Microstructure of Processed Cheese Containing Different Textural Starches. *Int. Dairy J.* **2020**, *105*, 104685. [CrossRef]
24. Ordiales, E.; Benito, M.J.; Martin, A.; Fernández, M.; Hernández, A.; De Guia Córdoba, M. Proteolytic Effect of *Cynara cardunculus* Rennet for Use in the Elaboration of 'Torta Del Casar' Cheese. *J. Dairy Res.* **2013**, *80*, 429–438. [CrossRef]
25. Crespo, A.; Martín, A.; Ruiz-Moyano, S.; Benito, M.J.; Rufo, M.; Paniagua, J.M.; Jiménez, A. Application of Ultrasound for Quality Control of "Torta Del Casar" Cheese Ripening. *J. Dairy Sci.* **2020**, *103*, 8808–8821. [CrossRef]
26. Martín, I.; Rodríguez, A.; Córdoba, J.J. Application of Selected Lactic-Acid Bacteria to Control *Listeria monocytogenes* in Soft-Ripened "Torta Del Casar" Cheese. *LWT* **2022**, *168*, 113873. [CrossRef]
27. Sanahuja, A.B.; de Almeida, R.P.; Marí, K.-A.I.; Lamadrid, M.C.; García, A.V.; Nadal, E.S. Sensory Attributes and Instrumental Chemical Parameters of Commercial Spanish Cured Ewes' Milk Cheeses: Insights into Cheese Quality Figures. *Foods* **2023**, *13*, 127. [CrossRef] [PubMed]
28. Tabla, R.; Roa, I. Use of Gaseous Ozone in Soft Cheese Ripening: Effect on the Rind Microorganisms and the Sensorial Quality. *LWT* **2022**, *170*, 114066. [CrossRef]
29. Bintsis, T. Yeasts in Different Types of Cheese. *AIMS Microbiol.* **2021**, *7*, 447. [CrossRef] [PubMed]
30. Merchán, A.V.; Ruiz-Moyano, S.; Hernández, M.V.; Martín, A.; Lorenzo, M.J.; Benito, M.J. Characterization of Autochthonal *Hafnia* spp. Strains Isolated from Spanish Soft Raw Ewe's Milk PDO Cheeses to Be Used as Adjunct Culture. *Int. J. Food Microbiol.* **2022**, *373*, 109703. [CrossRef]

31. Al-Gamal, M.S.; Ibrahim, G.A.; Sharaf, O.M.; Radwan, A.A.; Dabiza, N.M.; Youssef, A.M.; El-Ssayad, M.F.; El, M.F. The Protective Potential of Selected Lactic Acid Bacteria against the Most Common Contaminants in Various Types of Cheese in Egypt. *Heliyon* **2019**, *5*, e01362. [CrossRef]
32. Ordiales, E.; Martín, A.; Benito, M.J.; Fernández, M.; Casquete, R.; de Guía Córdoba, M. Influence of the Technological Properties of Vegetable Rennet (*Cynara cardunculus*) on the Physicochemical, Sensory and Rheological Characteristics of 'Torta Del Casar' Cheese. *Int. J. Dairy Technol.* **2014**, *67*, 402–409. [CrossRef]
33. Yates, M.D.; Drake, M.A. Texture Properties of Gouda Cheese. *J. Sens. Stud.* **2007**, *22*, 493–506. [CrossRef]
34. Urbach, G. Contribution of Lactic Acid Bacteria to Flavour Compound Formation in Dairy Products. *Int. Dairy J.* **1995**, *5*, 877–903. [CrossRef]
35. Delgado-Martínez, F.J.; Carrapiso, A.I.; Contador, R.; Ramírez, M.R. Volatile Compounds and Sensory Changes after High Pressure Processing of Mature "Torta Del Casar" (Raw Ewe's Milk Cheese) during Refrigerated Storage. *Innov. Food Sci. Emerg. Technol.* **2019**, *52*, 34–41. [CrossRef]
36. Sousa, M.J.; Malcata, F.X. Ripening of Ovine Milk Cheeses: Effects of Plant Rennet, Pasteurization, and Addition of Starter on Lipolysis. *Food Chem.* **1997**, *59*, 427–432. [CrossRef]
37. Ferreira, I.M.P.L.V.O.; Pinho, O.; Sampaio, P. Volatile Fraction of DOP "Castelo Branco" Cheese: Influence of Breed. *Food Chem.* **2009**, *112*, 1053–1059. [CrossRef]
38. Fröhlich-Wyder, M.T.; Bachmann, H.P.; Schmidt, R.S. Classification of Cheese Varieties from Switzerland Using Machine Learning Methods: Free Volatile Carboxylic Acids. *LWT* **2023**, *184*, 115095. [CrossRef]
39. Wang, J.; Yang, Z.J.; Wang, Y.D.; Cao, Y.F.; Wang, B.; Liu, Y. The Key Aroma Compounds and Sensory Characteristics of Commercial Cheddar Cheeses. *J. Dairy Sci.* **2021**, *104*, 7555–7571. [CrossRef] [PubMed]
40. Gan, H.H.; Yan, B.; Linforth, R.S.T.; Fisk, I.D. Development and Validation of an APCI-MS/GC–MS Approach for the Classification and Prediction of Cheddar Cheese Maturity. *Food Chem.* **2016**, *190*, 442–447. [CrossRef]
41. Chen, C.; Zhou, W.; Yu, H.; Yuan, J.; Tian, H. Characterization of Major Odor-Active Compounds Responsible for Nutty Flavor in Cheddar Cheese According to Chinese Taste. *Flavour. Fragr. J.* **2021**, *36*, 171–181. [CrossRef]
42. Curtin, Á.C.; McSweeney, P.L.H. Catabolism of Amino Acids in Cheese during Ripening. *Cheese Chem. Phys. Microbiol.* **2004**, *1*, 435–454. [CrossRef]
43. Jiao, J.; Zheng, Z.; Liu, Z.; You, C. Study of the Compositional, Microbiological, Biochemical, and Volatile Profile of Red-Veined Cheese, an Internal Monascus-Ripened Variety. *Front. Nutr.* **2021**, *8*, 649611. [CrossRef]
44. Curioni, P.M.G.; Bosset, J.O. Key Odorants in Various Cheese Types as Determined by Gas Chromatography-Olfactometry. *Int. Dairy J.* **2002**, *12*, 959–984. [CrossRef]
45. Xu, Z.; Chen, J.; Shi, X.; Wang, B.; Zheng, X.; Zheng, X. Characteristic Physicochemical Indexes and Flavor Compounds in Xinjiang Kazak Cheese during Ripening. *Food Biosci.* **2020**, *35*, 100586. [CrossRef]
46. Tekin, A.; Hayaloglu, A.A. Understanding the Mechanism of Ripening Biochemistry and Flavour Development in Brine Ripened Cheeses. *Int. Dairy J.* **2023**, *137*, 105508. [CrossRef]
47. Dahl, S.; Tavaria, F.K.; Xavier Malcata, F. Relationships between Flavour and Microbiological Profiles in Serra Da Estrela Cheese throughout Ripening. *Int. Dairy J.* **2000**, *10*, 255–262. [CrossRef]
48. Ordiales, E.; Martín, A.; Benito, M.J.; Hernández, A.; Ruiz-Moyano, S.; de Guía Córdoba, M. Role of the Microbial Population on the Flavor of the Soft-Bodied Cheese Torta Del Casar. *J. Dairy Sci.* **2013**, *96*, 5477–5486. [CrossRef] [PubMed]
49. Unno, R.; Suzuki, T.; Osaki, Y.; Matsutani, M.; Ishikawa, M. Causality Verification for the Correlation between the Presence of Nonstarter Bacteria and Flavor Characteristics in Soft-Type Ripened Cheeses. *Microbiol. Spectr.* **2022**, *10*, e02894-22. [CrossRef] [PubMed]
50. Delgado, F.J.; González-Crespo, J.; Cava, R.; García-Parra, J.; Ramírez, R. Characterisation by SPME–GC–MS of the Volatile Profile of a Spanish Soft Cheese P.D.O. Torta Del Casar during Ripening. *Food Chem.* **2010**, *118*, 182–189. [CrossRef]
51. Jia, R.; Zhang, F.; Song, Y.; Lou, Y.; Zhao, A.; Liu, Y.; Peng, H.; Hui, Y.; Ren, R.; Wang, B. Physicochemical and Textural Characteristics and Volatile Compounds of Semihard Goat Cheese as Affected by Starter Cultures. *J. Dairy Sci.* **2021**, *104*, 270–280. [CrossRef]
52. Cakmakci, S.; Gundogdu, E.; Hayaloglu, A.A.; Dagdemir, E.; Gurses, M.; Cetin, B.; Tahmas-Kahyaoglu, D. Chemical and Microbiological Status and Volatile Profiles of Mouldy Civil Cheese, a Turkish Mould-Ripened Variety. *Int. J. Food Sci. Technol.* **2012**, *47*, 2405–2412. [CrossRef]
53. Boltar, I.; Čanžek Majhenič, A.; Jarni, K.; Jug, T.; Bavcon Kralj, M. Volatile Compounds in Nanos Cheese: Their Formation during Ripening and Sesonal Variation. *J. Food Sci. Technol.* **2015**, *52*, 608–623. [CrossRef]

54. Teter, A.; Barłowska, J.; Król, J.; Brodziak, A.; Rutkowska, J.; Litwińczuk, Z. Volatile Compounds and Amino Acid Profile of Short-Ripened Farmhouse Cheese Manufactured from the Milk of the White-Backed Native Cow Breed. *LWT* **2020**, *129*, 109602. [CrossRef]
55. Niimi, J.; Eddy, A.I.; Overington, A.R.; Silcock, P.; Bremer, P.J.; Delahunty, C.M. Sensory Interactions between Cheese Aroma and Taste. *J. Sens. Stud.* **2015**, *30*, 247–257. [CrossRef]

foods

MDPI

Article

Application of *Ligilactobacillus salivarius* SP36, a Strain Isolated from an Old Cheese Seal, as an Adjunct Culture in Cheesemaking

Ramón Arias [1,*], Claudio Alba [2], Javier Calzada [3], Lorena Jiménez [1], Leónides Fernández [4], Marta Ávila [3], Mario Roman [5], Juan Roman [5], Juan M. Rodríguez [2] and Sonia Garde [3,*]

1 Centro Regional de Selección y Reproducción Animal (CERSYRA), Instituto Regional de Investigación y Desarrollo Agroalimentario y Forestal de la Junta de Comunidades de Castilla-La Mancha (IRIAF), CERSYRA Valdepeñas, 13300 Ciudad Real, Spain; ljimenez@quesomanchego.es

2 Department of Nutrition and Food Science, Complutense University of Madrid, 28040 Madrid, Spain; c.alba@ucm.es (C.A.); jmrodrig@ucm.es (J.M.R.)

3 Departamento de Tecnología de Alimentos, Instituto Nacional de Investigación y Tecnología Agraria y Alimentaria (INIA-CSIC), Carretera de La Coruña km 7, 28040 Madrid, Spain; jcalzada@inia.csic.es (J.C.); arribas@inia.csic.es (M.Á.)

4 Department of Galenic Pharmacy and Food Technology, Complutense University of Madrid, 28040 Madrid, Spain; leonides@ucm.es

5 QUALIAM s.l., Velazquez 119, 28003 Madrid, Spain; mroman@qualiam.es (M.R.); juanromanacacio@gmail.com (J.R.)

* Correspondence: rarias@jccm.es (R.A.); sgarde@inia.csic.es (S.G.); Tel.: +34-926276661 (R.A.); +34-913474900 (S.G.)

Abstract: Adjunct cultures originating from artisanal cheese environments may play an important role in recreating and developing traditional cheese flavours, thanks to their enzymatic activities, involved in different metabolic pathways that occur during cheese ripening. In this work, *Ligilactobacillus salivarius* SP36, a strain isolated from an old cheese seal, was added as an adjunct culture to the cheese's raw milk, and its effect on the microbiological, physical–chemical and sensory characteristics of the cheese was studied. The use of *L. salivarius* SP36 in cheese manufacturing had no significant ($p > 0.05$) effect on the cheese microbiota, gross composition (fat percentage, protein, total solids, moisture and NaCl concentration), colour or texture of the cheese. However, *L. salivarius* SP36 increased ($p < 0.01$) the formation of 25 volatile compounds, including 10 esters, 1 aldehyde, 8 alcohols and 6 carboxylic acids. In addition, cheeses made with *L. salivarius* SP36 received higher scores ($p < 0.01$) for aroma intensity and quality than control cheeses. *L. salivarius* SP36 proved to be a good candidate as an adjunct culture for cheesemaking, since it improved the cheese flavour by making it more intense and recovering typical sensorial notes of traditional cheeses.

Keywords: traditional cheese; *Ligilactobacillus salivarius*; adjunct culture; sensory properties

1. Introduction

The control on how the microbial communities evolve during cheese manufacturing and ripening is essential for modern cheesemaking [1,2]. As a consequence, cheese companies relies in the inoculation of milk with cultures composed of well-selected microorganisms, including starter cultures, when they are required to catalyse the fast initial acidification, and adjunct cultures, when they exert a noticeable influence on maturing activities, including the development of the sensorial properties that may characterize a certain type of cheese [3,4]. Starter cultures usually contain strains of *Lactococcus lactis* and/or *Streptococcus thermophilus*, while adjunct cultures may include a wide variety of species belonging to some of the genera in which the genus *Lactobacillus* was recently reclassified [5] and other lactic acid bacteria (LAB).

The microbiota of traditional raw-milk cheeses has gained a renewed interest in the search for LAB strains with relevant genomic and phenotypic traits for food and biotechnological applications [6–8]. In the dairy context, such strains could complement current commercial strains in order to recover or preserve the typical sensorial properties of the traditional cheeses or to provide novel dairy products [9–11]. Improvement of the organoleptic features of a given cheese may increase consumer's attraction and commercial value.

Many traditional cheeses have disappeared in recent decades, and many of those surviving at present are still manufactured by small familiar companies on a small scale for local or tourism-related sales. The number of those that do not employ starter cultures (and, therefore, which maturation relies on indigenous microorganisms, often applied using back-slopping approaches) is declining rapidly. In contrast, large-scale cheese companies employ pasteurized milk inoculated with starter cultures that repress the growth of the indigenous microorganisms that may arise from milk or from the dairy environment. Therefore, the biodiversity of the microbiota of industrial cheeses is reduced and often limited to or overwhelmingly dominated by the inoculated species and strains [12]. In turn, it leads to a loss of some of the organoleptic qualities linked to traditional cheeses. It has been repeatedly suggested that isolating and characterizing strains from naturally fermented traditional cheeses may help to provide new cultures that could be used by either small/traditional and large/industrial dairy companies to produce more safe or improved cheeses, respectively [13,14].

Recently, a *Ligilactobacillus salivarius* strain (*L. salivarius* SP36) was isolated from an old seal used to label the cheeses made by a producer living in a Pyrenean village [15]. In vitro assessment showed that the strain was safe and had some functional properties, including antimicrobial activity against cheese-related spoilage and foodborne microbes. The study of its genome revealed its potential for the production of organoleptic-related compounds. In this context, the objective of this work was to compare the performance of pressed cheese elaborated with a commercial starter culture and that of equivalent cheeses in which *L. salivarius* SP36 was added as an adjunct culture in addition to the starter culture.

2. Materials and Methods

2.1. Cheese Manufacture

Two batches of experimental cheeses were made in duplicate, using in the first type of preparation a commercial starter (CS) and in the second type the same commercial starter together with the strain *L. salivarius* SP36 as an adjunct culture (LS). The pressed cheeses were made with 80 litres of raw sheep bulk-tank milk, which was transported in less than 1 h and under refrigeration conditions (6 °C) from the farm to the CERSYRA cheese plant. In the case of the CS cheeses, the milk in agitation in the curd vat at 30 °C was inoculated with a commercial starter culture containing *Lactococcus lactis* subsp. *lactis*, *Lactococcus lactis* subsp. *cremoris*, *Lactococcus lactis* subsp. *lactis* biovar. *diacetylactis* and *Streptococcus thermophilus* (CHOOZIT MA4001 LYO 5 DCU, Danisco, France) at a dose of 4 DCU/80 L milk. In the case of the LS cheeses, the milk was inoculated with the same commercial starter culture at the same dose and at the same temperature (30 °C) but, also, with *Ligilactobacillus salivarius* SP36, a strain previously isolated from an old cheese seal [15] (50 g containing 9 log cfu/g in 80 L of milk). After 30 min, rennet was added (185 IMCU/mL, Laboratorios Arroyo, Santander, Spain), following the routine methodology of the cheese industry (30 °C, 40 min). Next, the curd was cut and mixed (30 °C, 10 min), and the temperature was gradually increased from 30 °C to 36 °C for 15 min. After placing the curd in the 3 kg cheese moulds, pressing was carried out by exerting a pressure of 2–3 kPa for 5–6 h at 15 °C and turning the cheeses every hour up to pH < 5.4. The cheeses were salted in a 20% NaCl brine pH = 5.2–5.4) for 15 h at 8 °C. The average cheese yield was 0.23 kg/L of milk. The cheeses were dried at 7 °C with a relative humidity of 90% for 10 d, and ripening was performed at 10 °C with a relative humidity of 80% for 240 d. In all the elaborations, there was a similar decrease in the weight (~750 g;

~20%) between 1 and 240 days after manufacture. Cheeses were analysed at 1, 60, 120, 180 and 240 d, classifying them as young (1 d), hard (60 and 120 d) and old cheeses (180 and 240 d) according to Spanish legislation.

2.2. Physico-Chemical and Microbiological Analyses of Milk

The physico-chemical characteristics (percentage of fat and protein) of the bulk-tank milk were analysed using MilkoScan equipment (Foss Electric, Hillerød, Denmark) and pH with pHmeter basic Crison (Barcelona, Spain). The microbiological composition of milk was assessed using PCA plates (Oxoid, Basingstoke, UK) for mesophilic bacteria counts and MRS plates (Oxoid) for lactic acid bacteria. Plates were incubated at 32 °C for 48 h.

2.3. Physico-Chemical and Microbiological Analyses of Cheeses

Cheeses were sampled at days 1, 60, 120, 180 and 240. The physico-chemical parameters (fat, protein, total solids and moisture, as well as NaCl concentration) were measured in duplicate with the near-infrared spectroscopy FOOD SCAN LAB (Foss Electric, Hillerød, Denmark), provided with ISISCAN 4.6 software. Each sample consisted of 50 g of crushed cheese collected from the central area of the cheese. At each sampling time, cheese aliquots were added to a sterile peptone solution and homogenized in a stomacher (IUL Instrument, Barcelona, Spain). Proper decimal dilutions of the homogenized samples were seeded in duplicate in MRS plates supplemented with L-cysteine (0.05%, *w*/*v*). MRS plates were incubated aerobically at 30 °C for 48 h for the selective enumeration of *L. lactis* and anaerobically at 37 °C for 48 h for the selective enumeration of *Lactobacillus* (*sensu lato* [5]). PCA plates were incubated at 42 °C for 48 h for the selective enumeration of *S. thermophilus*. Microbial counts were expressed in $\log_{10}$ cfu/g. Cheese pH was also measured at each sampling point.

Subsequently, a metataxonomic analysis of cheeses samples collected at days 1, 120 and 240 was performed. DNA was extracted as described previously [16]. A dual-barcoded 2-step PCR reaction was conducted to amplify a fragment of the V3–V4 hypervariable region of the bacterial 16S ribosomal RNA (rRNA) gene. Equimolar concentrations of the universal primers S-D-Bact-0341-b-S-17 (ACACTGACGACATGGTTCTACACC-TACGGGNGGCWGCAG) and S-D-Bact-0785-a-A-21 (TACGGTAGCAGAGACTTGGTCT-GACTACHVGGGTATCTAATCC) were used. Barcodes used for Illumina sequencing were appended to the 30 and 50 terminal ends of the PCR amplicons to allow for the separation of forward and reverse sequences. A bioanalyser (2100 Bioanalyzer, Agilent, Santa Clara, CA, USA) was used to determine the concentration of each sample. Barcoded PCR products from all samples were pooled at approximately equimolar DNA concentrations and run on a preparative agarose gel. The correctly sized band was excised and purified using a QIAEX II Gel Extraction Kit (Qiagen, Hilden, Germany) and then quantified with PicoGreen (BMG Labtech, Jena, Germany). Finally, 1 aliquot of pooled, purified and barcoded DNA amplicons was sequenced using the Illumina MiSeq pair-end protocol (Illumina Inc., San Diego, CA, USA) at the facilities of the Scientific Park of Madrid (Spain). The sequences analysed in this study are available in the BioSample database of the National Center for Biotechnology Information under accession number PRJNA1081834. The amplified fragments and results were taxonomically analysed using the Illumina™ software according to the manufacturer's guidelines and pipelines (version 2.6.2.3).

Taxonomic affiliation for individual amplicon sequence variants (ASVs) was ascertained through the q2 feature classifier, utilizing a naïve Bayes taxonomic classification schema against the SILVA 138.1 reference database. A matrix detailing ASVs specific to each respective sample was systematically generated. The subsequent normalization of bacterial taxa abundances was performed utilizing the total sum scaling (TSS) method, wherein each discrete ASV count was divided by the cumulative library size, thereby rendering the relative proportion of individual ASV counts per sample. For the examination of alpha diversity, both the Shannon and Richness diversity indices were employed, facilitated through the vegan package within R Statistical Computing Environment (Version 2.5.6).

2.4. Colour and Texture Analyses of Cheeses

The colour values of the cheeses were measured using the CIELAB colour space. Variables included lightness (L*), red/green value (a*) and blue/yellow value (b*), which were measured using a spectrophotometer CM-2300d (Konika Minolta Sensing, Inc., Tokyo, Japan). Texture profile analyses (TPAs) were performed to determine the values of hardness, adhesiveness, cohesiveness, gumminess, chewiness, springiness and resilience, following the method proposed by Romeith et al. [17]. A food texture analyser equipped with a 30 cm diameter cylindrical probe was used (Shimadzu, Kyoto, Japan). These internal cheese samples had a cylindrical shape with 11 mm of diameter and 15 mm of height and, immediately after their collection, were covered with plastic film to prevent desiccation and allowed to warm at room temperature (20 ± 2 °C) for 30 min before analysis. These analyses were carried out in quadruplicate.

2.5. Analysis of Volatile Compounds

Cheese pieces wrapped in aluminium foil were vacuum packed and frozen at −40 °C at 60, 120, 180 and 240 d of ripening, until analysis. Volatile compounds of cheeses were extracted by automated solid-phase microextraction (SPME) and analysed by gas chromatography–mass spectrometry (GC-MS) as described previously [18]. All determinations were carried out in duplicate. The detection, identification and semi-quantification of volatile compounds were carried out as described by Garde et al. [19]. Mass detection was performed in the scan mode, from 33 to 300 amu at 5.16 scan/s, and ionization by EI at 70 eV. Data were collected with the HP ChemStation program (G2070-90107), and volatile compounds were identified by comparison of spectra with the Wiley 275 library and by comparison of their retention times with authentic standards (Sigma Chemical Co., St. Louis, MO, USA). Relative abundances of compounds were expressed as percentages of their peak areas on the peak area of the internal standard cyclohexanone.

2.6. Sensory Analysis of the Cheeses

Two different sensorial analyses were performed during this work. The first one was carried on cheeses from 60 days of ripening, following the methodology of the ISO Standard N° 13299:2016 [20] and PDO Manchego Cheese Sensory Guide (Procedure PAS-02). On the other hand, a descriptive test was developed for cheeses based on the guidelines of hard and semi-hard cheeses given by Bérodier et al. [21]. Eight trained panelists evaluated the cheeses at 60, 120, 180 and 240 days of ripening for quality (overall acceptance) and intensity (overall intensity) of odour, aroma and taste, and for texture and colour quality on a 0–10-point scale. Panelists were also asked to assign a score on a 1- (very slight perception) to 7-point (very intense) scale to the taste attributes "sour", "bitter", "sweet", "salty" and "umami", and separately for odour and aroma attributes belonging to six families, namely, "lactic" (including the descriptors "milky", "buttery", "yogurt-like" and "cheesy"); "vegetal"; "fruity-flowery" (including the descriptors "fruity" and "nutty"); "toasted" (including the descriptors "caramel" and "toasted"); "animal" (including the descriptor "sheepy"); and "others" (including the descriptor "pungent"). Informed consent was obtained from the sensory participants, which were apprised about the samples, protocols and their rights as participants.

Prior to sensory evaluation, cheeses were held for 3 h at 20 to 22 °C. Then, they were cut in representative triangular slices (15–20 g) after removing the rind. Slices of the two cheeses per session from each of the two vats manufactured on the same day, coded with random three-digit numbers, were presented to panelists in random order. Bread and water were provided to cleanse the mouth between cheeses.

2.7. Statistical Analysis

Cheeses were manufactured in duplicate experiments on different days, and all determinations were performed in duplicate. Statistical treatment of data was performed by SPSS software (IBM SPSS Statistics version 29, IBM Corp., Armonk, NY, USA). Data were analysed by ANOVA using a general linear model, and comparison of means was carried out with a least significant difference (LSD) test.

3. Results and Discussion

3.1. Compositional and Microbiological Characteristics of Milk

The bulk-tank milk used had the following compositional characteristics: pH = 6.51, fat = 8.57% and protein = 5.85%. Counts on PCA and MRS agar plates were <5 log cfu/mL.

3.2. Microbiological Characteristics of the Cheeses

The populations of *L. lactis* and *S. thermophilus* decreased gradually over the ripening time, but no differences ($p > 0.05$) were found between the CS and the LS cheeses in relation to their counts throughout the ripening period (Table 1). The concentration of lactobacilli decreased significantly between day 1 and day 60 of ripening, from 8.72 to 6.54 log cfu/g while it remained stable from day 60 to the end of ripening, and no significant differences were found throughout the rest of ripening.

Table 1. Counts of starter microorganisms (*L. lactis* and *S. thermophilus*) and *Lactobacillus* (mean ± SD log cfu/g) in the ewes' milk cheeses made in this study.

Bacterial Species	Days	Type of Cheese	
		CS	LS
L. lactis	1	9.22 ± 0.02 [a]	9.27 ± 0.02 [a]
	60	8.58 ± 0.37 [b]	8.28 ± 0.03 [b]
	120	8.46 ± 0.02 [b]	8.42 ± 0.08 [b]
	180	8.02 ± 0.03 [c]	8.06 ± 0.08 [b]
	240	7.75 ± 0.26 [d]	7.77 ± 0.23 [c]
S. thermophilus	1	9.19 ± 0.16 [a]	8.83 ± 0.31 [a]
	60	8.57 ± 0.81 [b]	8.69 ± 0.12 [a]
	120	8.27 ± 0.13 [b]	8.16 ± 0.11 [a]
	180	7.76 ± 0.25 [b]	7.27 ± 0.02 [a]
	240	7.04 ± 0.23 [c]	6.70 ± 0.11 [a]
Lactobacillus	1	<3	8.72 ± 0.24 [a]
	60	<3	6.54 ± 0.43 [b]
	120	<3	6.47 ± 0.55 [b]
	180	<3	6.85 ± 0.45 [b]
	240	<3	6.85 ± 0.42 [b]

[a–d]: Different letters in the same column indicate significant statistical differences for days of ripening ($p < 0.05$). *Lactobacillus* counts correspond to *L. salivarius*.

In relation to the metataxonomic analysis of three CS cheeses at days 1, 120 and 240 and three LS cheeses at the same sampling times, the sequencing of the six cheese samples yielded 1,502,505 high-quality reads (median = 237,910.5 reads/sample, ranging from 200,787.0 to 284,112.8 reads/sample). The two diversity indices evaluated in this work (Shannon and Simpson indices) increased over time in both types of cheeses. For CS cheeses, the Shannon indices were 0.53, 0.66 and 0.99 at days 1, 120 and 240, respectively, while the values at the same sampling times where 0.60, 0.77 and 0.97 for LS cheeses. In relation to the Simpson index, the values were 0.33, 0.32 and 0.54 at days 1, 120 and 240, respectively, among the CS samples and 0.35, 0.36 and 0.49, respectively, among the LS ones. No differences were found when CS and LS samples were compared within each sampling time, but there was a significant increase for both diversity indices and both types of cheeses when samples of days 1 and 240 were compared ($p < 0.01$). It has been previously found that diversity in the microbiota of cheeses is a major driver in the organoleptic characteristics found in raw-milk cheeses [22].

The genera detected in the metataxonomic analysis of the samples are shown in Table 2. Among them, the genus Lactococcus was the most abundant in both cheese types although its concentration decreased from day 1 to day 240. The genus Streptococcus was the second genus in order of abundance in both cheese types at days 1 and 120, and its percentage also decreased throughout the ripening period. The genus *Ligilactobacillus* was present from the beginning in the LS cheeses and its relative abundance increased during ripening, probably because of the decrease in the relative abundances of genera *Lactococcus* and *Streptococcus*. In contrast, *Ligilactobacillus* sequences were sparsely detected in CS cheeses. Interistingly, *Lactiplantibacillus* sequences, probably with an environmental origin, were not detected on day 1 in any of the cheese types but its abundance increased over time, and in fact, it was the second most abundant genus in both cheese types at the end of the ripening period. *Lactiplantibacillus plantarum* and *Lactiplantibacillus pentosus* are common and often dominant members of the non-starter LAB communities in cheeses [12,23,24]. These microbes are present at low or even not detectable concentrations in the curd, but their levels increase notably during ripening [24], which is in agreement with the results of this study. Despite the great abundance of *Lactiplantibacillus* sequences in both cheeses, these microorganisms were not detected in the microbiological analyses, probably because they do not grow in the medium used and/or the conditions tested.

Table 2. Most abundant genera detected in the metataxonomic analysis performed for the ewes' milk cheeses made in this study.

	Type of Cheese					
	CS			LS		
Genera	Day 1	Day 120	Day 240	Day 1	Day 120	Day 240
Lactococcus	79.20	81.31	62.86	77.81	78.86	68.72
Streptococcus	20.60	10.51	7.27	20.37	13.45	9.64
Lactiplantibacillus	0.00	7.22	18.61	0.00	4.53	17.84
Ligilactobacillus	0.03	0.02	0.03	1.68	1.12	2.24
Minor genera	0.17	0.95	1.23	0.14	2.03	1.56

3.3. Compositional Characteristics of Cheeses

The physical–chemical characteristics of both cheese types were within the ranges established in the regulation of the differentiated quality figure for Manchego cheese [25]. No differences in the percentage of fat, protein, moisture and amount of NaCl were observed when both groups (LS and CS) were compared (Table S1). With respect to pH, the values were significantly higher in LS cheeses on 1 day (5.38 vs. 5.09) and on day 60 (5.23 vs. 5.16), but no differences were detected in the remaining sampling times (120, 180 and 240 days) ($p > 0.05$). There was no significant difference in colour parameters between both types of cheeses although the values of red index (a*) and yellow index (b*) were higher in the LS cheeses.

3.4. Textural Properties of Cheeses

The texture profiles of the cheeses at different stages of ripening are presented in Table S2. No significant differences were observed for these parameters between the CS and LS cheeses, although the evolution of descriptors was slightly less marked in the LS ones. The texture of the cheeses changed as the ripening process progressed, due to several biochemical events, as indicated by Lucey et al. [26]. Hardness of the cheeses was lower on day 1 than on day 60 of ripening for the two types of cheeses, with significant differences regarding LS cheeses, although the values tended to stabilize from then to day 240 of ripening. Adhesiveness also increased as cheese ripening progressed. In contrast, cohesiveness decreased with time. Gumminess of the cheeses decreased significantly as ripening progressed. This behaviour is similar to that described by Salvador et al. [27] in pressed cheeses. Chewiness showed significant differences among CS cheeses throughout ripening, with a maximum level at 60 days, then decreasing and increasing again at the end of ripening. On the contrary, the chewiness in LS cheeses evolved with a gradual decrease throughout the ripening process. Finally, springiness and resilience of cheeses shared similar values in the samples obtained up to 120 days of ripening in both cheese types but then they increased slightly until the end of ripening, with significant differences when CS cheeses at day 240 were compared with the same CS cheeses in the previous sampling times.

3.5. Analysis of Volatile Compounds

A total of 82 volatile compounds were identified by SPME/GC-MS in the cheeses, including 5 hydrocarbons, 5 sulphur compounds, 4 aldehydes, 9 ketones, 11 esters, 20 alcohols, 6 benzene compounds and 12 carboxylic acids. The use of *L. salivarius* SP36 as an adjunct culture significantly increased ($p < 0.01$) the formation of 25 of these volatile compounds (Tables 3–5).

Table 3. Esters [a] significantly affected by the use of *L. salivarius* SP36 as an adjunct culture in the ewes' milk cheeses made in this study.

	Time (d)	Type of Cheese	
		CS	LS
Ethyl acetate	60	4.44 ± 0.47 [a]	14.90 ± 9.58 [b]
	120	7.46 ± 0.44 [a]	18.09 ± 9.57 [b]
	180	11.79 ± 2.44 [a]	22.64 ± 6.97 [b]
	240	20.57 ± 6.10 [a]	52.27 ± 14.87 [b]
Ethyl propanoate	60	ND [a]	0.23 ± 0.02 [b]
	120	0.63 ± 0.26 [a]	1.23 ± 0.20 [b]
	180	1.54 ± 0.36 [a]	2.70 ± 0.29 [b]
	240	2.13 ± 0.21 [a]	4.98 ± 1.02 [b]
Ethyl butanoate	60	4.84 ± 0.67	7.23 ± 1.86
	120	10.49 ± 1.51 [a]	19.02 ± 4.13 [b]
	180	28.95 ± 8.12 [a]	37.56 ± 3.05 [b]
	240	33.93 ± 4.24 [a]	69.41 ± 4.60 [b]
Ethyl valerate	60	ND	ND
	120	ND	ND
	180	ND [a]	0.10 ± 0.01 [b]
	240	0.11 ± 0.02 [a]	0.28 ± 0.04 [b]
Ethyl hexanoate	60	1.83 ± 0.24	2.60 ± 0.77
	120	4.57 ± 0.84 [a]	7.17 ± 1.80 [b]
	180	11.35 ± 1.39 [a]	15.44 ± 3.24 [b]
	240	14.70 ± 2.14 [a]	28.50 ± 5.74 [b]
Ethyl octanoate	60	ND [a]	0.16 ± 0.07 [b]
	120	0.33 ± 0.06 [a]	0.61 ± 0.18 [b]
	180	1.07 ± 0.23	1.63 ± 0.73
	240	1.21 ± 0.13 [a]	2.30 ± 0.53 [b]
Propyl acetate	60	0.40 ± 0.25	0.78 ± 0.25
	120	3.12 ± 0.62 [a]	7.92 ± 3.08 [b]
	180	6.42 ± 2.99 [a]	20.35 ± 3.82 [b]
	240	9.12 ± 3.13 [a]	39.40 ± 13.75 [b]
Propyl butanoate	60	ND [a]	0.13 ± 0.07 [b]
	120	0.73 ± 0.16 [a]	1.87 ± 0.11 [b]
	180	2.85 ± 0.53 [a]	6.85 ± 0.57 [b]
	240	4.21 ± 0.75 [a]	12.12 ± 0.36 [b]
Butyl acetate	60	0.17 ± 0.08 [a]	0.90 ± 0.86 [b]
	120	0.63 ± 0.22 [a]	2.28 ± 1.78 [b]
	180	1.79 ± 0.50 [a]	5.13 ± 3.32 [b]
	240	2.88 ± 0.76 [a]	6.75 ± 2.81 [b]

Table 3. *Cont.*

	Time (d)	Type of Cheese CS	LS
1-Methylpropyl acetate	60	ND	ND
	120	1.48 ± 0.81 [a]	6.60 ± 5.46 [b]
	180	2.45 ± 1.13	10.91 ± 10.76
	240	4.34 ± 2.83	14.25 ± 11.82 [b]

[a] Results are mean (n = 4) ± SD of duplicate determinations in two cheese trials, expressed as relative abundances to internal standard cyclohexanone. Means in the same row followed by different letters differ significantly ($p < 0.01$). ND, not detected.

Table 4. Aldehydes and alcohols [a] significantly affected by the use of *L. salivarius* SP36 as an adjunct culture in the ewes' milk cheeses made in this study.

	Time (d)	Type of Cheese CS	LS
3-Methyl butanal	60	0.20 ± 0.16	0.33± 0.05
	120	0.47 ± 0.22 [a]	0.95 ± 0.18 [b]
	180	0.80 ± 0.39	1.03 ± 0.10
	240	0.63 ± 0.08 [a]	1.20 ± 0.16 [b]
Ethanol	60	125.02 ± 9.86	164.67 ± 18.85
	120	121.73 ± 14.42	145.50 ± 17.18
	180	138.59 ± 25.01	158.20 ± 17.02
	240	130.64 ± 25.84 [a]	221.30 ± 20.58 [b]
1-Propanol	60	7.48 ± 3.13 [a]	19.36 ± 11.65 [b]
	120	32.27 ± 2.81 [a]	65.70 ± 3.07 [b]
	180	56.64 ± 1.88 [a]	97.05 ± 4.73 [b]
	240	52.63 ± 6.49 [a]	116.35 ± 10.61 [b]
1-Butanol	60	14.35 ± 2.73	23.57 ± 14.35
	120	17.94 ± 3.08	22.34 ± 5.88
	180	37.46 ± 6.50	42.64 ± 4.78
	240	55.33 ± 14.07 [a]	87.83 ± 6.63 [b]
1-Hexanol	60	4.93 ± 1.63	17.91 ± 18.53
	120	4.04 ± 1.55 [a]	10.51 ± 7.82 [b]
	180	8.20 ± 1.54 [a]	13.38 ± 9.67 [b]
	240	8.88 ± 2.14 [a]	15.53 ± 7.83 [b]
1-Octanol	60	ND	ND
	120	0.10 ± 0.03 [a]	0.25 ± 0.16 [b]
	180	0.28 ± 0.05 [a]	0.54 ± 0.40 [b]
	240	0.37 ± 0.11 [a]	1.07 ± 0.79 [b]
2-Pentanol	60	6.10 ± 2.40 [a]	14.57 ± 2.96 [b]
	120	19.76 ± 2.42 [a]	33.29 ± 3.24 [b]
	180	51.34 ± 10.38	48.25 ± 11.11
	240	77.16 ± 21.24	87.72 ± 16.01
1,2-Propanediol	60	0.39 ± 0.26 [a]	1.56 ± 0.87 [b]
	120	0.62 ± 0.26 [a]	4.01 ± 2.12 [b]
	180	2.94 ± 1.19 [a]	8.17 ± 5.98 [b]
	240	2.16 ± 0.16 [a]	14.39 ± 2.86 [b]
2-Furanmethanol	60	ND [a]	0.31 ± 0.03 [b]
	120	ND [a]	0.33 ± 0.03 [b]
	180	0.09 ± 0.01 [a]	0.39 ± 0.04 [b]
	240	0.10 ± 0.02 [a]	0.46 ± 0.02 [b]

[a] Results are mean (n = 4) ± SD of duplicate determinations in two cheese trials, expressed as relative abundances to internal standard cyclohexanone. Means in the same row followed by different letters differ significantly ($p < 0.01$). ND, not detected.

Table 5. Carboxylic acids [a] significantly affected by the use of *L. salivarius* SP36 as an adjunct culture in the ewes' milk cheeses made in this study.

	Time (d)	Type of Cheese	
		CS	LS
Propanoic acid	60	5.64 ± 1.38 [a]	12.70 ± 2.05 [b]
	120	59.80 ± 11.92 [a]	91.62 ± 10.60 [b]
	180	130.20 ± 22.29	158.48 ± 35.51
	240	131.06 ± 41.38 [a]	186.09 ± 35.83 [b]
2-Methyl propanoic acid	60	1.44 ± 0.37 [a]	2.84 ± 1.66 [b]
	120	2.13 ± 0.54	2.51 ± 0.69
	180	3.54 ± 0.12	3.58 ± 0.28
	240	4.48 ± 1.87 [a]	6.22 ± 1.41 [b]
2-Methyl butanoic acid	60	0.37 ± 0.24	0.54 ± 0.33
	120	0.47 ± 0.13	0.61 ± 0.24
	180	1.22 ± 0.24	0.87 ± 0.03
	240	1.09 ± 0.50 [a]	1.70 ± 0.46 [b]
3-Methyl butanoic acid	60	0.59 ± 0.46	0.76 ± 0.40
	120	0.53 ± 0.15	0.74 ± 0.35
	180	1.34 ± 0.27	0.98 ± 0.05
	240	1.17 ± 0.50 [a]	1.88 ± 0.64 [b]
Pentanoic acid	60	0.76 ± 0.10	0.78 ± 0.08
	120	1.11 ± 0.20	1.27 ± 0.12
	180	2.26 ± 0.57	1.99 ± 0.09
	240	2.85 ± 0.79 [a]	3.69 ± 0.57 [b]
Hexanoic acid	60	79.00 ± 1.87	75.02 ± 7.90
	120	107.09 ± 19.96	110.46 ± 12.34
	180	208.83 ± 44.43	175.16 ± 12.97
	240	229.55 ± 64.23 [a]	293.97 ± 43.48 [b]

[a] Results are mean (n = 4) ± SD of duplicate determinations in two cheese trials, expressed as relative abundances to internal standard cyclohexanone Means in the same row followed by different letters differ significantly ($p < 0.01$).

L. salivarius favoured the formation of 6 ethyl-, 2 propyl-, 1 butyl- and 1 branched esters and their relative abundances increased significantly ($p < 0.001$) during LS cheese ripening (Table 3). The biosynthesis of aroma-active esters in dairy systems proceeds through two enzymatic mechanisms: esterification, a reaction in which esters are formed from alcohols and carboxylic acids, and alcoholysis, a transferase reaction in which fatty acyl groups from acylglycerols and acyl-CoA derivatives are directly transferred to alcohols. Evidence has been provided that esterases of lactic acid bacteria catalyse not only the hydrolysis of milk fat glycerides to release free fatty acids but also the synthesis of esters from glycerides and alcohols via alcoholysis in cheese [28]. The observed differences in ester concentration among cheeses could be related to the different lactic acid bacteria compositions; it has been described that *L. salivarius* strains isolated from human milk displayed esterase activities on most assayed substrates when tested for dairy technological properties [29]. Esters, which usually increase during cheese ripening, have a very low perception threshold, playing an important role in the aroma profiles of different cheese varieties [30–34]. Most esters encountered in cheese are described as having sweet, fruity and floral notes [30,34].

Cheeses made with the *L. salivarius* strain showed higher concentrations of 3-methyl butanal than control (CS) cheeses (Table 4). This branched-chain aldehyde originates from leucine by transamination, leading to an intermediate imide that can be decarboxylated, or by Strecker degradation, and has been identified as a potent odourant in several cheese varieties [30]. A high aminopeptidase activity has been described in *L. salivarius* strains, including Leu-pNA as a substrate [29]. 3-Methyl butanal presents a green malty odour, but at low concentration, the odour becomes fruity and rather pleasant.

The levels of 1-propanol, 1-hexanol, 1-octanol, 1,2-propanediol and 2-furanmethanol were significantly ($p < 0.01$) higher in cheeses made with *L. salivarius* SP36 throughout ripening (Table 4). However, the levels of ethanol and 1-butanol were only higher ($p < 0.01$) at the end of ripening, while the levels of 2-pentanol were higher ($p < 0.01$) at 60 and 120 d in cheeses made with *L. salivarius*. The relative abundances of all these alcohols increased ($p < 0.01$) with cheese age. Many metabolic pathways are involved in the biosynthesis of the alcohols commonly found in cheese, such as lactose metabolism, methyl ketone reduction, amino acid metabolism, as well as degradation of linoleic and linolenic acids [30]. In general, primary alcohols have a high perception threshold, so scarcely contribute to the aroma of cheese. However, they are the limiting factor in ester formation in hard cheeses [34]. With respect to 1,2-propanediol, Fröhlich-Wyder et al. [35] described the production of this compound when *Lactobacillus buchneri* and *Lactobacillus parabuchneri* were used as adjunct cultures in the manufacture of model cheese. On the other hand, 2-furanmethanol has been found to contribute to the nutty and roasted aroma of Parmigiano-Reggiano cheese [36].

The use of *L. salivarius* SP36 as an adjunct culture in cheese manufacturing also promoted the formation of six carboxylic acids at the end of ripening ($p < 0.01$) (Table 5). As with the rest of volatile compounds mentioned above, the concentrations of these acids increased ($p < 0.001$) with cheese age. During cheese ripening, most of the acids having between two and six carbon atoms originate from the degradation of lactose and amino acids, but they can also be derived from ketones, esters and aldehydes by oxidation [30]. Short-chain carboxylic acids have low perception thresholds and are major odourants of different cheese varieties. Propanoic acid has a typical vinegar odour, and pentanoic and hexanoic acids contribute significantly to the aroma of aged cheeses [31–33]. On the other hand, branched-chain carboxylic acids such as 2-methyl propanoic acid and 2- and 3-methyl butanoic acids are characteristic odour impact compounds of goat and sheep cheeses. They derived from valine, isoleucine and leucine, respectively [30].

To our knowledge, this is the first time that a *L. salivarius* strain has been used as an adjunct culture in the manufacture of sheep cheeses. In a previous work, the potential of two human milk *L. salivarius* strains in fresh model cheeses to develop a probiotic cheese was evaluated [37]. However, the effect of these strains on the volatile profile of cheeses was limited. More recently, the use of *L. salivarius* AR809 as an adjunct in the manufacture of Monascus-ripened cheese (a mould-ripened cheese) significantly promoted the formation of alcohols, acids and ketones but reduced the formation of certain ethyl esters [38].

3.6. Sensory Evaluation

The evaluation of the sensorial properties of the cheeses from 60 days of ripening was initially performed following the methodology of the ISO Standard N° 13299:2016 and the PDO Manchego Cheese Sensory Guide (Procedure PAS-02). This evaluation concluded that, in both types of cheeses, the external descriptors (shape, physical integrity, colour of the rind, uniformity of the engraving of the pleita, uniformity of the engraving of the flower, hardness of the rind, degree of rind distinction, uniformity of the colour of the cheese paste, colour of the cheese paste, distribution and size of the eyes, and integrity of the paste) as well as the internal descriptors (odour, texture and flavour) were within the ranges typical of Manchego cheese. To the appreciation of the trained panelists, both cheeses had similar sensorial properties overall (Figure S1), but LS cheeses had more complex flavours than CS ones. This is a logical finding since CS cheeses were made with a starter culture that is well tested by dairy companies, including those producing Manchego cheeses, and selected to provide cheese uniformity, which is an essential issue for consumers' reliability. Therefore, it is not strange that the addition of a *L. salivarius* strain with the ability to produce sensorial-related compounds led to the appreciation of "wilder" flavours; spicier touches; and greater firmness, microstructure and friability (Figure S1).

Subsequently, the cheeses were also evaluated and compared using a descriptive test based on the guidelines of Bérodier et al. [21]. The cheeses were evaluated for taste, odour and aroma quality and intensity, and for texture and colour quality, throughout ripening (Table 6). The taste intensity and aroma intensity, and texture quality ($p < 0.01$) of the cheeses increased significantly with cheese age ($p < 0.001$), whereas taste quality, odour quality, aroma quality and colour quality, and odour intensity remained stable. Odour intensity and quality scores were higher in cheeses made with the *L. salivarius* strain, but significant differences were not detected ($p > 0.01$). However, the panelists gave significantly higher aroma intensity scores to cheeses made with *L. salivarius* throughout ripening ($p < 0.01$) (Table 6). Aroma quality scores were also higher for cheeses made with *L. salivarius* and were significantly higher after 120 and 240 d of ripening ($p < 0.01$). The higher aroma intensity and aroma quality scores obtained for cheeses made with the *L. salivarius* strain are in accordance with the higher levels of volatile compounds detected in these cheeses (Tables 3–5).

Table 6. Sensory evaluation [a] of the cheeses made in this study.

	Time (d)	Type of Cheese	
		CS	LS
Odour intensity	60	5.54 ± 1.04	6.22 ± 1.07
	120	5.83 ± 1.55	6.75 ± 1.24
	180	5.89 ± 1.50	6.37 ± 1.65
	240	6.46 ± 1.61	6.58 ± 1.60
Odour quality	60	6.77 ± 1.12	7.16 ± 1.31
	120	6.89 ± 1.27	7.22 ± 1.31
	180	6.97 ± 1.00	7.41 ± 1.06
	240	7.04 ± 1.09	7.44 ± 1.27
Aroma intensity	60	6.11 ± 1.18 [a]	6.86 ± 1.26 [b]
	120	6.26 ± 1.04 [a]	7.24 ± 1.22 [b]
	180	6.19 ± 1.01 [a]	7.11 ± 1.34 [b]
	240	6.91 ± 1.31 [a]	7.40 ± 1.26 [b]
Aroma quality	60	6.92 ± 0.92	7.39 ± 1.29
	120	6.82 ± 1.21 [a]	7.78 ± 0.96 [b]
	180	7.19 ± 0.92	7.63 ± 1.30
	240	7.23 ± 1.01 [a]	8.03 ± 1.18 [b]
Taste intensity	60	6.05 ± 0.67 [a]	7.25 ± 1.06 [b]
	120	6.35 ± 1.07 [a]	7.17 ± 1.00 [b]
	180	6.69 ± 0.87 [a]	7.46 ± 1.09 [b]
	240	7.20 ± 1.29	7.49 ± 1.18
Taste quality	60	6.84 ± 1.06	7.61 ± 1.18
	120	6.49 ± 1.33 [a]	7.45 ± 1.16 [b]
	180	7.19 ± 0.81	7.60 ± 1.37
	240	7.27 ± 1.15	7.85 ± 1.36
Texture quality	60	7.11 ± 1.55	7.21 ± 1.42
	120	7.19 ± 1.22	6.81 ± 1.22
	180	7.00 ± 1.00	7.07 ± 1.08
	240	7.56 ± 1.26	7.63 ± 1.04
Colour quality	60	7.64 ± 1.86	7.79 ± 1.53
	120	8.09 ± 1.29	7.38 ± 1.54
	180	8.03 ± 1.32	7.60 ± 1.23
	240	7.69 ± 1.54	7.37 ± 1.29

[a] Results are mean (n = 16) ± SD from 8 trained panelists on a 0–10 point scale of duplicate determinations in two cheese trials. Means in the same row followed by different letters differ significantly ($p < 0.01$).

In addition, cheeses made with *L. salivarius* SP36 obtained significantly higher "sheepy", "pungent", "fruity" and "nutty" odour and aroma scores than control cheeses ($p < 0.001$). The higher "pungent" odour and aroma scores obtained by cheeses made with *L. salivarius* SP36 could be related to the higher levels of carboxylic acids found in these cheeses since many of them have been associated with "pungent" odours notes [30]. The increase in "fruity" and "nutty" aroma scores in cheeses including the *L. salivarius* strain could be associated with the higher levels of esters and furanmethanol found in these cheeses, respectively. Fruity and nutty flavour notes have been described for these compounds, respectively [30,36]. In a previous study, the use of two human milk *L. salivarius* strains in fresh cheeses did not have a relevant impact in the sensory quality and acceptance of the cheese, which was in agreement with their limited effect on their volatile profiles [37]. In contrast, the use of *L. salivarius* AR809 as an adjunct in the manufacture of Monascus-ripened cheese had a positive effect on sensory properties [38]. This strain significantly promoted the formation of several volatile compounds in cheese.

Taste intensity and quality scores were also higher for cheeses made with *L. salivarius* and were significantly ($p < 0.01$) higher for 60-, 120- and 180-day-old cheeses and 120-day-old cheeses, respectively (Table 6). Although some taste descriptors, such as "sour", "bitter", "sweet" or "salty", were not influenced by the use of *L. salivarius* SP36 in cheese manufacture, "umami" descriptor scores were significantly higher in cheeses made with the *L. salivarius* strain (4.37 vs. 3.82, $p < 0.01$). There is a large number of factors that may exert an influence on the sensorial properties of cheeses, but among them, their microbial composition has a paramount relevance [39]. Overall, changes in the parameters cited above may be closely related to the impact of the strain in cheese maturation, most probably related to the proteolytic and lipolytic activities of the strain and its ability to participate in the biosynthesis of volatile compounds. The analysis of the genome of *L. salivarius* SP3 has already revealed the presence of a wide variety of genes potentially involved in such properties [15]. They are relevant traits since the flavour-forming potential is one of the LAB traits with the highest interest to the dairy industry [40]. Work is in progress to further elucidate the metabolic potential of such a strain for dairy applications.

4. Conclusions

L. salivarius SP36 proved to be a good candidate to be added as an adjunct culture for cheesemaking, since its use increased the formation of a variety of volatile compounds identified as potent odourants in several cheese varieties. As a result, the use of *L. salivarius* SP36 improved the cheese flavour by making it more intense and recovering typical sensorial notes of semi-hard and hard traditional cheeses made with raw sheep's milk in Spain, such as the Manchego, Roncal, Zamorano and Idiazábal varieties. In addition, it had no negative effect on the rest of the cheese characteristics. Further studies would be very interesting to shed light on the metabolic pathways involved in the formation of volatile compounds, as well as to study *L. salivarius* SP36 functional properties in greater depth.

Supplementary Materials: The following supporting information can be downloaded at https://www.mdpi.com/article/10.3390/foods13142296/s1, Figure S1: External and internal sensory descriptors of both cheese types (CS and LS) at days 60, 120, 180 and 240 of ripening; Table S1: Physical–chemical and colour parameters (mean ± SD) depending on the batch and days of ripening; Table S2: Texture parameters (mean ± SD) depending on the batch and days of ripening.

Author Contributions: Conceptualization, R.A., L.F., J.M.R. and S.G.; methodology, C.A., J.C., L.F., M.R., J.R. and M.Á.; software, C.A. and J.C.; formal analysis, R.A., C.A., J.C., L.J., L.F., M.Á. and S.G.; resources, R.A., J.M.R. and S.G.; writing—original draft preparation, R.A., C.A., J.C., L.J., M.Á., J.M.R. and S.G.; writing—review and editing, R.A., C.A., J.C., L.J., M.Á., J.M.R. and S.G.; visualization, R.A., C.A., J.C., L.J., M.Á. and S.G.; supervision, R.A., J.M.R. and S.G. All authors have read and agreed to the published version of the manuscript.

Funding: This research received no external funding.

Data Availability Statement: The original contributions presented in the study are included in the article/Supplementary Materials; further inquiries can be directed to the corresponding authors.

Acknowledgments: The authors are grateful to the National Manchega Breeders Association (AGRAMA) and PDO Manchego Cheese.

Conflicts of Interest: Author Mario Roman and Juan Roman were employed by the company QUALIAM s.l. The remaining authors declare that the research was conducted in the absence of any commercial or financial relationships that could be construed as a potential conflict of interest.

References

1. Mayo, B.; Rodríguez, J.; Vázquez, L.; Flórez, A.B. Microbial interactions within the cheese ecosystem and their application to improve quality and safety. *Foods* **2021**, *10*, 602. [CrossRef] [PubMed]
2. Nam, J.H.; Cho, Y.S.; Rackerby, B.; Goddik, L.; Park, S.H. Shifts of microbiota during cheese production: Impact on production and quality. *Appl. Microbiol. Biotechnol.* **2021**, *105*, 2307–2318. [CrossRef] [PubMed]
3. Cogan, T.M.; Beresford, T.P.; Steele, J.; Broadbent, J.; Shah, N.P.; Ustunol, Z. Invited review: Advances in starter cultures and cultured foods. *J. Dairy Sci.* **2007**, *90*, 4005–4021. [CrossRef] [PubMed]
4. Alegría, A.; Alvarez-Martín, P.; Sacristán, N.; Fernández, E.; Delgado, S.; Mayo, B. Diversity and evolution of the microbial populations during manufacture and ripening of Casín, a traditional Spanish, starter-free cheese made from cow's milk. *Int. J. Food Microbiol.* **2009**, *136*, 44–51. [CrossRef] [PubMed]
5. Zheng, J.; Wittouck, S.; Salvetti, E.; Franz, C.M.A.P.; Harris, H.M.B.; Mattarelli, P.; O'Toole, P.W.; Pot, B.; Vandamme, P.; Walter, J.; et al. A taxonomic note on the genus *Lactobacillus*: Description of 23 novel genera, emended description of the genus *Lactobacillus* Beijerinck 1901, and union of *Lactobacillaceae* and *Leuconostocaceae*. *Int. J. Syst. Evol. Microbiol.* **2020**, *70*, 2782–2858. [CrossRef] [PubMed]
6. Montel, M.C.; Buchin, S.; Mallet, A.; Delbes-Paus, C.; Vuitton, D.A.; Desmasures, N.; Berthier, F. Traditional cheeses: Rich and diverse microbiota with associated benefits. *Int. J. Food Microbiol.* **2014**, *177*, 136–154. [CrossRef] [PubMed]
7. Flórez, A.B.; Vázquez, L.; Rodríguez, J.; Mayo, B. Phenotypic and safety assessment of the cheese strain *Lactiplantibacillus plantarum* LL441, and sequence analysis of its complete genome and plasmidome. *Int. J. Mol. Sci.* **2022**, *24*, 605. [CrossRef] [PubMed]
8. Apostolakos, I.; Paramithiotis, S.; Mataragas, M. Comparative genomic analysis reveals the functional traits and safety status of lactic acid bacteria retrieved from artisanal cheeses and raw sheep milk. *Foods* **2023**, *12*, 599. [CrossRef] [PubMed]
9. Albenzio, M.; Corbo, M.R.; Rehman, S.U.; Fox, P.F.; De Angelis, M.; Corsetti, A.; Sevi, A.; Gobbetti, M. Microbiological and biochemical characteristics of Canestrato Pugliese cheese made from raw milk, pasteurized milk or by heating the curd in hot whey. *Int. J. Food Microbiol.* **2001**, *67*, 35–48. [CrossRef]
10. Feutry, F.; Oneca, M.; Berthier, F.; Torre, P. Biodiversity and growth dynamics of lactic acid bacteria in artisanal PDO Ossau-Iraty cheeses made from raw ewe's milk with different starters. *Food Microbiol.* **2012**, *29*, 33–42. [CrossRef]
11. Psomas, E.; Sakaridis, I.; Boukouvala, E.; Karatzia, M.A.; Ekateriniadou, L.V.; Samouris, G. Indigenous lactic acid bacteria isolated from raw Graviera cheese and evaluation of their most important technological properties. *Foods* **2023**, *12*, 370. [CrossRef] [PubMed]
12. Martín-Platero, A.M.; Valdivia, E.; Maqueda, M.; Martín-Sánchez, I.; Martínez-Bueno, M. Polyphasic approach to bacterial dynamics during the ripening of Spanish farmhouse cheese, using culture-dependent and -independent methods. *Appl. Environ. Microbiol.* **2008**, *74*, 5662–5673. [CrossRef] [PubMed]
13. Terzić-Vidojević, A.; Veljović, K.; Tolinački, M.; Živković, M.; Lukić, J.; Lozo, J.; Fira, Đ.; Jovčić, B.; Strahinić, I.; Begović, J.; et al. Diversity of non-starter lactic acid bacteria in autochthonous dairy products from Western Balkan Countries—Technological and probiotic properties. *Food Res. Int.* **2020**, *136*, 109494. [CrossRef] [PubMed]
14. Sionek, B.; Okoń, A.; Łepecka, A.; Zielińska, D.; Jaworska, D.; Kajak-Siemaszko, K.; Neffe-Skocińska, K.; Trząskowska, M.; Karbowiak, M.; Szymański, P.; et al. The role of autochthonous *Levilactobacillus brevis* B1 starter culture in improving the technological and nutritional quality of cow's milk acid-rennet cheeses-industrial model study. *Foods* **2024**, *13*, 392. [CrossRef] [PubMed]
15. Alba, C.; Arroyo, R.; Fernández, L.F.; Narbad, A.; Rodríguez, J.M. Characterization of a *Ligilactobacillus salivarius* strain isolated from a cheese seal which was last used in 1936. *Foods* **2024**, *13*, 2005. [CrossRef] [PubMed]
16. Castro, I.; Alba, C.; Aparicio, M.; Arroyo, R.; Jiménez, L.; Fernández, L.; Arias, R.; Rodríguez, J.M. Metataxonomic and immunological analysis of milk from ewes with or without a history of mastitis. *J. Dairy Sci.* **2019**, *102*, 9298–9311. [CrossRef] [PubMed]
17. Romeith, E.A.; Michaelidou, A.; Biliaderis, C.G.; Zerfiridis, G.K. Low-fat White-brined cheese made from bovine milk and two commercial fat mimetic: Chemical, physical and sensory attributes. *Int. Dairy J.* **2002**, *12*, 525–540. [CrossRef]
18. Gómez-Torres, N.; Avila, M.; Delgado, D.; Garde, S. Effect of reuterin producing *Lactobacillus reuteri* coupled with glycerol on the volatile fraction, odour and aroma of semi-hard Ewe milk cheese. *Int. J. Food Microbiol.* **2016**, *232*, 103–110. [CrossRef] [PubMed]
19. Garde, S.; Carbonell, M.; Fernández-García, E.; Medina, M.; Nuñez, M. Volatile compounds in Hispánico cheese manufactured using a mesophilic starter, a thermophilic starter and bacteriocin-producing *Lactococcus lactis* subsp. *lactis INIA 415*. *J. Agric. Food Chem.* **2002**, *50*, 6752–6757. [CrossRef]
20. *ISO Standard No. 13299:2016*; Sensory Analysis—Methodology—General Guidance for Establishing a Sensory Profile. ISO: International Organization for Standardization: Geneva, Switzerland, 2016. Available online: https://www.iso.org/standard/58042.html (accessed on 20 June 2023).
21. Bérodier, F.; Lavanchy, P.; Zannoni, M.; Casals, J.; Herrero, L.; Aramo, C. A guide to the sensory evaluation of smell, aroma and taste of hard and semi-hard cheeses. *Lebensm.-Wiss. Technol.* **1997**, *30*, 653–664. [CrossRef]

22. Abriouel, H.; Martín-Platero, A.; Maqueda, M.; Valdivia, E.; Martínez-Bueno, M. Biodiversity of the microbial community in a Spanish farmhouse cheese as revealed by culture-dependent and culture-independent methods. *Int. J. Food Microbiol.* **2008**, *127*, 200–208. [CrossRef] [PubMed]
23. Sánchez, I.; Seseña, S.; Poveda, J.M.; Cabezas, L.; Palop, L. Genetic diversity, dynamics, and activity of Lactobacillus community involved in traditional processing of artisanal Manchego cheese. *Int. J. Food Microbiol.* **2006**, *107*, 265–273. [CrossRef] [PubMed]
24. Martín-Platero, A.M.; Maqueda, M.; Valdivia, E.; Purswani, J.; Martínez-Bueno, M. Polyphasic study of microbial communities of two Spanish farmhouse goats' milk cheeses from Sierra de Aracena. *Food Microbiol.* **2009**, *26*, 294–304. [CrossRef] [PubMed]
25. European Union. Commission Implementing Regulation (EU) No 129/2012 of 13 February 2012 approving minor amendments to the specification for a name entered in the register of protected designations of origin and protected geographical indications (Queso Manchego (PDO)). *Off. J. Eur. Union* **2012**, *43*, 1–5. Available online: https://eur-lex.europa.eu/eli/reg_impl/2012/129/oj (accessed on 11 March 2024).
26. Lucey, J.A.; Johnson, M.E.; Horne, D.S. Invited review: Perspectives on the basis of the rheology and texture properties of cheese. *J. Dairy Sci.* **2003**, *86*, 2725–2743. [CrossRef] [PubMed]
27. Salvador, A.; Igual, M.; Contreras, C.; Martínez-Navarrete, N.; Camacho, M.M. Effect of the inclusion of citrus pulp in the diet of goats on cheeses characteristics. *Small Rumin. Res.* **2014**, *121*, 361–367. [CrossRef]
28. Holland, R.; Liu, S.Q.; Crow, V.L.; Delabre, M.L.; Lubbers, M.; Bennett, M.; Norris, G. Esterases of lactic acid bacteria and cheese flavour: Milk fat hydrolysis, alcoholysis and esterification. *Int. Dairy J.* **2005**, *15*, 711–718. [CrossRef]
29. Cárdenas, N. Estudio de las Propiedades Tecnológicas de Bacterias Aisladas de Leche Materna: Aplicación para el Desarrollo de Alimentos Funcionales. Doctoral Thesis, Universidad Complutense de Madrid, Madrid, Spain, 2015.
30. Curioni, P.M.G.; Bosset, J.O. Key odorants in various cheese types as determined by gas chromatography-olfactometry. *Int. Dairy J.* **2002**, *12*, 959–984. [CrossRef]
31. Fernández-García, E.; Carbonell, M.; Núñez, M. Volatile fraction and sensory characteristics of Manchego cheese. 1. Comparison of raw and pasteurized milk cheese. *J. Dairy Res.* **2002**, *69*, 579–593. [CrossRef]
32. Fernández-García, E.; Carbonell, M.; Gaya, P.; Núñez, M. Evolution of the volatile components of ewes raw milk Zamorano cheese. Seasonal variation. *Int. Dairy J.* **2004**, *14*, 701–711. [CrossRef]
33. Fernández-García, E.; Gaya, P.; Medina, M.; Núñez, M. Evolution of the volatile components of raw ewes' milk Castellano cheese: Seasonal variation. *Int. Dairy J.* **2004**, *14*, 39–46. [CrossRef]
34. Liu, S.Q.; Holland, R.; Crow, V.L. Esters and their biosynthesis in fermented dairy products: A review. *Int. Dairy J.* **2004**, *14*, 923–945. [CrossRef]
35. Frohlich-Wyder, M.T.; Guggisberg, D.; Badertscher, R.; Wechsler, D.; Wittwer, A.; Irmler, S. The effect of *Lactobacillus buchneri* and *Lactobacillus parabuchneri* on the eye formation of semi-hard cheese. *Int. Dairy J.* **2013**, *33*, 120–128. [CrossRef]
36. Qian, M.; Reineccius, G. Identification of aroma compounds in Parmigiano-Reggiano cheese by gas chromatography/olfactometry. *J. Dairy Sci.* **2002**, *85*, 1362–1369. [CrossRef] [PubMed]
37. Cárdenas, N.; Calzada, J.; Peirotén, A.; Jiménez, E.; Escudero, R.; Rodríguez, J.M.; Medina, M.; Fernández, L. Development of a potential probiotic fresh cheese using two *Lactobacillus salivarius* strains isolated from human milk. *Biomed. Res. Int.* **2014**, *2014*, 801918. [CrossRef] [PubMed]
38. Yang, Y.; Xia, Y.; Li, C.; Wang, G.; Xiong, Z.; Song, X.; Zhang, H.; Wang, M.; Ai, L. Metabolites, flavor profiles and ripening characteristics of Monascus-ripened cheese enhanced by *Ligilactobacillus salivarius* AR809 as adjunct culture. *Food Chem.* **2024**, *436*, 137759. [CrossRef] [PubMed]
39. Smit, G.; Smit, B.A.; Engels, W.J. Flavour formation by lactic acid bacteria and biochemical flavour profiling of cheese products. *FEMS Microbiol. Rev.* **2005**, *29*, 591–610. [CrossRef]
40. Anastasiou, R.; Kazou, M.; Georgalaki, M.; Aktypis, A.; Zoumpopoulou, G.; Tsakalidou, E. Omics approaches to assess flavor development in cheese. *Foods* **2022**, *11*, 188. [CrossRef]

foods

MDPI

Article

Characterization of a *Ligilactobacillus salivarius* Strain Isolated from a Cheese Seal Which Was Last Used in 1936

Claudio Alba [1], Rebeca Arroyo [1], Leónides Fernández [2], Arjan Narbad [3] and Juan M. Rodríguez [1,*]

1 Department of Nutrition and Food Science, Complutense University of Madrid, 28040 Madrid, Spain; c.alba@ucm.es (C.A.); rebecaa@vet.ucm.es (R.A.)
2 Department of Galenic Pharmacy and Food Technology, Complutense University of Madrid, 28040 Madrid, Spain; leonides@ucm.es
3 Food Microbiome and Health Institute Strategic Programme, Quadram Institute Bioscience, Rosalind Franklin Road, Colney, Norwich NR4 7UQ, UK; arjan.narbad@quadram.ac.uk
* Correspondence: jmrodrig@ucm.es; Tel.: +34-913943749

Abstract: Cheesemaking played a pivotal role in the life of the Pyrenean villages where cheese was a most prized commodity and the subject of much local competition. In one of them (Sasa de Sobrepuerto), Mrs. Sebastiana Palacio decided in 1877 to label all the cheeses made in her household with a seal to differentiate them from those made by other local producers. The cheese seal was last used in 1936 and, since then, it has been kept under excellent storage conditions. Since well-preserved cheese seals are rare, and bacterial cells may survive desiccation for long periods, the objective of this work was to isolate and characterize any lactic acid bacteria that survived in the seal. Analysis of the milky crust material revealed the presence of sheep caseins. Culture-based analysis led to the isolation of a strain of *Bacillus licheniformis* and a strain of *Ligilactobacillus salivarius* (*L. salivarius* SP36). The latter was characterized in vitro for safety and dairy-related functional properties. Its genome encodes several genes involved in protein, peptide, and amino acid catabolism, and flavor. Overall, the phenotypic and genetic features of this strain support a high potential for being used as adjunct culture in cheesemaking.

Keywords: cheese; cheese seal; *Ligilactobacillus salivarius*; adjunct culture; genome mining

1. Introduction

Cheesemaking has been documented from, at least, the 6th millennium BC in Europe [1]. Such technology has been regarded as a milestone in human history due to the nutritional, cultural, and socio-economic impact of this activity [2]. The fact that cheese has a longer shelf life and is easier to transport than milk, together with its high nutritional value as a relevant source of proteins, fat, and calories, makes cheese as a very valuable food for surviving during cold seasons or periods of scarcity. It has been suggested that the introduction of this nutritious dairy product in the European diet decreased infant mortality rates among early farmers, enabling their demographic expansion [3].

Traditionally, cheesemaking was a key cultural and economic element in mountainous regions, such as the Spanish Pyrenees, where sheep constituted the main source of wealth until the 1950s–1960s of the last century. Cheese and cheesemaking played a pivotal role in the whole life cycle of the Pyrenean communities where "mountain cheese" was a most prized commodity and the subject of much local competition. Small changes in any of its properties, from its texture and shape to its flavor or cleanness, could lead to a particularly successful and highly prized cheese, highlighting the high skills of the farm where it was made. So, cheeses were a differential trait of each geographical zone, village, or even farm, a fact that is still evident under the different labels covering a wide variety of artisan cheeses in Europe: "Protected designation of origin" (PDO), Protected Geographical Indication (PGI), and Traditional Speciality Guaranteed (TSG) [4]. In other words, many cheeses are often heritage icons that remain deeply entwined with local and familiar culture and identity [5].

This was the case in a small region of the Spanish Pyrennees called "*Sobrepuerto*", where cheeses were highly appreciated for centuries in Aragón. Sheep flocks moved ("*trashumaban*") to lower lands between October/November and May/June. When they returned to the pastures around their villages in summer, the ewes' milk was used to make their then, famous cheeses. All the villages within the Sobrepuerto region were completely uninhabited in the 1950s–1960s of the 20th century. In one of them (Sasa de Sobrepuerto), one woman (Mrs. Sebastiana Palacio, born in 1853 in Sarsa de Surta, Huesca) decided in 1877 to label all the cheeses made in her house (Casa Juan Domingo) with a seal containing her initials since she was proud of her skills and hygienic procedures during cheesemaking. Such a practice lasted until 1936, when the last owners of the house (Mr. José Villlacampa and Mrs. Julia López) decided to move to a bigger village (Fiscal, Huesca), and to give up sheep farming after their ovine flock was requisitioned during the Spanish Civil War.

A few years ago, we discovered the existence of the cheese seal, which had been kept by a direct descendent (Mrs. Anabel Puyuelo, Fiscal), under excellent storage conditions (original box, cold and dark place). The seal was covered by a homogeneous dairy crust. Since such well-preserved cheese seals are rare, and vegetative bacterial cells may survive desiccation for long periods, particularly when protected by milky environments [6–9], the objectives of this work were, first, to submit the dairy crust to an enrichment step in order to know if culturable lactic acid bacteria remained in the seal and, if it was the case, then undertake a preliminary in vitro and in silico characterization of the isolated strain(s). In parallel, a metataxonomic analysis of this old biological material was also performed to explore which bacterial species might have been originally present in the seal.

2. Materials and Methods

2.1. In Search for the Seal

During an interview with Antonio Bellosta (born in 1914 in Naval, Huesca, Spain), who worked as a muleteer in the Spanish Pyrenees from the late 1920s to the early 1950s of the last century, he reported that he used to acquire cheeses, which were usually exchanged with olive oil (approximately 100 kg of cheeses with 20 L of olive oil), during his regular trips to the region called "*Sobrepuerto*". The cheeses were derived from milk collected from local ewes (Churra tensina breed) using rennet that had been obtained directly from the stomach of a lactating lamb. The cheeses were prepared via a back-slopping fermentation and matured for at least 6 months. Among the different local cheese producers, he remembered that the ones with the best organoleptic properties, and therefore, reaching the highest demand and economic value among his clients, were those made in Casa Juan Domingo (Sasa de Sobrepuerto, Huesca). In addition, they were uniquely labeled with a seal or stamp ("*a couple of letters*") on their upper surface. Some months later, and in an independent interview, Mrs. Angelines Villacampa (Casa Mallau, Susín, Huesca) informed that her mother (Mrs. Avelina Villacampa) was born in Casa Juan Domingo. Mrs. Villacampa said that, during her infancy, she used to spend summers in her mother's village where she witnessed cheesemaking by her grandmother, her mother, and other women living in the same house. She confirmed the existence of a cheese seal used to label all the cheese pieces produced in the household. In addition, she said that Mrs. Anabel Puyuelo, a daughter of her cousin Joaquín Puyuelo, kept the seal in her house in Fiscal (Huesca), the village where the family settled when they left Sasa de Sobrepuerto in 1936.

In the subsequent visit to Mrs. Puyuelo, she showed us the seal. It was a wooden oval, measuring 4.3 × 3.6 × 0.6 cm, which had been carved to carry the negative of two letters ("S.P.", after Sebastiana Palacio) (Figure 1). The external surface of the seal conserved a dairy crust. It has been conserved inside a crystal box in a dry and dark cupboard since it was last used in the summer of 1936.

Figure 1. The seal created in 1877 to label the cheeses made by Mrs. Sebastiana Palacio in Sasa de Sobrepuerto (Huesca, Spain). The seal was uninterruptedly used to label all the cheeses produced in Casa Juan Domingo since then and until 1936.

2.2. *Collection of Biological Material from the Seal*

The whitish milky crust present in the external surface of the seal was carefully removed using a scalpel and a sterile swab soaked in sterile peptone water. Then, the material was diluted in 3 mL of sterile peptone water.

2.3. *ELISA Detection of Ovine- and Bovine-Specific Caseins and β-Lactoglobulins*

The biological material obtained from the external surface of the seal was analyzed for the presence of either ovine- or bovine-specific caseins and β-lactoglobulins using a quantitative sandwich ELISA approach. In the case of caseins, the sheep casein alpha (CSN1) and the bovine casein alpha (CSN1) ELISA kits were employed (MyBioSource, Inc., San Diego, CA, USA). The sensitivity of the sheep casein-targeting kit was 5.0 pg/mL (detection range: 31.2–1000 pg/mL) while that of the cow casein-targeting kit was <7.5 pg/mL (detection range: 15.6–1000 pg/mL). In parallel, the sample was also assayed for the presence of either ovine- or bovine-specific β-lactoglobulins using the sheep beta-lactoglobulin and the bovine beta-lactoglobulin ELISA kits, respectively (MyBioSource, Inc.). In these cases, the sensitivity of the sheep-targeting kit was 0.1 mg/mL (detection range: 0.25–8 mg/mL) while that of the cow-targeting kit was 0.067 ng/mL (detection range: 156–10 ng/mL). To perform these analyses, an aliquot of the sample (0.5 mL) was centrifuged for 15 min at 1500× *g* and the supernatant was collected and immediately assayed.

2.4. *Culture of the Biological Material*

Proper dilutions of an aliquot (0.5 mL) of the seal-derived material were spread onto plates of the following media: MacConkey (MCK, BioMérieux, Marcy l'Etoile, France, isolation of enterobacteria), Columbia Nadilixic Acid (CNA, BioMérieux, streptococci, enterococci, staphylococci, and related Gram-positive bacteria), Sabouraud Dextrose Chloramphenicol (SDC, BioMérieux; yeasts), Polymyxin Pyruvate Egg yolk Mannitol Bromothymolblue Agar (PEMBA, BioMérieux; *Bacillus*), and De Man, Rogosa and Sharpe (MRS, Oxoid) supplemented with 0.05% (*w*/*v*) L-cysteine and bromophenol blue (MRS-Cys; isolation of lactic acid bacteria and bifidobacteria). CNA, MCK, SDC, and PEMBA plates were incubated aerobically at 32 °C for 48 h while MRS-Cys plates were incubated in anaerobiosis (85% nitrogen, 10% hydrogen, 5% carbon dioxide) at 37 °C for 48 h. A second aliquot (0.5 mL) was used for enrichment in Brain Heart Infusion (BHI) broth (25 mL) for 48 h at 32 °C. Then, serial dilutions of the enriched broth were spread on the same agar media described above and the plates were incubated under the same conditions. The isolates, including at least one representative of each colony morphology, were identified by both 16S rDNA sequencing [10] and MALDI-TOF mass spectrometry [11]. The only isolate that belonged to the lactic acid bacteria group was *Ligilactobacillus salivarius* SP36, which was submitted to the characterization of some properties related to the safety and functionality of cheesemaking.

2.5. In Vitro Characterization of L. salivarius SP36

The scheme used for the assessment of some features of *L. salivarius* SP36 was similar to that previously used by our group for the characterization of *L. salivarius* CECT 9145 and *L. salivarius* PS7, two strains belonging to the same species [12,13]. All the assays were carried out in triplicate. The antimicrobial activity of the strain against pathobionts and spoilage microbes that are prevalent in cheese environments was assayed, employing an overlay method [14,15]. The indicator strains included the bacteria *Clostridium tyrobutyricum* CECT 4011, *Staphylococcus aureus* CECT 976, *Staphylococcus aureus* CECT 5191, *Staphylococcus epidermidis* CECT 231, *Listeria monocytogenes CECT* 936, *Listeria monocytogenes CECT* 4031, *Salmonella enterica* subsp. *enterica* serotype Enteritidis *CECT* 4300, *Salmonella enterica* subsp. *enterica* serovar Typhi CECT 409, *Salmonella enterica* subsp. *enterica* serovar *Typhimurium CECT* 443, *Escherichia coli* CECT 4076, *Escherichia coli* CECT 4076 (O157:H7), *Klebsiella pneumoniae* CECT 142, *Klebsiella oxytoca* CECT 860 and *Proteus vulgaris* CECT 484, the molds *Aspergillus fumigatus* CECT 2071 and *Penicillium commune* CECT 20707, and the yeasts *Rhodotorula mucilaginosa* CECT 10359, *Kluyveromyces marxianus* CECT 10357, *Candida intermedia* CECT1431, and *Debaryomyces hansenii* CECT10360. All the indicator strains were provided by the Spanish Collection of Type Cultures (CECT, Burjassot, Spain). The plates overlaid with bacterial indicators were incubated at 32 or 37 °C (depending on the optimal temperature) for 48 h, while those overlaid with yeasts cells or fungal spores were incubated at 30 °C (yeasts, molds) for 48 h (yeasts) or 120 h (molds). Finally, the plates were examined for clear zones of inhibition (>2 mm) around the *L. salivarius* SP36 streaks.

Potential bacteriocinogenic activity against the bacterial indicator strains was assessed as described previously [15]. L- and D-lactic acid concentrations in the MRS supernatants of *L. salivarius* SP36 (obtained after an incubation at 37 °C for 16 h) were assayed enzymatically using the Roche Diagnostics (Mannheim, Germany) kit. Their pH values were also determined. The concentration of both lactic acid enantiomers and the pH were also quantified in UHT milk (pH 6.5) inoculated with the strain (approximately 10^6 cfu/mL) following the same procedure. Hydrogen peroxide production was qualitatively and quantitatively assayed using previously described methods [16,17], including the modifications proposed by Martín et al. [15].

Potential co-aggregation between *L. salivarius* SP36 and the bacterial and yeast indicator strains, its bacteriocinogenic activity against the bacterial indicator strains, and its capacity to adhere to intestinal (Caco-2 and HT-29) epithelial cells were assessed as described previously [15]. Adhesion of the probiotic strain to porcine mucin was assayed as described [18].

The capability of *L. salivarius* SP36 to survive after exposition to oral and gastrointestinal-like conditions was evaluated. For this purpose, the methodology proposed by Marteau et al. [19], and modified by Martín et al. [15], was selected.

The susceptibility of *L. salivarius* SP36 to antibiotics was tested using the E-test system (Biomerieux, Marcy l'Etoile, France) and the antibiotics (ampicillin, clindamycin, chloramphenicol, erythromycin, streptomycin, gentamicin, kanamycin, and tetracycline) and cut-off values proposed for this species by EFSA [20]. In addition, the strain was also examined for hemolysis when grown on horse blood agar medium [12], for biogenic amines biosynthesis (putrescine, cadaverine, tyramine, and histamine) [21], and for its potential for gastric mucin (HGM; Sigma, Kawasaki City, Japan) degradation [22].

2.6. Metataxonomic Analysis of the Milky Crust of the Seal

For the study of the seal microbiota, 1 mL of the material extracted from the milky crust was used for DNA extraction. This process followed a previously established method [23]. The extracted DNA was eluted in 22 μL of nuclease-free water and immediately stored at −20 °C until further analysis. The purity and concentration of the DNA were estimated using a NanoDrop 1000 spectrophotometer (NanoDrop Technologies, Inc., Rockland, DE, USA). To prevent contamination, a negative control was processed in parallel with the sample.

The V3-V4 hypervariable regions of the 16S rDNA were amplified by PCR using universal primers S-D-Bact-0341-b-S-17 (CCTACGGGNGGCWGCAG) and S-D-Bact-0785-a-A-21 (GACTACHVGGGTATCTAATCC) [24]. The PCR products were sequenced on the Illumina MiSeq system at the facilities of Parque Científico de Madrid (Tres Cantos, Spain). Barcodes were added to both the 3′ and 5′ terminal ends of the PCR amplicons to facilitate the separation of forward and reverse sequences in a subsequent PCR reaction. The DNA concentration of the PCR products was quantified using a 2100 Bioanalyzer system (Agilent, Santa Clara, CA, USA). After pooling the PCR products at equal molar ratios, the DNA amplicons were purified using a QIAEX II Gel Extraction Kit (Qiagen, Hilden, Germany). The DNA concentration was then quantified with PicoGreen (BMG Labtech, Jena, Germany). The pooled, purified, and barcoded DNA amplicons were finally sequenced using the Illumina MiSeq pair-end protocol following the manufacturer's guidelines.

Raw sequence data from the cheese seal sample were demultiplexed and filtered using Illumina MiSeq Reporter v2.6 analysis software. Microbiome bioinformatics was performed using QIIME 2 (2022.2), performing sequence denoising via DADA2 to enhance data accuracy. Taxonomic classifications were assigned using the q2-feature-classifier and the naïve Bayes classifier classify-sklearn, based on the SILVA database version 138.1. Further bioinformatic analysis was conducted in R (version 4.3.2), generating a table of Amplicon Sequence Variants (ASVs) counts. Bacterial taxa abundances were normalized utilizing the total sum scaling (TSS) method.

2.7. Whole Genome Sequencing and Mining of L. salivarius SP36

A bacterial culture of the *L. salivarius* SP36 was grown to a stationary phase over a period of 1–3 days and then harvested by centrifugation (1.5 mL at 13,000 rpm for 2 min). The pellet was resuspended in 1 mL of TE buffer (10 mM Tris-HCl, pH 8.0, 1 mM EDTA), centrifuged again, and the resultant pellet was frozen. For enzymatic lysis, the pellet was resuspended in 300 μL of TE buffer supplemented with 100 μL of lysozyme (25 mg/mL), 1–3 μL of mutanolysin (10 U/μL), and 1 μL of RNase A (10 mg/mL), and incubated at 37 °C for 30 min. DNA was extracted using the Maxwell® RSC Cultured Cells DNA Kit (Promega, Madison, WI, USA) according to the manufacturer's instructions. Libraries were created using the Nextera XT DNA Library Preparation Kit (Illumina, San Diego, CA, USA) and sequenced on a NextSeq 550 System (Illumina) producing 150 base pair (bp) paired end reads, as previously described [25]. Whole genome sequences of *L. salivarius* SP36 were deposited in GenBank under the accession number SAMN40261862.

The genomic sequence reads were processed using an integrated bioinformatics pipeline. Initial quality control was carried out using Fastp (Version 0.19.5) [26], which refined the read quality using adapter trimming and quality-based filtering. Taxonomic classification was performed with Kraken2 (Version 2.1.1 [27], utilizing the 'k2_nt_20230502' database to ensure precise taxonomic resolution at minimal confidence thresholds.

Subsequent de novo assembly of the quality-filtered reads was conducted using Shovill (Vr. 1.1.0) [28]. Genomic annotation was performed with RAST and Bakta, optimized for bacterial genomes to enhance annotation accuracy. Assembly integrity was evaluated with CheckM (Vr. 1.0.11) [29], while plasmid content was predicted with PlasFlow (Vr. 1.0) [30]. The assembled contigs were submitted to virulence factor analysis against the 'vfdb_full' database, antimicrobial resistance gene detection with ABRicate (Vr. 0.9.7) [31], and qualitative assessment through Quast (Vr. 5.0.2) [32]. Prophage regions were detected using Phigaro, a high-throughput prophage sequence application [33] while AntiSMASH (v. 7.0) was used for the mining of bacteriocin-related genes.

3. Results

3.1. ELISA Detection of Ovine- and Bovine-Specific Caseins and β-Lactoglobulins

In this study, the ELISA technique was used for the detection of ovine- and bovine-specific caseins and β-lactoglobulins in the milky crust present on the surface of the seal. Cow proteins and sheep β-lactoglobulin were below the detection limit of the assay. In contrast, the concentration of ovine casein was 229 pg/mL, a value that was within the detection range of the technique.

3.2. Isolation and In Vitro Characterization of L. salivarius SP36

Bacterial growth from the dairy crust was only achieved after the enrichment step in BHI broth. When this broth medium was cultured on agar plates, growth was only obtained in PEMBA and MRS-Cys plates. Only one type of colony was observed in each medium. The ones that grew on PEMBA plates were identified as *Bacillus licheniformis* while those grown on MRS-Cys plates were identified as *L. salivarius* (Figure 2).

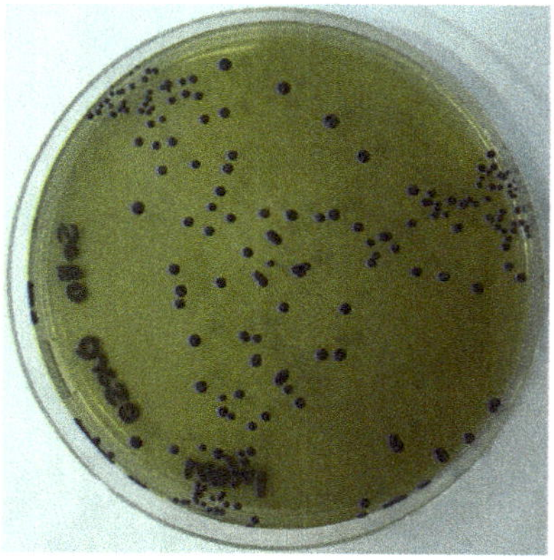

Figure 2. Colonies obtained on MRS-Cys agar plates after the BHI enrichment step.

L. salivarius SP36 showed antimicrobial activity against all the indicator microorganisms used in this work, except for *D. hansenii* CECT 10360 (no inhibition zone). The inhibition zone was large (>3 mm around the streak) against the *C. tyrobutyricum* strain and the Gram-negative bacterial indicators, medium (2–3 mm) against the remaining Gram-positive bacteria, and narrow (<2 mm) against the molds and the remaining yeasts. To elucidate the compound(s) responsible for the antimicrobial activity, the strain was screened for the production of bacteriocins, hydrogen peroxide, and lactate. Bacteriocin activity could not be detected against the spectrum of indicator organisms used in this study while it showed an ability to produce H_2O_2 (0.78 μg/mL ± 0.12). The L-lactic acid production by *L. salivarius* SP36 when grown in MRS broth for 16 h at 37 °C was 11.09 ± 0.87 mg/mL, which corresponded with a mean pH value of 3.94. In contrast, the amount of L-lactate detected in the milk cultures was much lower (0.21 ± 0.04 mg/mL, pH = 5.97). D-lactic acid was detected neither in the MRS supernatants nor in the milk cultures of the strain.

The strain displayed a co-aggregation ability against all the bacterial and yeast indicators, which was particularly intense against the *L. monocytogenes* strains and against the yeasts *R. mucilaginosa* CECT 10359 and *C. intermedia* CECT 1431. Its adherence to HT-29 and Caco-2 cells was high since it showed means of 371.2 ± 46.0 and 333 ± 39 adhered lactobacilli cells in 20 random microscopic fields, respectively. Adhesion of the strain to porcine mucin (12.3% ± 1.4 of retained fluorescence) was also high. The viability of the strain after exposition to oral and gastrointestinal-like conditions was ~63.7%.

The MIC values of *L. salivarius* SP36 for the antibiotics tested in this study using the E-test procedure revealed that the strain was sensitive to all of them, with MIC values that were acceptable according to the breakpoints proposed by EFSA [20]. The only exception was kanamycin, but this is an intrinsic feature of the *L. salivarius* species. Concerning other safety-related properties, the strain lacked hemolytic activity, and neither degraded gastric mucin nor produced biogenic amines.

3.3. Metataxonomic Analysis of the Milky Crust of the Seal

In the metataxonomic analysis of the sample collected from the seal, a total of 66,116 high-quality sequences were obtained. The analysis revealed a low diversity composition since most of the sequences (91.5%) belonged to only six genera. The most abundant genus was *Bacillus* (80.3% of the retrieved sequences) followed by, although at much lower percentages, *Lactobacillus* (3.6%), *Staphylococcus* (3.5%), *Dialister* (1.6%), *Ralstonia* (1.5%), and *Parvimonas* (1%) (Figure 3). The rest of the genera detected in the sample (minor genera) collectively accounted for 8.5% of the total sequences.

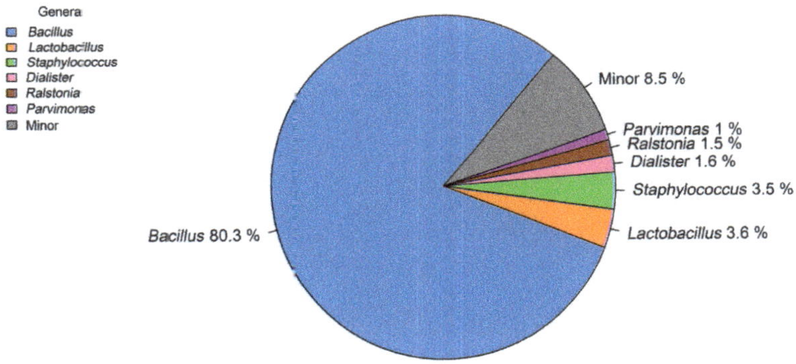

Figure 3. Pie chart with the relative abundance of microbial genera in the milky crust of the seal.

3.4. Genome Sequencing and Mining of L. salivarius SP36

The genome of *L. salivarius* SP36 comprises 1,964,896 base pairs, with a GC content of 32.83%, and includes 1968 coding sequences. These sequences are further categorized into 210 subsystems as analyzed by the RAST server. None of the genes found in its genome had a similarity with any known gene involved in virulence or pathogenicity. Some genes that may be involved in functionality, safety, stress resistance, and cheesemaking performance were detected in the genome of *L. salivarius* SP36 (Table 1).

Table 1. Genes detected in the genome of *L. salivarius* SP36 that may be involved in functionality, safety, stress resistance, and cheesemaking performance.

Category	Peptide or Protein-Encoding Genes	Brief Description
Bacteriocin production	*nisA*/*nisI* (nisin) *salA*/*salI* (salivaricin)	Antimicrobial peptides [34]
Lactate dehydrogenases	*ldhL* (L-lactate dehydrogenase)	Critical enzyme for lactic acid fermentation, affecting taste and shelf-life of fermented products [35].
Peroxide biosynthesis	*pox* (pyruvate oxidase) *nox* (NADH oxidase)	Generation of hydrogen peroxide, which can serve as a defense mechanism against pathogens [36].
Antibiotic resistance	*mate* (multi-antimicrobial extrusion protein) *gyrB* (DNA gyrase subunit B) *gyrA* (DNA gyrase subunit A)	Genes that may confer intrinsic resistance to some antibiotics [37].

Table 1. *Cont.*

Category	Peptide or Protein-Encoding Genes	Brief Description
Bile salt hydrolases	*bsh* (bile salt hydrolase)	Key enzyme for bacterial survival in the gastrointestinal tract by deconjugating bile acids [38].
Phage proteins	*int* (phage integrase) *portal* (phage portal protein)	Components of prophages that may impact bacterial genetics and the stability of bacterial populations in fermented foods [39].
Stress response (thermal stress)	*dnaK* (chaperone DnaK) *dnaJ* (chaperone DnaJ) *groEL* (chaperonin GroEL) *groES* (co-chaperonin GroES)	Proteins that assist in the proper folding of other proteins and stress recovery, essential for survival at high temperatures [40].
Stress response (oxidative stress)	*gshA* (glutamate-cysteine ligase) *grx* (glutaredoxin)	Enzymes that participate in the cellular response to oxidative stress by synthesizing antioxidants and repairing damaged proteins [40].
Stress response (osmotic stress)	*glpF* (glycerol uptake facilitator), proP (propanediol diffusion facilitator)	Proteins that help bacteria cope with high osmolarity environments by regulating the uptake of compatible solutes [40].
Proteases	*glpG* (Rhomboid protease) *clpX* (ATP-dependent Clp protease ATP-binding subunit)	Proteases that degrade misfolded proteins, influencing texture and flavor maturation in fermented products [41,42].
Peptidases	*pepT* (Peptidase T) *pepD* (Dipeptidase)	Enzymes that cleave peptides into amino acids, affecting the taste profile of fermented foods by releasing flavor precursors [41,42].
Aromatic pathways	*aroD* (3-dehydroquinate dehydratase) *qsuB* (quinate 5-dehydrogenase)	Enzymes involved in the synthesis of aromatic amino acids, precursors to a variety of flavor and aroma compounds in foods [43]

The strain harbors L-lactate dehydrogenase genes (*ldhL*) and may metabolize lactose predominantly through the Leloir pathway with alpha-galactosidase, a N-acetylgalactosamine permease IID component, galactokinase, galactose-1-phosphate uridylyltransferase, and aldose 1-epimerase as the featuring key enzymes. Additionally, a galactose operon repressor from the GalR-LacI family of transcriptional regulators, playing a crucial role in metabolic regulation, was also detected. The strain did not contain genes encoding caseinolytic proteinases, like serine proteases but, in contrast, the genome encodes a complex array of genes involved in the catabolism of other proteins and peptides, including various proteases, endo-, amino-, and carboxypeptidases, as well as enzymes responsible for amino acid catabolism and flavor (Table 1). Notable among these are the aromatic amino acid aminotransferase gamma and quinate 5-dehydrogenase I beta. Additionally, the genome features several genes related to lipolysis, including cyclic-di-AMP phosphodiesterase, phosphoesterase, thioesterase, and glycerophosphoryl diester phosphodiesterase.

Additionally, the search tool Antismash identified two regions potentially involved in bacteriocin production. The first region is classified as a class I lanthipeptide, showing a 63% similarity to the known nisin cluster, a member of the RiPPs (ribosomally synthesized and post-translationally modified peptides) within the lanthipeptide group of bacteriocins. However, the similarity rate with the nisin gene cluster was relatively low (63%, 63%, and 66% with respect to nisin Q, nisin A, and nisin Z, respectively), and two key genes for nisin biosynthesis (the one encoding the leader peptidase and the immunity gene) were lacking in the genome of our strain. The second region is characterized as a RiPP-like peptide, exhibiting a 66% similarity to the salivaricin CRL1328 α and β peptides. The strain also contains genetic determinants enabling the production of hydrogen peroxide, another compound with antimicrobial activity.

The presence of the *bsh* gene may contribute to its resistance when exposed to human gastrointestinal-like conditions while the presence of several genes involved in different types of stress may help to explain its survival for decades in the cheese seal.

The genome includes phage-related open reading frames scattered across the chromosome, alongside an entire prophage sequence detected by the Phigaro tool, encompassing a DNA segment of 23.8 kb identified as a *Siphoviridae* prophage.

4. Discussion

In this work, we isolated an *L. salivarius* strain from the biological remains that existed on the surface of an old cheese seal. This seal was used uninterruptedly during all the summers from 1877 to 1936. It was made of wood (with small pores and fine cracks) and remained at least 8 months a year covered by a dried milk layer. These factors favored the long-term survival of some of the microbes that inhabited and were able to remain alive within the desiccation-adapted biofilms [9]. It has been shown that some microbes, including foodborne pathogens, can resist drying processes and survive for long periods in low-moisture foods and environments, especially after exposure to non-lethal stress conditions, such as sub-lethal acidic or salty environments, which are present during cheese production, and, eventually, can grow if the dry food is rehydrated [6,7]. As an example, some species of the genus *Acinetobacter*, which are not particularly resistant to environmental conditions, can maintain their viability in desiccated powdered infant formula for two years, which is often the whole shelf-life of these products; *Acinetobacter* spp. are able to maintain their viability in desiccated infant formula for 2 years [8].

The culture-based assessment of the seal-derived sample led to the growth of isolates belonging to two bacterial species (*B. licheniformis* and *L. salivarius*). Interestingly, the metataxonomic analysis revealed that the genus *Bacillus* (80.3%) and the genus *Lactobacillus* (3.6%) were the two most abundant genera in the seal sample. It must be considered that *Lactobacillus* sequences in this metataxonomic analysis include all the genera in which this genus was reclassified in 2020 [44], including the genus *Ligilactobacillus*. The genus *Bacillus* includes spore-forming species and some of them (*B. licheniformis*, *B. cereus*, etc.) are often found in dairy farm environments (soil, feed, udders, and milk) [45,46]. They can also be frequently found in artisanal and commercial ewe's milk cheese [47,48]. The high resistance of *Bacillus* spp. to environmental conditions may explain why this was the most abundant genus in the metataxonomic analysis. In 2006, one strain of *B. licheniformis* and others of *Bacillus subtilis* were isolated from milk powder taken to Antarctica on Shackleton's expedition in 1907 [49].

In relation to the presence of *L. salivarius* in the seal sample, this species has been previously isolated from ovine biological samples, including sheep dung [50] and the rumen of a grazing lamb [51], and also from milk of different mammalian species [15,52–56]. Therefore, it seems to be a species that is particularly well adapted to milk and mammary environments. The resistance of *L. salivarius* strains to spray-drying depends largely on the strain and the process parameters [57] but some strains show high survival rates after drying (>94%) and they remain stable (>80% survival) for several months [58]. Anyway, this work is, to our knowledge, the first in which an isolate of this species, or any other lactic acid bacteria species, has been recovered after several decades (more than 80 years) from its last use. The fact that they have been conserved at a relatively constant temperature (approximately 12–16 °C) and in the dark throughout this period, together with the presence of several genes related to different types of environmental stress in the genome of *L. salivarius* SP36, may have helped them to maintain the viability of a small percentage of the original bacterial load.

The metataxonomic analysis also detected DNA sequences at an abundance of $\geq$1% from the genera *Staphylococcus*, *Ralstonia*, *Dialister*, and *Parvimonas*. The genus *Staphylococcus* is the most abundant one in the core microbiota of milk from either healthy or mastitic ewes [59,60]. So far, there are no reports describing the presence of *Ralstonia*, *Dialister*, or

Parvimonas DNA in ovine milk samples, although, *Ralstonia* sequences have been found in donkey and cow milk [61,62].

WGS of strains isolated from artisanal cheeses provides useful information for the selection of potential starter or adjunct cultures to recreate, at least to some extent, the organoleptic properties of the traditional cheeses [63]. In addition to safety-related issues, the production of antimicrobial substances and the metabolic and enzymatic activities responsible for acidification or sensorial properties are traits of particular interest to the dairy industry [64,65]. In this work, *L. salivarius* SP36 displayed noticeable inhibitory activity against dairy-related microbes that may cause spoilage or foodborne disease. Such activity seems to be due to the in vitro production of high amounts of L-lactate and small amounts of hydrogen peroxide. This strain harbors genes that are associated with the production of such compounds. In contrast, this activity was not due to the biosynthesis of bacteriocins. The strain contains gene information related to the bacteriocins nisin and salivaricin; however, it lacks some genetic elements that are essential for their expression or transport, and, therefore, the strain is unable to produce these antimicrobial peptides. In previous work, the analysis of the genome of *L. salivarius* CECT 5713 showed that it contained all the genes required for the biosynthesis of salivaricin Abp118, with almost a 100% identity with respect to the same genes present in the genome of the salivaricin-producing strain *L. salivarius* UCC118. The only significant difference was related to the *abpK* gene, which encodes the histidine kinase determinant involved in the *quorum sensing* system required for the regulation of bacteriocin synthesis in strain UCC118. More specifically, there is a four nucleotide deletion at the beginning of the *abpK* gene of CECT 5713, which introduces a frameshift mutation resulting in a premature stop codon, resulting in the absence of bacteriocinogenicity [39].

The strain lacks genes encoding caseinolytic proteinases, a fact that may explain its low growth and poor acidification when inoculated in milk, which may prevent its use as a starter culture. In contrast, the genome encodes a complex array of genes involved in protein and peptide degradation, amino acid catabolism, and flavor. Overall, these phenotypic and genetic features support the high potential of this strain for being used as adjunct culture in cheesemaking.

5. Conclusions

The results of this work show that old cheese-related biological materials can contain viable bacteria that can be recovered after an appropriate enrichment step. Such isolated strains were routinely employed through back-slopping before the generalization of the use of commercial starter cultures. They may display interesting functional and technological properties enabling their use as starter or adjunct cultures for local producers willing to recover traditional organoleptic properties. In this study, *L. salivarius* SP36 proved to be safe and to have some functional properties of interest for dairy companies. The analysis of its genome indicated that it is not a good candidate as a starter culture but an excellent one as an adjunct culture for cheesemaking.

Author Contributions: Conceptualization, L.F. and J.M.R.; methodology, R.A., C.A., A.N. and J.M.R.; software, C.A.; formal analysis, C.A. and J.M.R.; resources, A.N. and J.M.R.; data curation, C.A. and J.M.R.; writing—original draft preparation, C.A. and J.M.R.; writing—review and editing, C.A., A.N. and J.M.R.; supervision, L.F. and J.M.R.; funding acquisition, A.N. and J.M.R. All authors have read and agreed to the published version of the manuscript.

Funding: This research received no external funding.

Institutional Review Board Statement: Not applicable.

Informed Consent Statement: Not applicable.

Data Availability Statement: Whole genome sequences of *L. salivarius* SP36 were deposited in GenBank under the accession number SAMN40261862. The rest of the original contributions presented in the study are included in the article, further inquiries can be directed to the corresponding author.

Acknowledgments: We acknowledge Antonio Bellosta and Angelines Villacampa for providing us with the information required for finding the seal, and Anabel Puyuelo for allowing us to sample the seal. In addition, we are grateful to Ramón Arias, Lorena Jiménez, Sonia Garde, Marta Ávila, and Javier Calzada for critical reading of the manuscript.

Conflicts of Interest: The authors declare no conflicts of interest.

References

1. Salque, M.; Bogucki, P.I.; Pyzel, J.; Sobkowiak-Tabaka, I.; Grygiel, R.; Szmyt, M.; Evershed, R.P. Earliest evidence for cheese making in the sixth millennium BC in northern Europe. *Nature* **2013**, *493*, 522–525. [CrossRef]
2. Krausmann, F. Milk, manure and muscular power. Livestock and the industrialization of Agriculture. *Hum. Ecol.* **2004**, *32*, 735–773. [CrossRef]
3. McClure, S.B.; Magill, C.; Podrug, E.; Moore, A.M.T.; Harper, T.K.; Culleton, B.J.; Kennett, D.J.; Freeman, K.H. Fatty acid specific δ13C values reveal earliest Mediterranean cheese production 7200 years ago. *PLoS ONE* **2018**, *13*, e0202807. [CrossRef] [PubMed]
4. Dias, C.; Mendes, L. Protected Designation of Origin (PDO), Protected Geographical Indication (PGI) and Traditional Speciality Guaranteed (TSG): A bibiliometric analysis. *Food Res. Int.* **2018**, *103*, 492–508. [CrossRef]
5. Muchnik, J.; Biénabe, E.; Cerdan, C. Food identity/food quality: Insights from the "coalho" cheese in the Northeast of Brazil. *Anthropol. Food* **2005**, *4*. [CrossRef]
6. Bale, M.J.; Bennett, P.M.; Beringer, J.E.; Hinton, M. The survival of bacteria exposed to desiccation on surfaces associated with farm buildings. *J. Appl. Bacteriol.* **1993**, *75*, 519–528. [CrossRef] [PubMed]
7. Beuchat, L.; Komitopoulou, E. *Persistence and Survival of Pathogens in Dry Foods and Dry Food Processing Environments*; ILSI Europe Report Series; ILSI Europe: Brussels, Belgium, 2011; pp. 1–48.
8. Juma, N.A.; Manning, G.; Forsythe, S.J. Desiccation survival of *Acinetobacter* spp. in infant formula. *Food Control* **2016**, *68*, 162–166. [CrossRef]
9. Chen, G.; Lin, M.; Chen, Y.; Xu, W.; Zhang, H. Induction of a viable but nonculturable state, thermal and sanitizer tolerance, and gene expression correlation with desiccation-adapted biofilm and planktonic *Salmonella* in powdered infant formula. *J. Food Prot.* **2021**, *84*, 1194–1201. [CrossRef]
10. Martín, V.; Mediano, P.; Del Campo, R.; Rodríguez, J.M.; Marín, M. Streptococcal diversity of human milk and comparison of different methods for the taxonomic identification of streptococci. *J. Hum. Lact.* **2016**, *32*, NP84–NP94. [CrossRef]
11. Marín, M.; Arroyo, R.; Espinosa-Martos, I.; Fernández, L.; Rodríguez, J.M. Identification of emerging human mastitis pathogens by MALDI-TOF and assessment of their antibiotic resistance patterns. *Front. Microbiol.* **2017**, *8*, 1258. [CrossRef]
12. Martín, V.; Cárdenas, N.; Ocaña, S.; Marín, M.; Arroyo, R.; Beltrán, D.; Badiola, C.; Fernández, L.; Rodríguez, J.M. Rectal and vaginal eradication of *Streptococcus agalactiae* (GBS) in pregnant women by using *Lactobacillus salivarius* CECT 9145, a target-specific probiotic strain. *Nutrients* **2019**, *11*, 810. [CrossRef]
13. Cárdenas, N.; Martín, V.; Arroyo, R.; López, M.; Carrera, M.; Badiola, C.; Jiménez, E.; Rodríguez, J.M. Prevention of recurrent acute otitis media in children through the use of *Lactobacillus salivarius* PS7, a target-specific probiotic strain. *Nutrients* **2019**, *11*, 376. [CrossRef] [PubMed]
14. Magnusson, J.; Schnürer, J. *Lactobacillus coryniformis* subsp. coryniformis strain Si3 produces a broad-spectrum proteinaceous antifungal compound. *Appl. Environ. Microbiol.* **2001**, *67*, 1–5. [CrossRef] [PubMed]
15. Martín, R.; Jiménez, E.; Olivares, M.; Marín, M.L.; Fernández, L.; Xaus, J.; Rodríguez, J.M. *Lactobacillus salivarius* CECT 5713, a potential probiotic strain isolated from infant feces and breast milk of a mother-child pair. *Int. J. Food Microbiol.* **2006**, *112*, 35–43. [CrossRef] [PubMed]
16. Song, Y.L.; Kato, N.; Matsumiya, Y.; Liu, C.X.; Kato, H.; Watanabe, K. Identification of and hydrogen peroxide production by fecal and vaginal lactobacilli isolated from Japanese women and newborn infants. *J. Clin. Microbiol.* **1999**, *37*, 3062–3064. [CrossRef] [PubMed]
17. Yap, P.S.; Gilliland, S.E. Comparison of newly isolated strains of *Lactobacillus delbrueckii* subsp. lactis for hydrogen peroxide production at 5 degrees C. *J. Dairy Sci.* **2000**, *83*, 628–632. [CrossRef]
18. Cohen, P.S.; Laux, D.C. Bacterial adhesion to and penetration of intestinal mucus in vitro. *Methods Enzymol.* **1995**, *253*, 309–314. [CrossRef] [PubMed]
19. Marteau, P.; Minekus, M.; Havenaar, R.; Huis in't Veld, J.H. Survival of lactic acid bacteria in a dynamic model of the stomach and small intestine: Validation and the effects of bile. *J. Dairy Sci.* **1997**, *80*, 1031–1037. [CrossRef] [PubMed]
20. EFSA; Rychen, G.; Aquilina, G.; Azimonti, G.; Bampidis, V.; Bastos, M.L.; Bories, G.; Chesson, A.; Cocconcelli, P.S.; Flachowsky, G.; et al. Guidance on the characterisation of microorganisms used as feed additives or as production organisms. *EFSA J.* **2018**, *16*, e05206.
21. Bover-Cid, S.; Holzapfel, W.H. Improved screening procedure for biogenic amine production by lactic acid bacteria. *Int. J. Food Microbiol.* **1999**, *53*, 33–41. [CrossRef]
22. Zhou, J.S.; Gopal, P.K.; Hill, H.S. Potential probiotic lactic acid bacteria *Lactobacillus rhamnosus* (HN001), *Lactobacillus acidophilus* (HN017) and *Bifidobacterium lactis* (HN019) do not degrade gastric mucin in vitro. *Int. J. Food Microbiol.* **2001**, *63*, 81–90. [CrossRef] [PubMed]

23. Lackey, K.A.; Williams, J.E.; Meehan, C.L.; Zachek, J.A.; Benda, E.D.; Price, W.J.; Foster, J.A.; Sellen, D.W.; Kamau-Mbuthia, E.W.; Kamundia, E.W.; et al. What's normal? Microbiomes in human milk and infant feces are related to each other but vary geographically: The INSPIRE study. *Front. Nutr.* **2019**, *6*, 45. [CrossRef] [PubMed]
24. Klindworth, A.; Pruesse, E.; Schweer, T.; Peplies, J.; Quast, C.; Horn, M.; Glöckner, F.O. Evaluation of general 16S ribosomal RNA gene PCR primers for classical and next-generation sequencing-based diversity studies. *Nucleic Acids Res.* **2013**, *41*, e1. [CrossRef] [PubMed]
25. Palau, R.; Bloomfield, S.J.; Jenkins, C.; Greig, D.R.; Jorgensen, F.; Mather, A.E. *Yersinia enterocolitica* biovar 1A: An underappreciated potential pathogen in the food chain. *Int. J. Food Microbiol.* **2024**, *412*, 110554. [CrossRef] [PubMed]
26. Chen, S.; Zhou, Y.; Chen, Y.; Gu, J. fastp: An ultra-fast all-in-one FASTQ preprocessor. *Bioinformatics* **2018**, *34*, i884–i890. [CrossRef] [PubMed]
27. Wood, D.E.; Salzberg, S.L. Kraken: Ultrafast metagenomic sequence classification using exact alignments. *Genome Biol.* **2014**, *15*, R46. [CrossRef] [PubMed]
28. Seemann, T. Shovill: Faster SPAdes Assembly of Illumina Reads. Available online: https://github.com/tseemann/shovill (accessed on 14 March 2024).
29. Parks, D.H.; Imelfort, M.; Skennerton, C.T.; Hugenholtz, P.; Tyson, G.W. CheckM: Assessing the quality of microbial genomes recovered from isolates, single cells, and metagenomes. *Genome Res.* **2015**, *25*, 1043–1055. [CrossRef] [PubMed]
30. Krawczyk, P.S.; Lipinski, L.; Dziembowski, A. PlasFlow: Predicting plasmid sequences in metagenomic data using genome signatures. *Nucleic Acids Res.* **2018**, *46*, e35. [CrossRef] [PubMed]
31. Seemann, T. ABRicate: Mass Screening of Contigs for Antiobiotic Resistance Genes. Available online: https://github.com/tseemann/abricate (accessed on 14 March 2024).
32. Gurevich, A.; Saveliev, V.; Vyahhi, N.; Tesler, G. QUAST: Quality assessment tool for genome assemblies. *Bioinformatics* **2013**, *29*, 1072–1075. [CrossRef]
33. Starikova, E.V.; Tikhonova, P.O.; Prianichnikov, N.A.; Rands, C.M.; Zdobnov, E.M.; Ilina, E.N.; Govorun, V.M. Phigaro: High-throughput prophage sequence annotation. *Bioinformatics* **2020**, *36*, 3882–3884. [CrossRef]
34. Alvarez-Sieiro, P.; Montalbán-López, M.; Mu, D.; Kuipers, O.P. Bacteriocins of lactic acid bacteria: Extending the family. *Appl. Microbiol. Biotechnol.* **2016**, *100*, 2939–2951. [CrossRef] [PubMed]
35. Urbach, E.; Daniels, B.; Salama, M.S.; Sandine, W.E.; Giovannoni, S.J. The *ldh* phylogeny for environmental isolates of *Lactococcus lactis* is consistent with rRNA genotypes but not with phenotypes. *Appl. Environ. Microbiol.* **1997**, *63*, 694–702. [CrossRef]
36. Karaseva, O.; Ozhegov, G.; Khusnutdinova, D.; Siniagina, M.; Anisimova, E.; Akhatova, F.; Fakhrullin, R.; Yarullina, D. Whole genome sequencing of the novel probiotic strain *Lactiplantibacillus plantarum* FCa3L. *Microorganisms* **2023**, *11*, 1234. [CrossRef]
37. Bhukya, K.K.; Bhukya, B. Unraveling the probiotic efficiency of bacterium *Pediococcus pentosaceus* OBK05 isolated from buttermilk: An in vitro study for cholesterol assimilation potential and antibiotic resistance status. *PLoS ONE* **2021**, *16*, e0259702. [CrossRef] [PubMed]
38. Foley, M.H.; O'Flaherty, S.; Allen, G.; Rivera, A.J.; Stewart, A.K.; Barrangou, R.; Theriot, C.M. *Lactobacillus* bile salt hydrolase substrate specificity governs bacterial fitness and host colonization. *Proc. Natl. Acad. Sci. USA* **2021**, *118*, e2017709118. [CrossRef]
39. Langa, S.; Maldonado-Barragán, A.; Delgado, S.; Martín, R.; Martín, V.; Jiménez, E.; Ruíz-Barba, J.L.; Mayo, B.; Connor, R.I.; Suárez, J.E.; et al. Characterization of *Lactobacillus salivarius* CECT 5713. A strain isolated from human milk: From genotype to phenotype. *Appl. Microbiol. Biotechnol.* **2012**, *94*, 1279–1287. [CrossRef]
40. Derunets, A.S.; Selimzyanova, A.I.; Rykov, S.V.; Kuznetsov, A.E.; Berezina, O.V. Strategies to enhance stress tolerance in lactic acid bacteria across diverse stress conditions. *World J. Microbiol. Biotechnol.* **2024**, *40*, 126. [CrossRef] [PubMed]
41. Fernández, M.; Zúñiga, M. Amino acid catabolic pathways of lactic acid bacteria. *Crit. Rev. Microbiol.* **2006**, *32*, 155–183. [CrossRef]
42. Abdul Hakim, B.N.; Xuan, N.J.; Oslan, S.N.H. A Comprehensive review of bioactive compounds from lactic acid bacteria: Potential functions as functional food in dietetics and the food industry. *Foods* **2023**, *12*, 2850. [CrossRef]
43. Smit, G.; Smit, B.A.; Engels, W.J. Flavour formation by lactic acid bacteria and biochemical flavour profiling of cheese products. *FEMS Microbiol. Rev.* **2005**, *29*, 591–610. [CrossRef]
44. Zheng, J.; Wittouck, S.; Salvetti, E.; Franz, C.M.A.P.; Harris, H.M.B.; Mattarelli, P.; O'Toole, P.W.; Pot, B.; Vandamme, P.; Walter, J.; et al. A taxonomic note on the genus *Lactobacillus*: Description of 23 novel genera, emended description of the genus *Lactobacillus* Beijerinck 1901, and union of *Lactobacillaceae* and *Leuconostocaceae*. *Int. J. Syst. Evol. Microbiol.* **2020**, *70*, 2782–2858. [CrossRef] [PubMed]
45. Dréan, P.; McAuley, C.M.; Moore, S.C.; Fegan, N.; Fox, E.M. Characterization of the spore-forming *Bacillus cereus sensu lato* group and *Clostridium perfringens* bacteria isolated from the Australian dairy farm environment. *BMC Microbiol.* **2015**, *15*, 38. [CrossRef] [PubMed]
46. Knuth, R.M.; Stewart, W.C.; Taylor, J.B.; Bisha, B.; Yeoman, C.J.; Van Emon, M.L.; Murphy, T.W. Relationships among intramammary health, udder and teat characteristics, and productivity of extensively managed ewes. *J. Anim. Sci.* **2021**, *99*, skab059. [CrossRef] [PubMed]
47. Ozturkoglu-Budak, S.; Wiebenga, A.; Bron, P.A.; de Vries, R.P. Protease and lipase activities of fungal and bacterial strains derived from an artisanal raw ewe's milk cheese. *Int. J. Food Microbiol.* **2016**, *237*, 17–27. [CrossRef] [PubMed]

48. Tirloni, E.; Bernardi, C.; Celandroni, F.; Mazzantini, D.; Massimino, M.; Stella, S.; Ghelardi, E. Prevalence, virulence potential, and growth in cheese of *Bacillus cereus* strains isolated from fresh and short-ripened cheeses sold on the Italian market. *Microorganisms* **2023**, *11*, 521. [CrossRef] [PubMed]
49. Ronimus, R.S.; Rueckert, A.; Morgan, H.W. Survival of thermophilic spore-forming bacteria in a 90+ year old milk powder from Ernest Shackelton's Cape Royds Hut in Antarctica. *J. Dairy Res.* **2006**, *73*, 235–243. [CrossRef] [PubMed]
50. Balasubramanian, B.; Soundharrajan, I.; Al-Dhabi, N.A.; Vijayaraghavan, P.; Balasubramanian, K.; Valan Arasu, M.; Choi, K.C. Probiotic characteristics of *Ligilactobacillus salivarius* AS22 isolated from sheep dung and its application in corn-fox tail millet silage. *Appl. Sci.* **2021**, *11*, 9447. [CrossRef]
51. Carbonne, C.; Chadi, S.; Kropp, C.; Molimard, L.; Chain, F.; Langella, P.; Martin, R. *Ligilactobacillus salivarius* CNCM I-4866, a potential probiotic candidate, shows anti-inflammatory properties in vitro and in vivo. *Front. Microbiol.* **2023**, *14*, 1270974. [CrossRef]
52. Martín, R.; Delgado, S.; Maldonado, A.; Jiménez, E.; Olivares, M.; Fernández, L.; Sobrino, O.J.; Rodríguez, J.M. Isolation of lactobacilli from sow milk and evaluation of their probiotic potential. *J. Dairy Res.* **2009**, *76*, 418–425. [CrossRef]
53. Diaz, M.A.; Bik, E.M.; Carlin, K.P.; Venn-Watson, S.K.; Jensen, E.D.; Jones, S.E.; Gaston, E.P.; Relman, D.A.; Versalovic, J. Identification of *Lactobacillus* strains with probiotic features from the bottlenose dolphin (*Tursiops truncatus*). *J. Appl. Microbiol.* **2013**, *115*, 1037–1051. [CrossRef]
54. Quilodrán-Vega, S.; Albarracin, L.; Mansilla, F.; Arce, L.; Zhou, B.; Islam, M.A.; Tomokiyo, M.; Al Kassaa, I.; Suda, Y.; Kitazawa, H.; et al. functional and genomic characterization of *Ligilactobacillus salivarius* TUCO-L2 isolated from *Lama glama* milk: A promising immunobiotic strain to combat infections. *Front. Microbiol.* **2020**, *11*, 608752. [CrossRef] [PubMed]
55. Jobby, R.; Flora, Y.; Bora, A.; Jha, P.; Kawalkar, H.; Desai, N. Exploring probiotic activity of *Lactobacillus* sp. isolated from indigenous breeds of cattle milk and fecal samples in Bhatan village, MH., IN. *Curr. Microbiol.* **2020**, *77*, 1184–1190. [CrossRef] [PubMed]
56. Sobrino, O.J.; Alba, C.; Arroyo, R.; Pérez, I.; Sariego, L.; Delgado, S.; Fernández, L.; de María, J.; Fumanal, P.; Fumanal, A.; et al. Replacement of metaphylactic antimicrobial therapy by oral administration of *Ligilactobacillus salivarius* MP100 in a pig farm. *Front. Vet. Sci.* **2021**, *8*, 666887. [CrossRef] [PubMed]
57. Guerrero Sanchez, M.; Passot, S.; Campoy, S.; Olivares, M.; Fonseca, F. *Ligilactobacillus salivarius* functionalities, applications, and manufacturing challenges. *Appl. Microbiol. Biotechnol.* **2022**, *106*, 57–80. [CrossRef] [PubMed]
58. Buahom, J.; Siripornadulsil, S.; Sukon, P.; Sooksawat, T.; Siripornadulsil, W. Survivability of freeze- and spray-dried probiotics and their effects on the growth and health performance of broilers. *Vet. World.* **2023**, *16*, 1849–1865. [CrossRef] [PubMed]
59. Esteban-Blanco, C.; Gutiérrez-Gil, B.; Marina, H.; Pelayo, R.; Suárez-Vega, A.; Acedo, A.; Arranz, J.J. The milk microbiota of the Spanish churra sheep breed: New insights into the complexity of the milk microbiome of dairy species. *Animals* **2020**, *10*, 1463. [CrossRef] [PubMed]
60. Lysitsas, M.; Spyrou, V.; Billinis, C.; Valiakos, G. Coagulase-negative staphylococci as an etiologic agent of ovine mastitis, with a focus on subclinical forms. *Antibiotics* **2023**, *12*, 1661. [CrossRef] [PubMed]
61. Zhou, M.; Huang, F.; Du, X.; Wang, C.; Liu, G. Microbial quality of donkey milk during lactation stages. *Foods* **2023**, *12*, 4272. [CrossRef]
62. Zhang, J.; Liu, X.; Usman, T.; Tang, Y.; Mi, S.; Li, W.; Yang, M.; Yu, Y. Integrated analysis of transcriptome and milk metagenome in subclinical mastitic and healthy cows. *Anim. Biosci.* **2024**, *37*, 709–717. [CrossRef]
63. Flórez, A.B.; Vázquez, L.; Rodríguez, J.; Mayo, B. Phenotypic and safety assessment of the cheese strain *Lactiplantibacillus plantarum* LL441, and sequence analysis of its complete genome and plasmidome. *Int. J. Mol. Sci.* **2022**, *24*, 605. [CrossRef]
64. Apostolakos, I.; Paramithiotis, S.; Mataragas, M. Comparative genomic analysis reveals the functional traits and safety status of lactic acid bacteria retrieved from artisanal cheeses and raw sheep milk. *Foods* **2023**, *12*, 599. [CrossRef] [PubMed]
65. Luo, R.; Liu, C.; Li, Y.; Liu, Q.; Su, X.; Peng, Q.; Lei, X.; Li, W.; Menghe, B.; Bao, Q.; et al. Comparative genomics analysis of habitat adaptation by *Lactobacillus kefiranofaciens*. *Foods* **2023**, *12*, 1606. [CrossRef] [PubMed]

foods

MDPI

Article

Biocontrol of *L. monocytogenes* with Selected Autochthonous Lactic Acid Bacteria in Raw Milk Soft-Ripened Cheese under Different Water Activity Conditions

José M. Martín-Miguélez [1], Jurgen Robledo [2], Irene Martín [1], Cristina Castaño [1], Josué Delgado [1] and Juan J. Córdoba [1,*]

1 Higiene y Seguridad Alimentaria, Instituto Universitario de Investigación de Carne y Productos Cárnicos (IProCar), Facultad de Veterinaria, Universidad de Extremadura, 10003 Cáceres, Spain; jmmm@unex.es (J.M.M.-M.); iremartint@unex.es (I.M.); cristinacs@unex.es (C.C.); jdperon@unex.es (J.D.)

2 Laboratorio Hidromante S.L., C. Isaac Peral, 15. Pol. Ind. Sepes, 10600 Plasencia, Spain; jurgen@hidromante.com

* Correspondence: jcordoba@unex.es

Abstract: The effect of selected autochthonous Lactic Acid Bacteria (LAB) against *Listeria monocytogenes* was evaluated in two elaborations of soft-ripened cheese performed under high and low relative humidity (RH) elaborations, to achieve a_w ranging from 0.97 to 0.94 in ripened cheeses. Two selected autochthonous strains of *Lacticaseibacillus casei* 31 and 116 were used. In each elaboration, 8 batches were physicochemically and microbiologically evaluated throughout the ripening process. The a_w and pH decreased during ripening to final values ranging from 0.944 to 0.972 a_w and 5.0 to 5.3 pH, respectively. LAB was the only microbial group that increased throughout the ripening in high and low RH elaborations. In batches that were uninoculated with LAB strains, *L. monocytogenes* was either maintained at the initial inoculation level or showed a slight reduction by the end of the ripening process. However, in LAB-inoculated batches in the two elaborations, steady decreases of *L. monocytogenes* were observed throughout maturation. *L. casei* 31 alone or in combination with strain 116 provoked reductions of 2 to 4 log CFU/g in *L. monocytogenes* over 60 days of ripening, which could be enough as a strategy for biocontrol to deal with the usual contamination by *L. monocytogenes* during cheese processing.

Keywords: *Lacticaseibacillus casei*; *L. monocytogenes* control; cheeses

1. Introduction

Listeria monocytogenes is a ubiquitous Gram-positive bacterium usually transmitted to humans through food due to its wide distribution in food processing facilities, causing gastroenteritis, abortion, or even death, depending on the host's susceptibility [1,2]. This pathogen is a persistent bacterium able to form biofilms on surfaces in food processing facilities and to adapt to stressful conditions, such as temperatures from −0.4 to 45 °C, 4.6–9.5 pH, and salt concentrations over 20%, making it a cause of public health and economic burden [3,4]. Its ability to survive under a wide range of physicochemical parameters, even after decontamination technologies have been applied, proves how listeriosis is a food safety hazard, with ready-to-eat (RTE) foods, such as soft-ripened cheeses, being the most important vehicle of transmission to humans [5,6].

Most cheeses present physicochemical characteristics that allow *L. monocytogenes* to survive and grow, causing listeriosis outbreaks regardless of the country of study [7–9]. The hazard of the presence of *L. monocytogenes* in soft-ripened cheeses made with raw milk is higher than in those whose elaboration includes a pasteurization step, due to the possible presence of this pathogen in raw milk [10,11].

Microbial control has traditionally been achieved in this kind of soft-ripened cheese by salting, acidification, drying, and chemical preservation [12]. Nowadays, consumers concern about the presence of chemical compounds among food ingredients, have promoted the development of more sustainable preservation methods, such as high hydrostatic pressure, cold plasma, pulsed light, ultrasound, bioactive films, natural antimicrobial agents such as plant extracts, or biocontrol by microbial protective cultures [13,14]. The last method is recommended due to its low environmental impact and easiness of application [15].

"Torta del Casar" is one of the traditional soft-ripened cheeses made from raw milk, for which Protected Designations of Origin (P.D.O.) regulations forbid any practice or addition of substances external to the product itself; yet, at the same time, its physicochemical characteristics may allow the development of pathogenic microorganisms such as *L. monocytogenes* [16]. In addition, in the elaboration of this kind of soft-ripened cheese that is usually made under traditional conditions, the a_w can vary between 0.98 and 0.94 throughout the ripening, resulting in a_w values ranging from 0.97 to 0.94 in the final product due to differences in relative humidity (RH) during the traditional process, which usually varies from 90% to 70% RH. These differences can be critical for the persistence and even growth of *L. monocytogenes*. Biocontrol with selected protective cultures of lactic acid bacteria (LAB) has been shown to be an effective method for reducing this pathogenic bacterium in "Torta del Casar" soft-ripened cheese [17]. In other P.D.O. cheeses from Europe, the use of LAB as protective cultures has proved to be an efficient strategy to achieve a reduction of around 2 log CFU/g of *L. monocytogenes* [18,19]. However, the efficiency in the reduction of *L. monocytogenes* with selected LAB strains has always been demonstrated in cheeses reaching a_w values below 0.94 at the end of processing [7]. Nevertheless, the efficiency of selected autochthonous LAB against *L. monocytogenes* in soft-ripened cheese with higher a_w values (0.97 to 0.94) in the final product has not been evaluated yet, even though these variations can be found in traditional soft-ripened cheeses and can be critical to controlling *L. monocytogenes*.

This work aimed to evaluate selected LAB for the biocontrol of *L. monocytogenes* in the traditional soft-ripened cheese "Torta del Casar" under different relative humidity conditions during ripening. In addition, the physicochemical and microbiological modifications in the LAB-inoculated cheeses throughout the ripening process were evaluated.

2. Materials and Methods

2.1. Microbial Cultures

L. monocytogenes strain S7-2 (serotype 4b) belonging to the National Institute of Agricultural and Food Research and Technology (INIA) Culture Collection (Madrid, Spain) was utilized throughout the experiment carried out. The LAB strains used, "31" and "116" (*Lacticaseibacillus casei*), were isolated from "Torta del Casar" cheese and selected by their anti-*L. monocytogenes* activity in cheese-based agar in vitro models [20]. *L. monocytogenes* and LAB were sub-cultured twice into 10 mL of brain heart infusion (BHI; Conda Pronadisa, Madrid, Spain) and Man Rogosa Sharpe (MRS; Conda Pronadisa), respectively, from frozen stock cultures (−80 °C) of the same media containing 20% (*w*/*v*) of glycerol (Thermo Fisher Scientific, Waltham, MA, USA).

At the end of the incubation period, microbial cells were centrifugated at 10,621 g for 10 min; the supernatants were discarded; and cells were dissolved in phosphate buffered saline (PBS; 0.32 g of sodium dihydrogen phosphate (Scharlau Chemie S.A., Barcelona, Spain), 1.09 g of disodium hydrogen phosphate (Scharlau Chemie S.A.), 9 g of NaCl (Thermo Fisher Scientific), 1 L of distilled water), achieving a concentration of ≈11 log CFU/mL for LAB and ≈4–5 log CFU/mL in the case of *L. monocytogenes*. Serial dilutions were carried out to verify the inoculation load; LAB was incubated on MRS agar at 30 °C and *L. monocytogenes* on Agar *Listeria* Ottaviani and Agosti (ALOA; Conda Pronadisa) at 37 °C for 48 h. MRS agar and ALOA are selective media for LAB and *L. monocytogenes*, respectively. Moreover, the initial counts (CFU/g) were assessed at the beginning of the maturation time.

2.2. Preparation of "Torta del Casar" Cheeses

"Torta del Casar" cheeses were elaborated in the facilities of a P.D.O. manufacturer in Casar de Cáceres (Extremadura region, Spain) to bring this study close to the commercial scale. Two separate elaborations were performed at two different times of the year (winter and summer) to evaluate the heterogeneity in relative humidity conditions during ripening and consequently in the a_w of the product throughout the processing of an artisanal product such as "Torta del Casar" and to study the use of LAB in several scenarios. In each production, four batches were produced: A (control without LAB added), B (with LAB 116), C (with LAB 31), and D (LAB 116 + 31). The material utilized during processing was the same material that the facility uses to elaborate their commercial products. Fifty liters of raw ewe milk collected from several farms were used for each batch with the addition of 10 mL of PBS with a LAB concentration of 11 log CFU/mL to reach 6 log CFU/mL in the milk to be processed. Thirty fresh pieces of approximately 500 g of cheese from each of the above-mentioned batches were made under the employees' supervision, following their usual routine for "Torta del Casar" which involves cheese curdling for 1 h at 28 °C with vegetable coagulant obtained from flowers of *Cynara cardunculus* L., pressing for 1 h, and salting in brine for 2 h. Hygienic measures were taken to avoid contamination between batches following the industry-established protocol and using the facility's disinfection materials.

The elaborated fresh cheeses were transported on the same day of production to the Faculty of Veterinary of the University of Extremadura under refrigeration conditions (3 °C $\pm$ 0.5) for inoculation with *L. monocytogenes* and further ripening. Cheeses from every batch were inoculated the same day in a laminar flow cabinet (Azbil Telstar Technologies S.L.U., Barcelona, Spain) with a 4 log CFU/mL PBS concentration of *L. monocytogenes* in the center of the product (in a cube of 16 cm^2 of surface and 6 cm deep [the entire volume of fresh cheese], and $\approx$100 g of weight) according to Martín et al. [17]. A concentration of 4–5 log CFU/g was intended to be achieved on the initial day of ripening. The four original batches (A, B, C, and D) were subdivided into batches without inoculation by *L. monocytogenes* (A, B, C, and D) for physicochemical analyses and four additional batches inoculated with *L. monocytogenes*: E (from batch A only inoculated with *L. monocytogenes*), F (from batch B inoculated with LAB 116 + *L. monocytogenes*), G (from batch C inoculated with LAB 31 + *L. monocytogenes*), H (from batch D inoculated with LAB 116 + 31 + *L. monocytogenes*). Fifteen cheeses belonging to each one of the 8 batches were ripened for 60 days (minimum ripening time of processing according to P.D.O. regulations) at the pilot plant of the Faculty of Veterinary (Extremadura region, Spain).

The conditions followed by maturation in the two processes were slightly different in terms of relative humidity (RH), following the usual fluctuations of this parameter in the traditional ripening of this kind of soft-ripened cheese. Thus, maturation 1 (named Low RH) was processed as follows: 35 days at 6 °C and 90% RH, 10 days at 8 °C and 70% RH, 10 days at 9 °C and 70% RH, and 5 days at 10 °C and 70% RH. The conditions for maturation 2 (named High RH) were: 35 days at 6 °C and 90% RH, 10 days at 8 °C and 80% RH, 10 days at 9 °C and 80% RH, and 5 days at 10 °C and 80% RH.

In each maturation, 3 cheeses of each batch (A, B, C, D, E, F, G, and H) were taken for physicochemical and microbiological analyses in the different sampling days (0, 15, 30, and 60 days of ripening). Samples were taken in a laminar flow cabinet (Azbil Telstar Technologies) from the central cube of the cheeses, with 16 cm^2 of surface, 6 cm depth, and $\approx$100 g in weight. This is the part of the cheese that was inoculated with *L. monocytogenes* as previously described.

2.3. Microbiological Analysis

Microbiological analyses were performed for each elaboration on days 0, 15, 30, and 60, by taking the central cube comprising 16 cm^2 (except for the outer layers) and homogenizing the cheese in Stomacher homogenizer bags (Interscience, Saint-Germain-en-Laye, France). Then, 25 g of cheese from batches A, B, C, and D was removed to confirm the absence of natural contamination of *L. monocytogenes* according to the International Organization for Standardization (ISO) 11290-1 standard [21]. In addition, 10 g of each homogenized cheese from every batch were analyzed with 90 mL of 1% (*w*/*v*) peptone water (Conda Pronadisa); subsequently, decimal dilutions of peptone water were prepared, and agar plates from different media according to the microbial group studied were used. Plate count agar (PCA; Conda Pronadisa), mannitol salt agar (MSA; Conda Pronadisa), and MRS were utilized to sample total viable counts, Gram-positive catalase-positive cocci (GC+), and LAB, respectively, after incubation at 30 °C for 72 h, except in the case of MRS, which was incubated for 48 h under microaerophilic conditions using a plastic bag to provide an oxygen-deprived environment. Malt Extract Agar (MEA; 20 g of malt extract (Conda Pronadisa), 20 g of glucose (Labbox Labware S.L., Barcelona, Spain), 1 g of bacteriological peptone (Conda Pronadisa), 20 g of bacteriological agar (Conda Pronadisa), 1 L of distilled water) was used to evaluate yeasts after being incubated at 25 °C for 48 h. Violet red bile glucose agar (VRBG; Conda Pronadisa) was incubated at 37 °C for 24 h to determine *Enterobacteriaceae*. *L. monocytogenes* quantification was carried out just in batches E, F, G, and H on ALOA, which allowed the differentiation of *L. monocytogenes* from other *Listeria* spp. after 48 h at 37 °C, by their development of green color colonies surrounded by the characteristic opaque halo. The counts were performed in plates with more than 10 colonies and less than 300 [22], and they were expressed at CFU/g.

2.4. Physicochemical Analysis

Physicochemical analyses were conducted according to the Association of Official Agricultural Chemists [23]. They were performed for each production on days 0, 15, 30, and 60 for batches A, B, C, and D, by taking the homogenized cheese used for microbial analyses. The samples were homogenized in ultrapure water to be measured with a pH meter (Hach Lange Spain S.L.U., Barcelona, Spain) previously calibrated with standard solutions (Scharlau Chemie S A.). The a_w was measured with a Novasina Lab Master Water activity meter model a_w SPRINT-TH 300 (Novasina AG, Lachen, Switzerland). Moisture content was calculated through the dehydration of samples at 105 °C with the addition of previously dehydrated washed sea sand (Scharlau Chemie S.A.).

2.5. Characterization of LAB Strains by Pulsed-Field Gel Electrophoresis Typing

To evaluate the implantation of LAB 116 and 31 during ripening, LAB-inoculated batches sampled at the end of the ripening time and inoculated onto MRS plates, grew on this medium and were selected from each production. Fifty percent of the colonies on each MRS plate were isolated and inoculated into MRS broth to get a higher cell concentration after incubation at 30 °C for 48 h. LAB 116 and 31 were also inoculated into MRS broth from their frozen stock cultures and incubated under the same conditions to be used as standards. LAB strains were characterized by pulsed-field gel electrophoresis (PFGE) according to Martín et al. [20].

2.6. Statistical Analysis

Different statistical analyses were performed using IBM SPSS Statistics v.22 (IBM Co., New York, NY, USA, EEUU), segmenting by batch and analyzing days as the dependent variable, and vice versa. Shapiro–Wilk and Levene's tests were performed to evaluate normality distribution and homogeneity of variances, respectively. Normal and homogeneous variables were analyzed with the HSD Tukey test; the T3 Dunnet was used to study the statistical differences between normal and non-homogeneous variables; and the Kruskal-Wallis test was carried out on non-normal variables.

3. Results

3.1. Physicochemical Changes during Ripening

The evolution of the moisture content, a_w, and pH throughout the ripening process in the different batches and the two elaborations are shown in Table 1. The moisture content in batches produced under Low RH conditions decreased in all the analyzed batches from initial values on day 0 of ripening, which ranged from 57–55%, to 32–39% by the end of maturation. However, in the High RH elaboration, the decrease in humidity content only reached 43–45% in all batches by the end of the processing. At days 30 and 60 of ripening, the humidity content was significantly ($p \leq 0.05$) higher for High RH than for the Low RH elaboration for all the analyzed batches (Table 1).

Table 1. Evolution of moisture content (%), water activity (a_W), and pH for low and high relative humidity soft-ripened cheese elaborations.

	Batches	Days of Ripening			
		0	15	30	60
1. Low relativity humidity maturation					
Moisture content (%)	A	55.34 ± 0.99 *b4	50.75 ± 2.18 3	40.24 ± 0.70 *2	32.77 ± 1.05 *a1
	B	52.37 ± 1.22 *a3	48.99 ± 2.87 3	40.14 ± 0.69 *2	33.75 ± 0.65 *a1
	C	57.85 ± 1.25 c23	47.23 ± 1.25 *3	43.24 ± 0.72 *2	39.35 ± 0.22 *b1
	D	55.12 ± 0.78 b4	48.65 ± 2.57 *3	38.73 ± 1.83 *2	33.61 ± 1.11 *a1
a_W	A	0.980 ± 0.002 2	0.980 ± 0.002 b3	0.949 ± 0.002 *1	0.949 ± 0.003 *a1
	B	0.977 ± 0.001 *3	0.977 ± 0.001 ab4	0.951 ± 0.004 *2	0.944 ± 0.001 *ab1
	C	0.982 ± 0.005 2	0.982 ± 0.005 ab3	0.950 ± 0.007 *1	0.962 ± 0.002 b1
	D	0.979 ± 0.001 2	0.981 ± 0.006 a2	0.942 ± 0.005 *1	0.952 ± 0.010 ab1
pH	A	6.93 ± 0.11 *b2	5.33 ± 0.29 1	4.95 ± 0.13 *1	5.16 ± 0.04 1
	B	6.69 ± 0.05 *a3	5.27 ± 0.02 2	4.86 ± 0.04 *1	5.15 ± 0.23 12
	C	6.61 ± 0.12 a2	5.21 ± 0.10 1	4.82 ± 0.25 1	5.18 ± 0.24 1
	D	6.65 ± 0.04 a2	5.31 ± 0.22 1	4.94 ± 0.10 *1	4.99 ± 0.03 *1
2. High relative humidity maturation					
Moisture content (%)	A	56.96 ± 0.44 *3	50.02 ± 0.77 a2	51.80 ± 1.34 *2	43.25 ± 1.10 *a1
	B	56.18 ± 1.45 *3	51.47 ± 0.90 a2	52.72 ± 0.91 *23	45.78 ± 0.23 *a1
	C	58.39 ± 1.11 3	51.47 ± 0.66 *a2	54.14 ± 1.93 *123	47.78 ± 0.66 *a1
	D	56.49 ± 1.14 2	53.90 ± 1.27 *b2	53.69 ± 1.28 *2	45.98 ± 1.03 *b1
a_W	A	0.976 ± 0.024 23	0.981 ± 0.004 *3	0.974 ± 0.004 *2	0.963 ± 0.010 *b1
	B	0.982 ± 0.004 *2	0.976 ± 0.005 *2	0.975 ± 0.007 *12	0.962 ± 0.005 *ab1
	C	0.980 ± 0.012 123	0.982 ± 0.003 *3	0.976 ± 0.001 *2	0.967 ± 0.007 ab1
	D	0.979 ± 0.001 2	0.983 ± 0.001 2	0.974 ± 0.001 *1	0.972 ± 0.008 b1
pH	A	7.09 ± 0.29 *bc2	5.68 ± 0.25 2	5.36 ± 0.28 *b2	5.37 ± 0.16 1
	B	7.35 ± 0.20 *c3	5.82 ± 0.30 2	5.31 ± 0.10 *ab1	5.24 ± 0.23 1
	C	6.87 ± 0.12 ab4	5.40 ± 0.19 3	5.12 ± 0.16 a1	5.39 ± 0.04 2
	D	6.84 ± 0.18 a3	5.66 ± 0.13 1	5.21 ± 0.05 *a2	5.35 ± 0.04 *2

A (uninoculated control), B (inoculated with LAB 116), C (inoculated with LAB 31), and D (inoculated with LAB 116 and 31). Values are expressed as mean ± standard deviation. Means with different letters as superscripts (a–c) in the same column indicate significant differences ($p \leq 0.05$) among batches from the same sampling day and maturation. Means with different numbers as superscripts (1–4) in the same row indicate significant differences ($p \leq 0.05$) among sampling days of the same batch and elaboration. Means with an asterisk as a superscript (*) in the same column indicate significant differences ($p \leq 0.05$) between elaborations (low and high relative humidity maturation) of the same batch and sampling day.

The a_W values decreased during ripening to reach values that ranged from 0.944–0.962 in finished cheeses under Low RH maturation and values ranging from 0.962 to 0.972 in the final cheeses under High RH elaboration. Only in A and B batches, the a_W values were significantly ($p \leq 0.05$) lower for Low RH than for the High RH elaboration (Table 1).

The pH of cheeses decreased during the first 30 days of ripening from initial values around 7 to levels below 5 in all batches of the Low RH elaboration and below 5.4 in all batches of the High RH maturation. From day 30 to 60 of ripening, the pH slightly increased ($p \leq 0.05$) for the Low RH elaboration, reaching values around 5.0–5.1, but remained around 5.3 in High RH ripening (Table 1). However, no significant differences ($p \leq 0.05$) were found in the final pH values between batches from each elaboration except for batch D, which displayed higher ($p \leq 0.05$) values for the High RH elaboration than for the Low RH elaboration.

3.2. Microbiological Changes during Ripening

The levels of all analyzed microbiological groups, except *L. monocytogenes*, throughout the ripening process in Low and High RH elaborations are displayed in Table 2. Total viable counts (TVC) increased ($p \leq 0.05$) from values around 6 or 7 log CFU/g at day 0 to levels close to 8 or 9 log CFU/g at day 15 of ripening in all batches of both maturations. In the remaining days of ripening, the levels of TVC remained at the same level or even decreased in all batches of both elaborations, except for B, F, and H batches, which showed slight increases ($p \leq 0.05$) to levels ranging from 9.18 to 9.52 log CFU/g (Table 1). By the end of ripening, there were no significant differences between batches of either elaboration, but significantly ($p \leq 0.05$) higher TVC levels were found in C, E, F, G, and H batches of the High RH elaboration.

Enterobacteriaceae counts increased ($p \leq 0.05$) in the first 15 days of elaboration to reach levels ranging from 7 to 8 log CFU/g in all batches of both Low and High RH elaborations (Table 1). Then, levels of *Enterobacteriaceae* decreased at day 60 in both elaborations. In Low RH maturation, levels of *Enterobacteriaceae* at days 15, 30, and 60 of ripening were significantly higher in A and E batches (without LAB) than in most of the LAB-inoculated ones (Table 2).

Table 2. Evolution of total viable counts (TVC), *Enterobacteriaceae* (EB), Gram-positive catalase-positive cocci (GC+), lactic acid bacteria (LAB), and yeasts from low and high relative humidity soft-ripened cheese elaborations.

Microbial Group	Batches	Days of Ripening			
		0	15	30	60
		1. Low relative humidity maturation (log CFU/g)			
TVC	A	7.49 ± 0.09 ^{*c1}	9.68 ± 0.09 ^{*c3}	8.75 ± 0.15 ^{de2}	8.85 ± 0.01 ^{2}
	B	7.30 ± 0.06 ^{*bc1}	9.14 ± 0.04 ^{a2}	7.58 ± 0.20 ^{*a1}	8.66 ± 0.91 ^{12}
	C	6.58 ± 0.13 ^{ab1}	9.05 ± 0.18 ^{abc3}	7.92 ± 0.17 ^{*ab2}	8.22 ± 0.26 ^{*2}
	D	6.19 ± 0.13 ^{*a1}	9.26 ± 0.15 ^{abc3}	8.19 ± 0.09 ^{*bc2}	8.81 ± 0.21 ^{3}
	E	7.49 ± 0.09 ^{*c1}	9.53 ± 0.07 ^{*bc3}	9.06 ± 0.08 ^{*e2}	8.91 ± 0.04 ^{*2}
	F	7.30 ± 0.06 ^{*bc1}	8.78 ± 0.32 ^{abc3}	7.85 ± 0.22 ^{*ab12}	7.94 ± 0.17 ^{*2}
	G	6.58 ± 0.13 ^{ab1}	9.23 ± 0.04 ^{*ab3}	8.29 ± 0.19 ^{*bcd2}	8.03 ± 0.10 ^{*2}
	H	6.19 ± 0.13 ^{*a1}	9.21 ± 0.13 ^{*abc3}	8.17 ± 0.25 ^{*bc2}	8.23 ± 0.13 ^{*2}
EB	A	7.05 ± 0.08 ^{*c1}	8.65 ± 0.17 ^{*c2}	8.11 ± 0.03 ^{*c2}	7.01 ± 0.12 ^{*bc1}
	B	6.82 ± 0.12 ^{*bc12}	7.16 ± 0.03 ^{*ab2}	6.13 ± 0.98 ^{*a12}	6.97 ± 0.02 ^{*b1}
	C	6.01 ± 0.09 ^{*ab}	7.48 ± 0.05 ^{*abc}	7.01 ± 0.44 ^{ab}	5.65 ± 0.79 ^{abc}
	D	5.70 ± 0.04 ^{a1}	7.64 ± 0.11 ^{bc2}	7.38 ± 0.08 ^{bc12}	6.01 ± 0.32 ^{*abc12}
	E	7.05 ± 0.08 ^{*c1}	8.56 ± 0.12 ^{*c4}	8.13 ± 0.09 ^{*c3}	7.28 ± 0.02 ^{*c2}
	F	6.82 ± 0.12 ^{*bc2}	7.11 ± 0.12 ^{*a2}	6.83 ± 0.22 ^{*ab2}	6.36 ± 0.07 ^{a1}
	G	6.01 ± 0.09 ^{*ab}	7.36 ± 0.09 ^{*ab1}	7.23 ± 0.11 ^{*abc1}	6.26 ± 0.11 ^{a1}
	H	5.70 ± 0.04 ^{a1}	7.69 ± 0.23 ^{bc3}	7.28 ± 0.08 ^{*abc23}	5.96 ± 0.47 ^{abc12}
GC+	A	6.19 ± 0.16 ^{*c2}	5.52 ± 0.13 ^{b1}	5.41 ± 0.30 ^{c12}	6.29 ± 0.82 ^{*ab12}
	B	5.36 ± 0.13 ^{b}	5.54 ± 0.18 ^{*b}	5.23 ± 0.21 ^{c}	5.88 ± 0.47 ^{*ab}
	C	4.87 ± 0.04 ^{*a1}	4.52 ± 0.06 ^{*a2}	4.55 ± 0.28 ^{ab12}	6.65 ± 0.24 ^{*b3}
	D	4.85 ± 0.13 ^{a2}	4.67 ± 0.15 ^{a12}	4.33 ± 0.08 ^{a1}	6.12 ± 0.13 ^{*ab3}
	E	6.19 ± 0.16 ^{*c2}	5.29 ± 0.15 ^{b1}	5.01 ± 0.09 ^{bc1}	6.53 ± 0.41 ^{*ab12}
	F	5.36 ± 0.13 ^{b}	5.37 ± 0.13 ^{*b}	5.36 ± 0.22 ^{c}	6.50 ± 0.49 ^{*ab}
	G	4.87 ± 0.04 ^{*a2}	4.38 ± 0.08 ^{a1}	4.80 ± 0.05 ^{*abc2}	5.29 ± 0.22 ^{*a3}
	H	4.85 ± 0.13 ^{a1}	4.60 ± 0.18 ^{a1}	4.67 ± 0.05 ^{*ab1}	5.71 ± 0.37 ^{*ab2}

Table 2. *Cont.*

Microbial Group	Batches	Days of Ripening			
		0	15	30	60
		1. Low relative humidity maturation (log CFU/g)			
LAB	A	7.12 ± 0.17 ^{*a1}	8.49 ± 0.00 ^{*bc2}	8.81 ± 0.06 ^{*b3}	8.84 ± 0.08 ^{a23}
	B	7.56 ± 0.06 ^{*b1}	8.29 ± 0.03 ^{ab2}	8.93 ± 0.20 ^{a2}	9.37 ± 0.49 ^{bc12}
	C	7.55 ± 0.03 ^{b1}	8.31 ± 0.04 ^{ab2}	8.89 ± 0.05 ^{a3}	9.23 ± 0.02 ^{*c4}
	D	7.52 ± 0.03 ^{*ab1}	8.30 ± 0.10 ^{ab2}	8.77 ± 0.07 ^{ab3}	8.90 ± 0.06 ^{*ab3}
	E	7.12 ± 0.17 ^{*a1}	8.62 ± 0.11 ^{*c2}	8.96 ± 0.03 ^{*ab2}	8.82 ± 0.05 ^{*a2}
	F	7.56 ± 0.06 ^{*b1}	8.42 ± 0.12 ^{*bc2}	8.74 ± 0.06 ^{*ab3}	8.92 ± 0.07 ^{*ab3}
	G	7.55 ± 0.03 ^{b1}	8.33 ± 0.11 ^{ab2}	8.69 ± 0.03 ^{a2}	9.08 ± 0.04 ^{bc3}
	H	7.52 ± 0.03 ^{*ab1}	8.22 ± 0.06 ^{a2}	8.97 ± 0.29 ^{ab3}	8.98 ± 0.10 ^{*abc3}
Yeasts	A	7.29 ± 0.20 ^{*b1}	9.47 ± 0.20 ^{*cd3}	8.45 ± 0.10 ^{c2}	8.90 ± 0.07 ^{3}
	B	7.26 ± 0.20 ^{*b1}	9.11 ± 0.17 ^{*abc2}	7.29 ± 0.12 ^{*a1}	9.01 ± 0.15 ^{2}
	C	6.27 ± 0.14 ^{a1}	8.96 ± 0.18 ^{*ab3}	7.28 ± 0.50 ^{*a2}	9.11 ± 0.08 ^{3}
	D	5.81 ± 0.23 ^{a1}	9.21 ± 0.10 ^{*abcd3}	7.58 ± 0.16 ^{*a2}	8.22 ± 1.09 ^{123}
	E	7.29 ± 0.20 ^{*b1}	9.48 ± 0.06 ^{*d4}	8.33 ± 0.12 ^{*b2}	8.74 ± 0.13 ^{*3}
	F	7.26 ± 0.20 ^{b}	8.68 ± 0.24 ^{a}	7.64 ± 0.36 ^{*abc}	8.75 ± 1.11
	G	6.27 ± 0.14 ^{a1}	9.35 ± 0.13 ^{bcd4}	7.65 ± 0.08 ^{*ab2}	8.97 ± 0.02 ^{*3}
	H	5.81 ± 0.23 ^{a1}	9.42 ± 0.08 ^{*cd4}	7.71 ± 0.06 ^{*abc2}	9.05 ± 0.06 ^{*3}
		2. High relative humidity maturation (log CFU/g)			
TVC	A	6.33 ± 0.02 ^{*a1}	9.42 ± 0.08 ^{*3}	8.65 ± 0.06 ^{ab2}	9.03 ± 0.37 ^{3}
	B	6.53 ± 0.06 ^{*c1}	8.90 ± 0.26 ^{2}	8.96 ± 0.03 ^{*abc2}	9.18 ± 0.08 ^{3}
	C	6.39 ± 0.07 ^{ab1}	9.00 ± 0.16 ^{2}	9.00 ± 0.03 ^{*bc2}	8.92 ± 0.29 ^{*2}
	D	6.48 ± 0.05 ^{*bc1}	9.17 ± 0.21 ^{2}	8.99 ± 0.10 ^{*abc2}	9.14 ± 0.08 ^{2}
	E	6.33 ± 0.02 ^{*aa1}	8.92 ± 0.05 ^{*3}	8.59 ± 0.04 ^{*a2}	9.15 ± 0.13 ^{*3}
	F	6.53 ± 0.06 ^{*c1}	8.77 ± 0.05 ^{2}	8.99 ± 0.06 ^{*abc3}	9.27 ± 0.04 ^{*4}
	G	6.39 ± 0.07 ^{ab1}	8.90 ± 0.17 ^{*2}	9.11 ± 0.05 ^{*c23}	9.36 ± 0.11 ^{*3}
	H	6.48 ± 0.05 ^{*bc1}	8.90 ± 0.07 ^{*2}	9.07 ± 0.06 ^{*c3}	9.52 ± 0.11 ^{*4}
EB	A	5.59 ± 0.07 ^{*1}	7.89 ± 0.39 ^{*2}	8.01 ± 0.05 ^{*b2}	4.92 ± 0.12 ^{*abcd1}
	B	5.69 ± 0.29 ^{*2}	7.83 ± 0.15 ^{*3}	7.76 ± 0.21 ^{*ab3}	3.30 ± 0.27 ^{*a1}
	C	5.71 ± 0.09 ^{*2}	7.85 ± 0.12 ^{*3}	7.45 ± 0.23 ^{a3}	4.69 ± 0.48 ^{abc1}
	D	5.57 ± 0.33 ^{12}	7.22 ± 1.11 ^{23}	7.51 ± 0.25 ^{a3}	4.17 ± 0.69 ^{*ab1}
	E	5.59 ± 0.07 ^{*1}	7.71 ± 0.12 ^{*3}	7.55 ± 0.15 ^{*a3}	6.09 ± 0.24 ^{*cde2}
	F	5.69 ± 0.29 ^{*1}	7.90 ± 0.13 ^{*2}	7.78 ± 0.04 ^{*ab2}	6.24 ± 0.29 ^{e1}
	G	5.71 ± 0.09 ^{*1}	7.86 ± 0.12 ^{*2}	7.75 ± 0.15 ^{*ab2}	6.16 ± 0.37 ^{de1}
	H	5.57 ± 0.33 ^{1}	7.56 ± 0.58 ^{23}	7.79 ± 0.13 ^{*ab3}	5.51 ± 0.04 ^{bcde12}
GC+	A	4.67 ± 0.26 ^{*a123}	5.48 ± 0.16 ^{3}	5.11 ± 0.18 ^{bc3}	4.23 ± 0.51 ^{*ab1}
	B	5.31 ± 0.11 ^{b2}	5.18 ± 0.02 ^{*2}	5.13 ± 0.08 ^{c2}	4.13 ± 0.08 ^{*ab1}
	C	4.59 ± 0.07 ^{*a2}	4.83 ± 0.07 ^{*2}	4.59 ± 0.07 ^{abc2}	3.85 ± 0.09 ^{*a1}
	D	4.98 ± 0.22 ^{ab2}	4.92 ± 0.16 ^{2}	4.49 ± 0.15 ^{a2}	3.85 ± 0.12 ^{*a1}
	E	4.67 ± 0.26 ^{*a}	5.31 ± 0.21	4.51 ± 0.50 ^{a}	4.24 ± 0.56 ^{*a1}
	F	5.31 ± 0.11 ^{ab}	5.01 ± 0.05 *	4.78 ± 0.40 ^{abc}	4.85 ± 0.25 ^{*ab}
	G	4.59 ± 0.07 ^{*b}	4.44 ± 0.51	4.28 ± 0.10 ^{*ab}	3.89 ± 0.38 ^{*ab}
	H	4.98 ± 0.22 ^{ab3}	4.81 ± 0.08 ^{23}	4.23 ± 0.12 ^{*ab1}	4.48 ± 0.12 ^{*b12}
LAB	A	5.73 ± 0.08 ^{*a1}	8.06 ± 0.02 ^{*ab2}	7.83 ± 0.33 ^{*ab2}	8.43 ± 0.47 ^{a2}
	B	7.68 ± 0.03 ^{*b1}	8.39 ± 0.11 ^{bd2}	8.67 ± 0.11 ^{bc3}	9.21 ± 0.15 ^{bc4}
	C	7.62 ± 0.07 ^{ab1}	8.46 ± 0.10 ^{d2}	8.72 ± 0.09 ^{c3}	8.84 ± 0.27 ^{*ab23}
	D	7.62 ± 0.02 ^{*ab1}	8.39 ± 0.05 ^{d2}	8.54 ± 0.33 ^{abc123}	9.14 ± 0.08 ^{*abc3}
	E	5.73 ± 0.08 ^{*a1}	7.76 ± 0.10 ^{*ac2}	7.58 ± 0.10 ^{*a2}	8.54 ± 0.21 ^{*a3}
	F	7.68 ± 0.03 ^{*b1}	8.17 ± 0.07 ^{*abcd2}	8.59 ± 0.06 ^{*abc3}	9.22 ± 0.03 ^{*bc4}
	G	7.62 ± 0.07 ^{ab1}	8.03 ± 0.29 ^{abcd1}	8.67 ± 0.14 ^{c2}	9.20 ± 0.10 ^{bc3}
	H	7.62 ± 0.02 ^{*ab1}	8.26 ± 0.03 ^{cd2}	8.72 ± 0.12 ^{c2}	9.37 ± 0.10 ^{*c3}

Table 2. *Cont.*

Microbial Group	Batches	Days of Ripening			
		0	15	30	60
2. High relative humidity maturation (log CFU/g)					
Yeasts	A	$6.07 \pm 0.13^{*1}$	$8.64 \pm 0.07^{*b2}$	8.58 ± 0.16^{a2}	9.06 ± 0.18^{ab3}
	B	$6.32 \pm 0.11^{*1}$	$8.35 \pm 0.07^{*a2}$	$9.09 \pm 0.14^{*c3}$	9.08 ± 0.11^{ab3}
	C	6.10 ± 0.07^{1}	$8.40 \pm 0.10^{*a2}$	$8.89 \pm 0.15^{*abc3}$	8.90 ± 0.17^{a3}
	D	6.25 ± 0.18^{1}	$8.40 \pm 0.09^{*a2}$	$8.94 \pm 0.01^{*bc3}$	9.11 ± 0.08^{ab3}
	E	$6.07 \pm 0.13^{*1}$	$8.68 \pm 0.03^{*b2}$	$8.71 \pm 0.12^{*ab2}$	$9.09 \pm 0.11^{*ab3}$
	F	$6.32 \pm 0.11^{*1}$	8.44 ± 0.01^{a2}	$9.02 \pm 0.08^{*bc3}$	9.28 ± 0.12^{b4}
	G	6.10 ± 0.07^{1}	8.44 ± 0.02^{a2}	$8.92 \pm 0.07^{*bc3}$	$9.35 \pm 0.04^{*b4}$
	H	6.25 ± 0.18^{1}	$8.25 \pm 0.08^{*a2}$	$8.93 \pm 0.10^{*bc3}$	$9.39 \pm 0.14^{*b4}$

A (uninoculated control), B (inoculated with LAB 116), C (inoculated with LAB 31), D (inoculated with LAB 116 and 31), E (inoculated with *L. monocytogenes*), F (inoculated with *L. monocytogenes* and LAB 116), G (inoculated with *L. monocytogenes* and LAB 31) and H (inoculated with *L. monocytogenes*, LAB 116 and 31). Values are expressed as mean ± standard deviation. Means with different letters as superscripts ($^{a–e}$) in the same column indicate significant differences ($p \leq 0.05$) among batches from the same sampling day and maturation. Means with different numbers as superscripts ($^{1–4}$) in the same row indicate significant differences ($p \leq 0.05$) among sampling days of the same batch and elaboration. Means with an asterisk as a superscript (*) in the same column indicate significant differences ($p \leq 0.05$) between elaborations (low and high relative humidity maturation) of the same batch and sampling day.

Gram-positive catalase-positive cocci (GC+) showed levels generally ranging from 4 to 6 log CFU/g in all batches of the two elaborations, showing increases in the first 15 days of processing (Table 2). From day 15 to the end of ripening, there were no increases in this microbial group, and even some decreases were observed. At the end of ripening, counts of GC+ were lower ($p \leq 0.05$) than those found on the rest of the days of processing for most of the batches. The levels of GC+ at day 60 of maturation were significantly ($p \leq 0.05$) lower in all batches of High RH elaboration in comparison with Low RH maturation (Table 2).

LAB levels increased ($p \leq 0.05$) in most batches of Low and High RH elaborations at days 15, 30, and 60 of ripening (Table 2). At the end of ripening, the highest counts were always found in LAB-inoculated batches (B, C, D, F, G, and H) of both Low and HR elaborations. In these batches, levels of LAB were higher than 7 log CFU/g throughout the ripening period. At day 60 of maturation, LAB counts of the High RH maturation showed significantly higher levels in D, E, F, and H batches than in the same batches of the Low RH elaboration.

Yeast counts increased ($p \leq 0.05$) at day 15 in all batches of the two elaborations (Table 2). In the next days of ripening, steady increases until day 60 were found only for the High RH elaboration. At the end of ripening, levels of yeasts were higher than 8 log CFU/g in all batches and elaborations.

Levels of *L. monocytogenes* throughout the ripening are displayed in Table 3 for both elaborations (Low and High RH). In the E batch (uninoculated with LAB), during the first 30 days of ripening, *L. monocytogenes* showed a significant ($p \leq 0.05$) increase under Low RH maturation, while under High RH maturation, levels of *L. monocytogenes* remained at inoculation levels in the first 15 days of ripening and decreased to less than 1 log CFU/g at day 30 of ripening. In any case, a slight decrease of 1 log CFU/g of *L. monocytogenes* since the initial day was only observed at the end of ripening in batch E of the High RH elaboration (Table 3). On the contrary, in the LAB-inoculated batches of both Low and High RH elaborations, continuous reductions of *L. monocytogenes* counts compared to initial inoculation levels were observed throughout the ripening time. At the end of ripening, the reduction ($p \leq 0.05$) of *L. monocytogenes* in the LAB-inoculated batches was around 1.5–2 log CFU/g in Low RH batches and 3.5–4 log CFU/g in High RH ones (Table 3). The highest reductions of this pathogen in the two elaborations were observed in batch G (inoculated with LAB 31) and especially in batch H (inoculated with LAB 116 and 31).

Table 3. Evolution of *Listeria monocytogenes* counts (log CFU/g) for low and high relative humidity soft-ripened cheese elaborations at 0, 15, 30, and 60 days of ripening.

Batches	Days of Ripening			
	0	15	30	60
	1. Low relative humidity maturation (log CFU/g)			
E	$4.09 \pm 0.15^{*ab1}$	$4.30 \pm 0.27^{*b1}$	$5.73 \pm 0.08^{*b2}$	4.97 ± 0.45^{b12}
F	$4.05 \pm 0.18^{*ab2}$	$3.75 \pm 0.19^{*a12}$	$2.87 \pm 0.04^{*a1}$	3.61 ± 1.21^{ab12}
G	$4.16 \pm 0.08^{*b3}$	$3.82 \pm 0.05^{*a2}$	$2.88 \pm 0.15^{*a1}$	2.03 ± 0.69^{a12}
H	$3.89 \pm 0.18^{*a3}$	$3.71 \pm 0.15^{*a23}$	$2.69 \pm 0.13^{*a1}$	2.37 ± 0.22^{a1}
	2. High relative humidity maturation (log CFU/g)			
E	$5.56 \pm 0.10^{*2}$	$5.21 \pm 0.20^{*b12}$	$4.58 \pm 0.05^{*b1}$	4.55 ± 0.05^{b1}
F	$5.42 \pm 0.06^{*2}$	$4.46 \pm 0.19^{*a1}$	$4.17 \pm 0.26^{*ab1}$	2.94 ± 1.14^{ab12}
G	$5.56 \pm 0.08^{*3}$	$4.50 \pm 0.02^{*ab2}$	$3.78 \pm 0.33^{*ab23}$	2.11 ± 0.57^{ab1}
H	$5.51 \pm 0.05^{*4}$	$4.37 \pm 0.09^{*a3}$	$3.46 \pm 0.26^{*a2}$	1.44 ± 0.83^{a1}

E (inoculated with *L. monocytogenes*), F (inoculated with *L. monocytogenes* and LAB 116), G (inoculated with *L. monocytogenes* and LAB 31) and H (inoculated with *L. monocytogenes*, LAB 116 and 31). Values are expressed as mean ± standard deviation. Means with different letters as superscripts (^{a,b}) in the same column indicate significant differences ($p \leq 0.05$) among batches from the same sampling day and maturation. Means with different numbers as superscripts ($^{1-4}$) in the same row indicate significant differences ($p \leq 0.05$) among sampling days for the same batch and elaboration. Means with an asterisk as a superscript (*) in the same column indicate significant differences ($p \leq 0.05$) between elaborations (low and high relative humidity maturation) of the same batch and sampling day.

3.3. Characterization of LAB Strains by Pulsed-Field Gel Electrophoresis Typing

The PFGE profiles obtained from the LAB isolates of inoculated batches were identified by comparison with the standard LAB 116 and 31 PFGE profiles. The analyzed LAB isolates showed similar restriction patterns to the original inoculated 116 and 31 bacteria on the final day of ripening. The 116 LAB-inoculated cheeses showed a 75% of similarities and the 31 LAB-inoculated a 66%.

4. Discussion

Raw milk soft-ripened cheeses represent a consumer hazard as they have not undergone a pasteurization process [16] and may allow the development of *L. monocytogenes* during processing and commercialization due to their physicochemical characteristics. LAB have been proposed as an effective strategy to inhibit the growth or even decrease levels of *L. monocytogenes* in ripened products [15]. However, there is doubt about whether LAB can control this pathogenic bacterium under conditions of soft-ripened cheese made with raw milk, in which a_w during ripening and in the final product is, in most cases, in the range of 0.94–0.97.

In the present work, two different elaborations of cheeses were carried out to evaluate the efficiency of selected autochthonous LAB in controlling *L. monocytogenes* over a wide range of moisture content and a_w, including values of the latter at which control of this pathogenic bacterium can be more challenging. The two maturations allowed cheeses with different moisture content in the range of 32 to 46%, with higher values being observed for High HR maturation than for Low HR maturation during ripening and in final products. In all cases, the evolution of moisture content throughout the ripening period and the final content of moisture were in the reported range for this kind of cheese [24–26]. Although the a_w decreased in both elaborations as expected, the decrease was slower under High RH than under Low RH maturation. This allowed us to obtain a wide range of a_w values (from 0.94 to 0.97) in the cheeses, throughout the ripening process and in the final products. Higher values were observed in High RH than in the Low RH elaboration. The a_w values obtained were also in the range reported for this kind of cheese [24–26] and were not sufficient to control *L. monocytogenes* since they never reached values below 0.92 [27].

The pH decreased during the first 30 days of ripening in the two maturations followed by values that ranged between 4.82 and 5.36 but, in any case, reached values $\leq$4.4, which are reported as being able to inhibit *L. monocytogenes* by themselves [27]. From day 30 until day 60 of ripening, pH increased slightly until reaching final values of 5.0–5.3 in the two elaborations, probably due to the accumulation of products derived from the proteolysis [17]. The final pH values cannot control *L. monocytogenes* but, in combination with a_w ($a_w \leq 0.94$ and pH ≤ 5.4), would inhibit this pathogenic bacterium [27], although these conditions were not found in any of the elaborations and batches analyzed.

The differences in moisture content, a_w, and pH between batches throughout the ripening process are mainly attributed to the different conditions of relative humidity during ripening and not to the LAB inoculation since there were no consistent differences in the above physicochemical parameters between LAB-inoculated and uninoculated batches. Thus, the addition of selected autochthonous LAB strains 116 and 31 did not provoke any modification of physicochemical parameters that influence the sensorial characteristics of soft-ripened cheeses that have been reported in previous works [17].

TVC, *Enterobacteriaceae,* and GC+ only showed significant increases in all analyzed batches in the first 15 days of the two elaborations studied. Similar increases have also been reported at the beginning of the maturation of traditional soft-ripened cheeses such as "Torta del Casar" and "Queso de la Serena" when a_w is above 0.96 [24]. This could also explain the higher counts of TVC found under High RH maturation, in which the moisture content and a_w were higher [25]. Levels over 7 log CFU/g of *Enterobacteriaceae* have also been observed in "Torta del Casar" cheeses in the second week of ripening, with an increase close to 2 log CFU/g relative to initial levels [24]. This microbial group has also been found in high numbers in raw milk soft-ripened cheeses at the end of maturation and has been reported to contribute to flavor development [26,28]. GC+ levels were always lower than 7 log CFU/g in all batches and elaborations analyzed. Remarkably, GC+ displayed levels around 6 log CFU/g with a moisture content between 30–40% under Low RH maturation and around 3.5–4.5 log CFU/g with a moisture content close to 45% under High RH elaboration. The GC+ usually found during the ripening of soft-ripened cheeses are more adapted to the values of a_w found under Low RH elaboration [17].

LAB and yeasts were the only microbial groups that showed increases in most batches at days 15, 30, and 60 of ripening of both elaborations; but in the case of yeasts, these increases were only observed under High RH maturation. Changes in moisture and a_w reduction over the course of ripening do not appear to affect the growth of either microbial group [17], except under Low RH maturation, in which the growth of yeasts seems to be more limited in comparison with High RH elaboration. It is worth noting that the LAB reached levels above 7 log CFU/g throughout the ripening period, and by the end of the processing, there were always higher LAB concentrations in the inoculated batches (B, C, D, F, G, and H) of both Low and High HR elaborations. However, inoculation by selected autochthonous LAB was not reflected in a greater decrease in pH in the LAB-inoculated batches, probably due to *L. casei* belonging to heterofermentative LAB, which can be positive for maintaining the sensorial characteristics of ripened cheeses. No negative effects have been described in soft-ripened cheeses matured with similar levels of LAB [29,30].

PFGE analysis of the LAB strains in the final products confirmed the correct implantation of *L. casei* 116 and *L. casei* 31 LAB in their respective batches, since these strains were the majority found in the corresponding ripened cheeses, which also indicates that these strains were present throughout the ripening time.

Regarding *L. monocytogenes*, the levels of this pathogen in the E batch (uninoculated with LAB) increased or were maintained at initial inoculation levels during the first 15 days of ripening. This can be explained because the conditions of a_w (around 0.98) and pH (between 5.8–5.3) do not inhibit *L. monocytogenes* and even promote its growth [31]. During the second half of maturation (from day 30 to 60) of the E batch, this pathogenic bacterium either remained at the inoculation level or showed a slight reduction of about 1 log CFU/g, probably due to the cheese values of a_w and pH ($a_w \leq 0.94$ and pH ≤ 5.4) that do not favor the growth of *L. monocytogenes* [27]. However, in LAB-inoculated batches (F, G, and H) of both Low and High RH elaborations, continuous reductions of *L. monocytogenes* throughout the ripening time were observed, even though, in the first days of ripening, the a_w and pH conditions could promote growth of this pathogen [31]. Complete maturation of LAB-inoculated cheeses was able to reduce *L. monocytogenes* to the range of 1.51 to 4.07 log CFU/g (Figure 1).

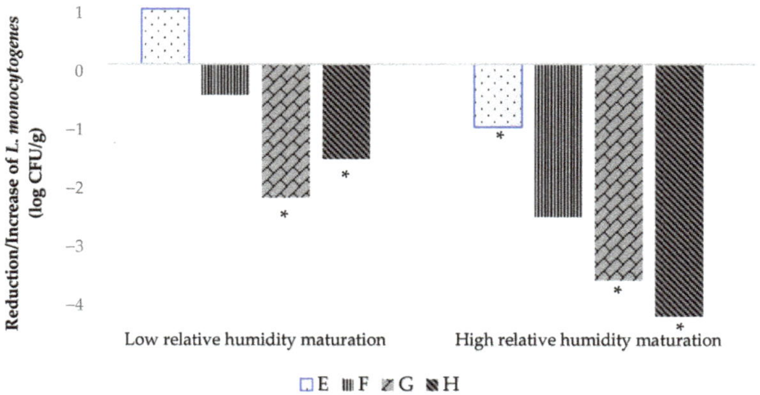

Figure 1. Evolution of *L. monocytogenes* counts during 60 days of ripening for low and high RH elaborations. E (inoculated with *L. monocytogenes*), F (inoculated with *L. monocytogenes* and LAB 116), G (inoculated with *L. monocytogenes* and LAB 31) and H (inoculated with *L. monocytogenes*, LAB 116 and 31). Values are expressed as mean and standard deviation in box and whisker plots. Values with an asterisk as a superscript (*) indicate significant differences ($p \leq 0.05$) with day 0 of the same elaboration.

The reduction in *L. monocytogenes* found in LAB 116- and 31-inoculated batches contrasts with that of batches not inoculated with LAB, in which this microbial group from contamination, mainly from raw milk, were at similar levels to the inoculated batches during ripening but in which the reduction of this pathogen was very low or absent. Thus, the inoculated LAB 116 and 31 exerted an antagonistic effect against *L. monocytogenes*, probably due to competition for nutrients and the production of compounds like bacteriocins that have been reported for *L. casei* and for the 116 strain used in the present work [32–34]. It deserves to be highlighted that LAB 31 alone or in combination with strain 116 provoke reductions of *L. monocytogenes* from 2 to 4 log CFU/g over 60 days of ripening (Figure 1). Several LAB strains of dairy origin have achieved reduction of *L. monocytogenes* in different types of cheeses elaborated in Brazil [35,36], Italy [37,38], and Spain [17]. In addition, reductions of about 1 log CFU/g of *L. monocytogenes* in soft-ripened cheese "Torta del Casar" inoculated with LAB have been previously reported [17]. However, in the present work, it is demonstrated that by using selected autochthonous LAB in soft-ripened cheeses, reductions of *L. monocytogenes* of up to 4 log CFU/g can be achieved during ripening under various a_w values, including those that promote the growth of this pathogenic bacterium. This is of utmost importance to ensure microbial safety concerning *L. monocytogenes* in this kind of soft-ripened cheese, in which the conditions of a_w and pH during part of the ripening time and in the final products are usually above those ($a_w \leq 0.94$ and pH ≤ 5.4) that do not allow growth of this pathogenic bacterium [27]. Even under these conditions, the utilization of selected autochthonous LAB guarantees the elimination of *L. monocytogenes*. The possible contamination of this bacterium could reach the product through the raw milk, equipment, utensils, or food handlers, usually at levels lower than 2 log CFU/g [39]. This is of great relevance to minimizing the risk of listeriosis associated with the consumption of soft-ripened cheeses and to avoid non-compliance with the microbiological criteria of RTE foods in industry and throughout their shelf-life in the EU [27]. In addition, the presence of selected LAB strains could help to control the microbial contamination of raw milk soft-ripened cheeses [17,40]. In the present work, the levels of *Enterobacteriaceae* during ripening under Low RH maturation were significantly lower in the LAB-inoculated batches.

Thus, the utilization of selected autochthonous LAB strains could be a good and sustainable strategy for traditional cheese industries to improve the safety of soft-ripened cheeses without affecting their physicochemical characteristics. However, the present study has applied certain selected autochthonous LAB to one kind of soft-ripened cheese, "Torta del Casar", and further studies must be performed with more selected autochthonous LAB strains obtained from other kinds of cheeses to expand the application of LAB for the biocontrol of *L. monocytogenes* in raw milk cheeses.

5. Conclusions

This work demonstrates that the utilization of selected autochthonous LAB as protective cultures in soft-ripened cheeses provoked reductions of *L. monocytogenes* of 1.5–4 log CFU/g during ripening under different conditions of a_w, including those that promote the growth of this pathogenic bacterium, without affecting the physicochemical characteristics of the final products. The combined utilization of *L. casei* 31 and *L. casei* 116 caused the most consistent reduction of *L. monocytogenes* counts, by about 4 log CFU/g, over the course of the industrial ripening process, independently of the physicochemical characteristics of soft-ripened cheeses, which are proved to be variable according to the artisanal elaboration of the product. The reduction achieved by the mix of the above LAB may be enough to deal with the natural contamination of *L. monocytogenes*. This biocontrol strategy could be included in the HACCP framework as a preventive measure, to reinforce the high safety standards required for artisanal soft-ripened cheeses made with raw milk.

6. Patents

Martín, I., Rodríguez, A and Córdoba, J.J, inventor 2021. The new strain of *Lacticaseibacillus casei* 116 with antagonist effect against *Listeria monocytogenes* to be used as protective culture in ripened cheese. No. P202131120.

Author Contributions: Conceptualization, J.J.C., I.M. and J.D.; methodology, J.M.M.-M., J.R., I.M. and C.C.; software analysis, J.M.M.-M.; formal analysis, J.M.M.-M. and C.C.; investigation J.M.M.-M., J.R., I.M. and J.J.C.; data curation, J.M.M.-M.; writing—original draft preparation, J.M.M.-M. and J.J.C.; writing—review and editing, J.M.M.-M., J.R., I.M, C.C., J.D. and J.J.C.; supervision, J.J.C. and I.M.; project administration, J.J.C. and J.R.; funding acquisition, J.J.C., J.R. and I.M. All authors have read and agreed to the published version of the manuscript.

Funding: This research was funded by Junta de Extremadura and FEDER grants IDA1-21-0130-3 and GR18056. I. Martín is the recipient of the Margarita Salas MS-13 grant funded by Ministerio de Universidades and European Union NextGenerationEU. This work was supported by grant 2022/00254/001 funded by MCIN/AEI/10.13039/50110001103.

Institutional Review Board Statement: Not applicable.

Informed Consent Statement: Not applicable.

Data Availability Statement: Data is contained within the article.

Acknowledgments: We thank the Torta del Casar Protected Designation of Origin, the Dña Francisca S.L. cheese factory (Casar de Cáceres, Spain), and Laboratory Hidromante S.L. (Plasencia, Spain).

Conflicts of Interest: Author Jurgen Robledo was employed by the company Laboratorio Hidromante S.L. This author and the remaining authors (José M. Martín-Miguélez, Irene Martín, Cristina Castaño, Josué Delgado and Juan J. Córdoba) declare that the research was conducted in the absence of any commercial or financial relationships that could be construed as a potential conflict of interest.

References

1. Chaves, R.D.; Kumazawa, S.H.; Khaneghah, A.M.; Alvarenga, V.O.; Hungaro, H.M.; Sant'Ana, A.S. Comparing the susceptibility to sanitizers, biofilm-forming ability, and biofilm resistance to quaternary ammonium and chlorine dioxide of 43 *Salmonella enterica* and *Listeria monocytogenes* strains. *Food Microbiol.* **2024**, *117*, 104380. [CrossRef] [PubMed]
2. Carpentier, B.; Cerf, O. Review—Persistence of *Listeria monocytogenes* in food industry equipment and premises. *Int. J. Food Microbiol.* **2011**, *145*, 1–8. [CrossRef] [PubMed]
3. Ferreira, V.; Wiedmann, M.; Teixeira, P.; Stasiewicz, M.J. *Listeria monocytogenes* persistence in food-associated environments: Epidemiology, strain characteristics, and implications for public health. *J. Food Prot.* **2014**, *77*, 150–170. [CrossRef] [PubMed]
4. Osek, J.; Wieczorek, K. *Listeria monocytogenes*—How this pathogen uses its virulence mechanisms to infect the hosts. *Pathogens* **2022**, *11*, 1491. [CrossRef] [PubMed]
5. Kurpas, M.; Wieczorek, K.; Osek, J. Ready-to-eat meat products as a source of *Listeria monocytogenes*. *J. Vet. Res.* **2018**, *62*, 49–55. [CrossRef] [PubMed]
6. Bucur, F.I.; Grigore-Gurgu, L.; Crauwels, P.; Riedel, C.U.; Nicolau, A.I. Resistance of *Listeria monocytogenes* to stress conditions encountered in food and food processing environments. *Front. Microbiol.* **2018**, *9*, 417209. [CrossRef] [PubMed]
7. Gérard, A.; El-Hajjaji, S.; Niyonzima, E.; Daube, G.; Sindic, M. Prevalence and survival of *Listeria monocytogenes* in various types of cheese—A review. *Int. J. Dairy Technol.* **2018**, *71*, 825–843. [CrossRef]
8. Palacios, A.; Otto, M.; Flaherty, E.; Boyle, M.M.; Malec, L.; Holloman, K.; Low, M.; Wellman, A.; Newhart, C.; Gollarza, L.; et al. Multistate outbreak of *Listeria monocytogenes* infections linked to fresh, soft hispanic-style cheese—United States, 2021. *Morb. Mortal. Wkly. Rep.* **2022**, *71*, 709. [CrossRef]
9. Nüesch-Inderbinen, M.; Bloemberg, G.V.; Müller, A.; Stevens, M.J.A.; Cernela, N.; Kollöffel, B.; Stephan, R. Listeriosis Caused by Persistence of *Listeria monocytogenes* serotype 4b sequence type 6 in cheese production environment. *Emerg. Infect. Dis.* **2021**, *27*, 284. [CrossRef]
10. Gopal, N.; Hill, C.; Ross, P.R.; Beresford, T.P.; Fenelon, M.A.; Cotter, P.D. The prevalence and control of *Bacillus* and related spore-forming bacteria in the dairy industry. *Front. Microbiol.* **2015**, *6*, 167862. [CrossRef]
11. Costanzo, N.; Ceniti, C.; Santoro, A.; Clausi, M.T.; Casalinuovo, F. Foodborne pathogen assessment in raw milk cheeses. *Int. J. Food Sci.* **2020**, *2020*, 3616713. [CrossRef] [PubMed]
12. Jan, A.; Sood, M.; Sofi, S.; Norzom, T. Non-thermal processing in food applications: A review. *Int. J. Food Sci. Nutr.* **2017**, *2*, 171–180.
13. Moula Ali, A.M.; Sant'Ana, A.S.; Bavisetty, S.C.B. Sustainable preservation of cheese: Advanced technologies, physicochemical properties and sensory attributes. *Trends Food Sci. Technol.* **2022**, *129*, 306–326. [CrossRef]
14. El-Saber Batiha, G.; Hussein, D.E.; Algammal, A.M.; George, T.T.; Jeandet, P.; Al-Snafi, A.E.; Tiwari, A.; Pagnossa, J.P.; Lima, C.M.; Thorat, N.D.; et al. Application of natural antimicrobials in food preservation: Recent views. *Food Control* **2021**, *126*, 108066. [CrossRef]
15. Delgado, J.; Álvarez, M.; Cebrián, E.; Martín, I.; Roncero, E.; Rodríguez, M. Biocontrol of pathogen microorganisms in ripened foods of animal origin. *Microorganisms* **2023**, *11*, 1578. [CrossRef]
16. Official Journal of the European Union Publication of an Amendment Application Pursuant to Article 50(2)(a) of Regulation (EU) No 1151/2012 of the European Parliament and of the Council on Quality Schemes for Agricultural Products and Foodstuffs. Available online: https://eur-lex.europa.eu/legal-content/ES/TXT/?uri=uriserv:OJ.C_.2015.235.01.0005.01.SPA&toc=OJ:C:2015:235:TOC (accessed on 30 September 2023).
17. Martín, I.; Rodríguez, A.; Córdoba, J.J. Application of selected lactic-acid bacteria to control *Listeria monocytogenes* in soft-ripened "Torta del Casar" cheese. *LWT* **2022**, *168*, 113873. [CrossRef]
18. Morandi, S.; Silvetti, T.; Battelli, G.; Brasca, M. Can lactic acid bacteria be an efficient tool for controlling *Listeria monocytogenes* contamination on cheese surface? The Case of Gorgonzola Cheese. *Food Control* **2019**, *96*, 499–507. [CrossRef]
19. Meloni, M.P.; Piras, F.; Siddi, G.; Cabras, D.; Comassi, E.; Lai, R.; McAuliffe, O.; De Santis, E.P.L.; Scarano, C. Comparison of activity of commercial protective cultures and thermophilic lactic acid bacteria against *Listeria monocytogenes*: A new perspective to improve the safety of Sardinian PDO cheeses. *Foods* **2023**, *12*, 1182. [CrossRef]
20. Martín, I.; Rodríguez, A.; Alía, A.; Martínez, R.; Córdoba, J.J. Selection and characterization of lactic acid bacteria with activity against *Listeria monocytogenes* from traditional RTE ripened foods. *LWT* **2022**, *163*, 113579. [CrossRef]
21. *ISO 11290-1:2017*; Microbiology of the Food Chain—Horizontal Method for the Detection and Enumeration of Listeria monocytogenes and of Listeria spp.—Part 1: Detection Method. International Organization for Standardization: Geneva, Switzerland, 2017.
22. *ISO 7218:2007*; Microbiology of Food and Animal Feeding Stuffs—General Requirements and Guidance for Microbiological Examinations. International Organization for Standardization: Geneva, Switzerland, 2007.
23. AOAC Association of Official Analytical Chemists. *Official Methods of Analysis of the AOAC International*; AOAC Association of Official Analytical Chemists: Washington, DC, USA, 2000.
24. Merchán, A.V.; Ruiz-Moyano, S.; Hernández, M.V.; Martín, A.; Lorenzo, M.J.; Benito, M.J. Characterization of autochthonal *Hafnia* spp. strains isolated from Spanish soft raw ewe's milk PDO cheeses to be used as adjunct culture. *Int. J. Food Microbiol.* **2022**, *373*, 109703. [CrossRef]

25. Weragama, D.; Weerasingha, V.; Jayasumana, L.; Adikari, J.; Vidanarachchi, J.K.; Priyashantha, H. The physicochemical, microbiological, and organoleptic properties and antioxidant activities of cream cheeses fortified with dried curry leaves (*Murraya Koenigii* L.) powder. *Food Sci. Nutr.* **2021**, *9*, 5774–5784. [CrossRef] [PubMed]
26. Gonçalves, M.T.P.; Benito, M.J.; de Guía Córdoba, M.; Egas, C.; Merchán, A.V.; Galván, A.I.; Ruiz-Moyano, S. Bacterial Communities in Serpa cheese by culture dependent techniques, 16S RRNA Gene Sequencing and High-Throughput Sequencing Analysis. *J. Food Sci.* **2018**, *83*, 1333–1341. [CrossRef] [PubMed]
27. European Commission. Commission Regulation (EC) No 1441/2007 of 5 December 2007 Amending Regulation (EC) No 2073/2005 on Microbiological Criteria for Foodstuffs (Text with EEA Relevance). Available online: https://eur-lex.europa.eu/eli/reg/2007/1441/oj (accessed on 22 September 2023).
28. Crespo, A.; Martín, A.; Ruiz-Moyano, S.; Benito, M.J.; Rufo, M.; Paniagua, J.M.; Jiménez, A. Application of ultrasound for quality control of "Torta del Casar" cheese ripening. *J. Dairy Sci.* **2020**, *103*, 8808–8821. [CrossRef] [PubMed]
29. Fusco, V.; Quero, G.M.; Poltronieri, P.; Morea, M.; Baruzzi, F. Autochthonous and probiotic lactic acid bacteria employed for production of "advanced traditional cheeses". *Foods* **2019**, *8*, 412. [CrossRef] [PubMed]
30. Cardinali, F.; Ferrocino, I.; Milanović, V.; Belleggia, L.; Corvaglia, M.R.; Garofalo, C.; Foligni, R.; Mannozzi, C.; Mozzon, M.; Cocolin, L.; et al. Microbial communities and volatile profile of Queijo de Azeitão PDO cheese, a traditional Mediterranean thistle-curdled cheese from Portugal. *Food Res. Int.* **2021**, *147*, 110537. [CrossRef] [PubMed]
31. Lomonaco, S.; Decastelli, L.; Nucera, D.; Gallina, S.; Manila Bianchi, D.; Civera, T. *Listeria monocytogenes* in Gorgonzola: Subtypes, diversity and persistence over time. *Int. J. Food Microbiol.* **2009**, *128*, 516–520. [CrossRef]
32. Deegan, L.H.; Cotter, P.D.; Hill, C.; Ross, P. Bacteriocins: Biological tools for bio-preservation and shelf-life extension. *Int. Dairy J.* **2006**, *16*, 1058–1071. [CrossRef]
33. Parada, J.L.; Caron, C.R.; Medeiros, A.B.P.; Soccol, C.R. Bacteriocins from lactic acid bacteria: Purification, properties and use as biopreservatives. *Braz. Arch. Biol. Technol.* **2007**, *50*, 512–542. [CrossRef]
34. Martín, I.; Barbosa, J.; Pereira, S.I.A.; Rodríguez, A.; Córdoba, J.J.; Teixeira, P. Study of lactic acid bacteria isolated from traditional ripened foods and partial characterization of their bacteriocins. *LWT* **2023**, *173*, 114300. [CrossRef]
35. Campagnollo, F.B.; Margalho, L.P.; Kamimura, B.A.; Feliciano, M.D.; Freire, L.; Lopes, L.S.; Alvarenga, V.O.; Cadavez, V.A.P.; Gonzales-Barron, U.; Schaffner, D.W.; et al. Selection of indigenous lactic acid bacteria presenting anti-listerial activity, and their role in reducing the maturation period and assuring the safety of traditional Brazilian cheeses. *Food Microbiol.* **2018**, *73*, 288–297. [CrossRef]
36. Ortolani, M.B.T.; Yamazi, A.K.; Moraes, P.M.; Viçosa, G.N.; Nero, L.A. Microbiological quality and safety of raw milk and soft cheese and detection of autochthonous lactic acid bacteria with antagonistic activity against *Listeria monocytogenes*, *Salmonella* spp., and *Staphylococcus aureus*. *Foodborne Pathog. Dis.* **2010**, *7*, 175–180. [CrossRef] [PubMed]
37. Panebianco, F.; Giarratana, F.; Caridi, A.; Sidari, R.; De Bruno, A.; Giuffrida, A. Lactic acid bacteria isolated from traditional Italian dairy products: Activity against *Listeria monocytogenes* and modelling of microbial competition in soft cheese. *LWT* **2021**, *137*, 110446. [CrossRef]
38. Dalzini, E.; Cosciani-Cunico, E.; Monastero, P.; Bernini, V.; Neviani, E.; Bellio, A.; Decastelli, L.; Losio, M.N.; Daminelli, P.; Varisco, G. *Listeria monocytogenes* in Gorgonzola cheese: Study of the behaviour throughout the process and growth prediction during shelf life. *Int. J. Food Microbiol.* **2017**, *262*, 71–79. [CrossRef] [PubMed]
39. Ricci, A.; Allende, A.; Bolton, D.; Chemaly, M.; Davies, R.; Fernández Escámez, P.S.; Girones, R.; Herman, L.; Koutsoumanis, K.; Nørrung, B.; et al. *Listeria monocytogenes* contamination of ready-to-eat foods and the risk for human health in the EU. *EFSA J.* **2018**, *16*, e05134. [CrossRef]
40. Falardeau, J.; Trmčić, A.; Wang, S. The occurrence, growth, and biocontrol of *Listeria monocytogenes* in fresh and surface-ripened soft and semisoft cheeses. *Compr. Rev. Food Sci. Food Saf.* **2021**, *20*, 4019–4048. [CrossRef]

foods

MDPI

Article

Manufacture of a Potential Antifungal Ingredient Using Lactic Acid Bacteria from Dry-Cured Sausages

Tiago de Melo Nazareth *, Jorge Calpe, Carlos Luz, Jordi Mañes and Giuseppe Meca

Department of Food Science and Toxicology, Faculty of Pharmacy, University of Valencia, Ave. Vicent Andrés Estellés s/n, 46100 Burjassot, Spain

* Correspondence: tiago@uv.es; Tel.: +34-96-354-4959

Abstract: The growing interest in functional foods has fueled the hunt for novel lactic acid bacteria (LAB) found in natural sources such as fermented foods. Thus, the aims of this study were to isolate, identify, characterize, and quantify LAB's antifungal activity and formulate an ingredient for meat product applications. The overlay method performed a logical initial screening by assessing isolated bacteria's antifungal activity in vitro. Next, the antifungal activity of the fermented bacteria-free supernatants (BFS) was evaluated by agar diffusion assay against six toxigenic fungi. Subsequently, the antifungal activity of the most antifungal BFS was quantified using the microdilution method in 96-well microplates. The meat broth that showed higher antifungal activity was selected to elaborate on an ingredient to be applied to meat products. Finally, antifungal compounds such as organic acids, phenolic acids, and volatile organic compounds were identified in the chosen-fermented meat broth. The most promising biological candidates belonged to the *Lactiplantibacillus plantarum* and *Pediococcus pentosaceus*. *P. pentosaceus* C15 distinguished from other bacteria by the production of antifungal compounds such as nonanoic acid and phenyl ethyl alcohol, as well as the higher production of lactic and acetic acid.

Keywords: *Pediococcus pentosaceus*; *Lactiplantibacillus plantarum*; antifungal activity; organic acids; phenolic acids; volatile organic compounds

1. Introduction

Due to their high nutritional content, meat and its derivatives are an important food category in many people's diets. Consuming processed meats such as sausages, hot dogs, and luncheon meats has grown mainstream [1]. Between 1998 and 2018, global meat consumption rose by 58%, reaching a total of 360 million tonnes [2].

Dry-cured meat products are a type of traditional food that are manufactured and eaten across the globe. Their market dominance is well-known, owing to customers' stringent expectations for high-quality and safe foodstuffs. Consumption of these fungus-infected foods increases the risk of exposure to mycotoxins; this is a worldwide public health concern [3].

Dry-cured meats mostly comprise muscle tissue, and their surface physical-chemical characteristics, such as low water activity (A_w), neutrality to low pH, high salt content, and protein content, influence the microbial flora that grows on their exterior layers [4]. Although dry-fermented sausages have low A_w and high salt content, alterations of these properties influence the metabolism by facilitating mycotoxin biosynthesis [5].

Several species are recognized for producing mycotoxins, including *Aspergillus*, *Alternaria*, *Claviceps*, *Fusarium*, and *Penicillium* [6]. In certain environmental and substrate conditions over meat products, *Aspergillus flavus*, *A. parasiticus*, *A. ochraceus*, *P. nordicum*, *P. polonicum*, and *P. verrucosum* produce mycotoxins, posing a risk to consumers through their growth in salami, dry-cured hams, and other meat products [7,8].

Penicillium nordicum is the most important ochratoxin A (OTA) producing species frequently detected in cold-chain protein foods such as dry-cured ham, salami, and salted fish.

In particular, this fungus grows well at temperatures of about 15 °C and salt concentrations of more than 5% NaCl [9].

The consumption of cereals contaminated with mycotoxins may result in the accumulation of these toxins in the organs of both animals and humans, thereby increasing the risk of various acute and chronic diseases. This is due to the carcinogenic, mutagenic, genotoxic, teratogenic, neurotoxic, and estrogenic effects associated with mycotoxins [10]. Mycotoxin contamination of these products may occur at any stage along the manufacturing chain, from animals contaminated in feed through the end product's manufacture or storage. To make matters worse, despite having been detected in dry-cured meats and cheese worldwide, most nations do not regulate mycotoxins in this kind of food. Although the EU has published a new amending regulation and concluded that additional monitoring for OTA occurrence is required before setting maximum levels, this mycotoxin will soon be regulated (Commission Regulation (EC) No 2022/1370).

Various physical and chemical methods have been developed to manage the growth of fungi and their toxins. However, an effective strategy for reducing the occurrence of mycotoxins remains elusive. Furthermore, certain molds have developed resistance to chemical treatments and preservatives. Thus, reducing the prevalence of these molds in food production is of utmost importance, and significant efforts are being made to develop safe and efficient methods for this purpose. Biopreservation, which involves controlling the growth of one organism using natural substances, has garnered considerable interest in the past decade as a promising solution [11]. The quest to make healthier meat products has prompted studies to minimize saturated fat, salt, and cholesterol. Better composition of unsaturated fatty acids and integration of postbiotics, probiotics, and prebiotics were also encouraged. These functional additives may benefit human health while improving meat products' nutrition [12]. Thus, contemporary customers desire high-quality, safe, minimally processed, and chemical-free foodstuffs.

Lactic Acid Bacteria (LAB) are either naturally present in meals or introduced as pure cultures to a variety of dietary items. LAB have a GRAS classification (generally regarded as safe), and it is estimated that fermented foods make up 25% of the European diet and 60% of the diet in many developing nations [13]. LAB are often used as starting cultures in the production of acidophilus milk, yogurt, buttermilk, cottage cheeses, hard cheeses, and soft cheeses, among other dairy products [14]. The cohabitation of LAB and yeast is also crucial for the success of other biotechnological applications, such as the production of sourdough bread [15]. Ancient traditions of employing LAB in food and animal feed, together with a new understanding of the favorable health benefits of probiotic LAB use, imply that they might serve as viable alternatives to chemical preservatives.

Regarding the antifungal activity of bioprotective cultures, it is often the consequence of the synergistic impact of many compounds since organic acids are not the only known active molecules. Antifungal action may require other molecules, such as fatty acids [16], reuterin [17], cyclic dipeptides, and proteinaceous substances [18], among others.

Phenolic acids are the most representative subgroup of phenolic compounds; they are important for fermented products because of their relationship with the sensory characteristics of foods. LAB have the ability to metabolize and release bound phenolic acids in their free form, and some of them have demonstrated a broad spectrum of antimicrobial activity, making them an important preservative to consider for foods [19]. For instance, salicylic and gallic acid have shown antifungal potential against postharvest fungi such as *P. expansum* and *F. graminearum* [20,21].

The use of probiotics in meat products is regarded as an attractive strategy for enhancing their healthfulness, as the fermentation carried out by probiotics can generate health-improving compounds, typically through the hydrolysis of polysaccharides, proteins, and fats, as well as biologically active compounds such as peptides, organic acids, and conjugated linoleic acid [22]. In addition, fermented sausages are significant matrices for probiotic delivery since they may be taken without heat treatment, which increases the survival rates of bacteria and fungi.

Therefore, a continuous survey of OTA-producing strains, especially during ripening, could support safety assurance in dry-cured meat production. Against this background, the objective of this study was to isolate LAB with antifungal properties from traditional dry-cured sausages and to develop an antifungal ingredient with potential application in the manufacture of dry-cured meat products. Additionally, the antifungal properties of the ingredient were characterized, and its responsible metabolites were identified using various analytical techniques, including high-performance liquid chromatography with diode array detection (HPLC-DAD), Quadrupole Time-of-Flight (QTOF) mass spectrometry, and Headspace Solid-Phase Microextraction (HS-SPME).

2. Materials and Methods

2.1. Chemicals

Each analyte had at least 95% of purity. Sigma-Aldrich (St. Louis, MO, USA) provided the phenolic acids 1,2–dihydroxybenzene, 3–(4–hydroxy–3–methoxyphenyl) propionic, benzoic acid, caffeic acid, gallic acid, p–Coumaric acid, sinapic acid, syringic acid, vanillic acid, and vanillin. Bachem (Bubendorf, Switzerland) provided the DL–3–phenyllactic acid (PLA). MP Biomedicals (Santa Ana, CA, USA) provided the ferulic acid. The lactic acid and acetic acid were purchased from Sigma-Aldrich (St. Louis, MO, USA).

Solvents suitable for liquid chromatography (≥99.9% purity), including acetonitrile (ACN), ethyl acetate, formic acid, and methanol, were provided by VWR Chemicals (Radnor, PA, USA). Sigma-Aldrich supplied ammonium formate (≥99.5%), C18, magnesium sulfate ($MgSO_4$), sodium chloride (NaCl), and potato dextrose broth (PDB) (P6685). Liofilchem (Roseto degli Abruzzi, Teramo, Italy) provided de Man, Rogosa Sharpe broth (MRSb) (Oxoid CM359), and MRS agar (MRSa) (Oxoid CM361), plate count agar (PCA) (Oxoid CM0463) and potato dextrose agar (PDA) (Oxoid CM0139). Water was deionized with <18 MΩ/cm using a Milli-Q purification system (Millipore Corp., Bedford, MA, USA).

2.2. Sampling Isolation, Cultivation, and Preliminary Characterization of LAB Isolates

Sampling, isolation, cultivation, and preliminary characterization LABs were collected from various Spanish homemade-fermented products obtained from local shops in the Valencia region: "longaniza de payés", "fuet", "longaniza de Pascua", and "longaniza classic". Samples (20 g) were added to 180 mL of sterile physiological saline solution and were homogenized for 5 min. Simultaneously, bacteria were isolated by scraping a swab dipped in 0.1% peptone water (Liofilchem, Roseto degli Abruzzi, Teramo, Italy; Oxoid LP 0037) over the food surface. The swabs were then submerged in tubes containing MRSb to recover the bacteria, and the tubes were kept anaerobic at 37 °C for LAB culture. Appropriate dilutions were plated onto MRSa and cultures anaerobically for 48 h at 37 °C. Colonies were purified on MRSa after being randomly chosen from MRSa plates containing 15 to 300 colonies [23]. Microorganisms were first evaluated microscopically for Gram reaction and morphology as well as catalase production. Only Gram-positive and catalase-negative isolates were investigated, and the strains were maintained at −80 °C in MRSb containing 20% glycerol.

Biochemical and physiological tests were used to describe the microorganisms. The bacteria were evaluated for their capacity to grow at different temperatures (15, 30, and 45 °C), at different pH values (3, 5, and 6), and MRSb supplemented with NaCl (3, 6.5, and 10%).

2.3. Fungal Culture and Inoculum Preparation

Fungal strains purchased from the Spanish Type Culture Collection (Valencia, Spain) were *Aspergillus parasiticus* CECT 2681, *Penicillium commune* CECT 20767, *P. griseofulvum* CECT 2605–T, and *P. nordicum* CECT 2320. *A. flavus* ITEM 8111 was acquired from the Institute of Sciences for Food Production's microbial culture collection (Bari, Italy). The VTT Culture Collection provided *P. verrucosum* VTT D–01847 (Espoo, Finlandia). The fungal strains were stored at −80 °C in sterile glycerol (30% *v*/*v*).

The fungal strains were recovered by adding 1 mL of the glycerinated solution to 9 mL PDB media. After incubating the infected PDB for 72 h at 25 °C, aliquots were plated on PDA Petri plates to harvest spores. The experiments were performed with these spores.

2.4. Screening of Antifungal Properties by Overlay Method

The first screening of antifungal activity was performed using bacterial isolates grown overnight in 10 mL of MRSb at 37 °C. Then, the overlay assay was used to determine the antifungal activity of LAB isolates [24]. Bacterial suspensions (10 µL) were inoculated on MRSa plates and incubated at 37 °C for 48 h. Afterward, the agar plates were filled with 10 mL of soft (7% agar) PDA containing 10^6 conidia/mL and were incubated at 30 °C. The inhibition growth zone was determined 48 h later and measured on a cm scale. The experiments were conducted in triplicate.

2.5. Identification of Isolate by Matrix-Assisted Laser Desorption/Ionization Time-of-Flight Mass Spectrometry (MALDI-TOF MS) Fingerprinting

LAB was identified by extracting isolated cultures according to Maier et al. [25]. The MALDI-TOF MS was a Microflex L20 (Bruker Daltonics, Billerica, MA, USA) mass spectrophotometer equipped with an N_2 laser. The sample spectra were acquired in positive linear ion mode, with a voltage acceleration of 20 kV, and the mass ranged from 2000 to 20,000 Da. Moreover, the spectra were obtained in triplicate as suggested by MALDI Biotyper Realtime Classification (RTC). The identification of the samples was the spectra of the largest log score organism. Finally, the results were compared with the MBT 7854 y MBT 7311_RUO database (Bruker Daltonics).

2.6. Preparation of the Bacteria-Free Supernatant (BFS) and Development of a Postbiotic Antifungal Ingredient

Meat broth (MB) was formulated by modifying the composition of the MRSb, as plotted in Supplementary Material Table S1. Meat extract and yeast extract were substituted with freeze-dried pork loin at different concentrations: 2.00 (MB2), 4.00 (MB4), 8.00 (MB8), and 10.00 (MB10) g/L. The ingredients were mixed and dissolved in distilled water for 10 min, and the pH was adjusted to 6. Then, the media were autoclaved at 121 °C for 21 min. Afterward, the isolated LAB were defrosted and incubated in MRSb at 37 °C for 12 h to reach the exponential growth phase. Subsequently, the LAB were placed at 10^7 CFU/mL in 50 mL of the MB and incubated at 37 °C for 48 h. After fermentation, the LAB were mechanically separated by centrifugation at $4000\times g$ for 10 min to separate the BFS. Finally, BFS were kept at −80 °C for 24 h and freeze-dried to obtain the postbiotic antifungal ingredient [26]. Freeze-dried BFS of fermented MRSb was carried out as a control group.

2.7. Determination of Sensitivity of Food-Borne Toxigenic Fungi to Bacteria Isolates

2.7.1. In Vitro Antifungal Activity of Samples by Agar Diffusion Method

The agar diffusion test was used initially to determine the sensitivity of food spoilage and toxin-producing fungi to the BFS. The method was previously described by Fredua-Agyeman et al. [27] with some modifications.

Petri dishes containing PDA were infected with collected fungal conidial grown on PDA plates. The conidia of the fungal strains described in Section 2.3 were scraped off the agar and distributed on another PDA plate using sterile cotton swabs. Then, using sterile pipette tips, 10 mm diameter wells were created, and each well was filled with 100 µL of freeze-dried BFS resuspended in sterile water at a concentration of 250 g/L. A sterile water solution containing freeze-dried MRSb was elaborated as a control. After that, the PDA plates were incubated at 25 °C for 72 h to enable either fungal growth or BFS diffusion through the agar. Finally, the antifungal activity was determined by measuring the inhibition zone around the well on a cm scale.

2.7.2. Microdilution Susceptibility Test to Freeze-Dried BFS

To establish the Minimum Inhibitory Concentration (MIC) and the Minimum Fungicidal Concentration (MFC), the microdilution test was carried out in 96-well microplates following the CLSI M38-A2 [28] criteria, with a few changes. Firstly, the microplates were filled with 100 µL of PDB. Next, plates were filled with PDB containing freeze-dried BFS at a concentration of 340 g/L and then serially diluted 2-fold to reach the final concentration of 0.33 g/L. A volume of 100 µL of PDB containing 5×10^3 conidia/mL of the toxigenic fungi described by Section 2.3 was added to each well. The positive control consisted of contaminated PDB with fungal spores, whereas the negative control used PDB that had not been infected. Then, the microplates were incubated for 72 h at 25 °C. Each dose was tested in four replicates, and the experiment was carried out two times ($n = 8$).

Three parameters were determined: MIC50, MIC100, and MFC. After the incubation period, the MIC50 was established as the minimum concentration at which the BFS inhibited 50% of the fungal colonies, whereas the MIC100 was defined as the minimum concentration of BFS at which no growth was seen in the microplate. The MFC was then determined by reculturing 10 µL of the concentration matching the MIC100 and higher concentrations assessed on PDA plates. After 72 h of incubation at 25 °C, the MFC concentration that prevented fungal growth was determined. The plates were rechecked after seven days of incubation.

2.8. *Extraction and Quantification of Organic Acids*

The organic acids were extracted from the fermented MRSb and MB10, according to Özcelik et al. [29]. First, 5 mL of the BFS was aliquoted into a separated tube and treated with 1 mL of metaphosphoric acid. After 2 min of homogenization, the tubes were centrifuged at 4000× *g* for 10 min. The BFS was filtered using a membrane filter (0.45 µm). Finally, the BFS was diluted in water: formic acid (0.1%) and injected into an HPLC system.

The system applied for the chromatographic determination was an Agilent 1100 (Santa Clara, CA, USA) HPLC equipped with an autosampler, binary pump, and vacuum degasser. Results were expressed in g/L. The column used as the stationary phase was a Rezex™ ROA-Organic Acid H+ (8%) (150 × 7.8 mm, Ea Phenomenex®, Torrance, CA, USA). The mobile phases consisted of water as solvent A and ACN as solvent B, both acidified with 0.1% formic acid and eluted using the gradient (5% B at 0 min, 95% B at 30 min, and 5% at 35 min). Each analysis began with a 3-min equilibration of the column. The flow rate was set at 0.3 mL/min, and the sample volume was set to 20 µL. The photo diode array detector (DAD) (Agilent G1315B) was applied at 210 nm to monitor the chromatogram.

Individual stock solutions of each examined organic acid were produced by dissolving the compounds in the mobile phase. At least six-point calibration curves were created for acetic and lactic acid. The concentrations of organic acids were determined by comparing the peak area to the calibration curve standard area. Results were expressed in g/L.

2.9. *Extraction and Quantification of Phenolic Acids*

The phenolic acids of the fermented MRSb and the MB10 were purified using the QuEChERS extraction described by Brosnan et al. [30]. In 50-mL tubes, a solution of 4 g $MgSO_4$, 1 g NaCl, and 10 mL ethyl acetate with 1% formic acid (*v*/*v*) were added, followed by the addition of 10 mL of the BFS. For 1 min, the tubes were vortexed and cooled on ice. Next, the solution was centrifuged at 4 °C and 4000× *g* for 10 min. The supernatant was combined with 900 mg of $MgSO_4$ and 150 mg of C18 and vortexed for 1 min before centrifuging as previously described. Finally, the samples were filtered with a 0.22 µm pore filter and evaporated under nitrogen flow.

The phenolic acids analysis was performed in an Agilent 6450 Ultra High-Definition Accurate Mass QTOF-MS equipment, coupled to an Agilent Dual Jet Stream Electrospray Ionization (ESI) interface in negative ionization mode under the conditions described by Denardi-Souza et al. [31]. MassHunter Qualitative Analysis software version B.08.00 was used to handle data integration and elaboration. The results were expressed in µg/L.

2.10. Volatile Organic Compounds (VOCs) Determination

Volatile Organic Compounds (VOCs) emitted by the bacteria after fermentation in MRSb and ME10 media were determined by head-space solid phase microextraction (HS-SPME) technique and subsequent analysis in a gas chromatograph coupled to a mass spectrometer (GC-MS) following the methodology of Luz et al. [32] with minor modifications.

For this purpose, 10 mL of the BFS, obtained as described in Section 2.6, was placed in a 20 mL glass vial. The VOCs were extracted for 45 min at 50 °C with a divinylbenzene/carbon wide range/polydimethylsiloxane (DVB/C-WR/PDMS) coated fiber (80 μm × 10 mm) (Agilent Technologies, Santa Clara, CA, USA). Then, the fiber was desorbed for 10 min at 250 °C in splitless mode on an Agilent 7890A gas chromatograph coupled to an Agilent 7000A triple quadrupole mass spectrometer equipped with an electron impact (EI) source. The column used for chromatographic separation was an HP-5MS (30 m × 0.25 mm, 0.25 μm) (Agilent Technologies). The temperature ramp was programmed as follows: 40 °C held for 2 min and raised to 160 °C at 6 °C/min; then, it was ramped up to 250 °C at 10 °C/min and held for 4 min. The carrier gas was helium (99.99%) at a flow rate of 2.5 mL/min. Compound detection was performed in Full Scan mode in an m/z range of 40–450 Da.

The compounds were identified by comparison of their mass spectra with those recorded in the NIST 09 library. In addition, linear retention indices (LRI) were calculated based on the retention time of a solution of alkanes (C8–C20) tested under the same conditions as the samples and compared with the existing literature.

2.11. Statistical Analysis

For statistical analysis, GraphPad Prism version 3.0 software was employed. The differences ($p \leq 0.05$) between the groups of the organic acids and phenolic acids composition were analyzed by a one-way ANOVA followed by the Tukey post hoc test for multiple comparisons.

3. Results and Discussion

3.1. Isolation, Screening, and Identification of Antifungal Bacteria Strains

A total of 102 bacteria were isolated from different handmade dry Spanish sausages. Among these, 42 isolates were classified as LAB since they were gram-positive, could grow in micro-aerobic conditions, and lacked catalase activity (Table 1). Further characterization of bacteria strains highlighted that these isolates grew in the following conditions: temperature ranging from 15 to 45 °C; pH ranging from 3.5 to 6; and growth in MRS-NaCl at 3, 6.5, and 10%. Therefore, these bacteria were selected, and after the characterization, a first screening was performed using the overlay approach to assess the antifungal properties of the 42 bacteria isolated; they were assayed against six spoilage fungal strains in vitro that commonly contaminate cured sausage products. As presented in Table 2, only 14 bacteria isolates (C11, C12, C13, C15, C20, C28, C56, C58, C60, C66, C69, C71, C72, and C79) demonstrated inhibitory activity against all fungi examined, with varying degrees of inhibition depending on the fungal strain tested. In particular, *P. griseofulvum* CECT 2605-T and *P. commune* CECT 20767 were the most susceptible strains, with inhibition zones ranging from 1.5–3.0 cm and 1.0–3.0 cm, respectively. In contrast, *Aspergillus* strains (*A. parasiticus* CECT 2681 and *A. flavus* ITEM 8111) were the most resistant to the LAB, with inhibition zones ranging from 0.3–1.7 cm. An example of overlay methodology is given in Figure 1. These findings agree with those obtained by previous authors [33], which noticed that *Aspergillus* strains are more resistant to LAB than other fungal genera.

Table 1. Characterization of LAB isolated from dry-cured sausages.

	Preliminary Characterization			Growth at Temperature	Growth in pH	Growth in MRS-NaCl (%)	
Bacteria	Morphology	Gram *	Catalase	15 to 45 °C	3.5 to 6	3 to 6.5	10
C1	Coccoid	P	-	+	+	+	-
C4	Coccoid	P	-	+	+	+	-
C5	Coccoid	P	-	+	+	+	-
C6	Coccoid rod	P	-	+	+	+	-
C11	Coccoid	P	-	+	+	+	-
C12	Coccoid	P	-	+	+	+	-
C13	Coccoid	P	-	+	+	+	-
C14	Coccoid rod	P	-	+	+	+	+
C15	Coccoid	P	-	+	+	+	+
C16	Rod-shaped	P	-	+	+	+	+
C17	Rod-shaped	P	-	+	+	+	+
C19	Coccoid	P	-	+	+	+	+
C20	Rod-shaped	P	-	+	+	+	+
C27	Rod-shaped	P	-	+	+	+	+
C28	Coccoid	P	-	+	+	+	+
C29	Rod-shaped	P	-	+	+	+	+
C30	Rod-shaped	P	-	+	+	+	+
C31	Coccoid	P	-	+	+	+	+
C35	Coccoid	P	-	+	+	+	-
C36	Coccoid	P	-	+	+	+	+
C37	Coccoid	P	-	+	+	+	+
C38	Coccoid	P	-	+	+	+	-
C39	Coccoid	P	-	+	+	+	+
C44	Coccoid	P	-	+	+	+	+
C45	Coccoid	P	-	+	+	+	+
C46	Coccoid rod	P	-	+	+	+	-
C47	Coccoid	P	-	+	+	+	+
C48	Coccoid	P	-	+	+	+	+
C50	Coccoid	P	-	+	+	+	+
C56	Coccoid	P	-	+	+	+	+
C57	Coccoid	P	-	+	+	+	+
C58	Coccoid	P	-	+	+	+	+
C59	Coccoid	P	-	+	+	+	-
C60	Rod-shaped	P	-	+	+	+	+
C66	Coccoid	P	-	+	+	+	+
C68	Coccoid	P	-	+	+	+	+
C69	Coccoid	P	-	+	+	+	+
C70	Coccoid	P	-	+	+	+	-
C71	Coccoid	P	-	+	+	+	+
C72	Coccoid	P	-	+	+	+	+
C79	Coccoid	P	-	+	+	+	+
C80	Coccoid	P	-	+	+	+	+

* P means positive; + signifies visual growth, and - means no visual growth.

Table 2. Antifungal activity of LAB by overlay technique.

Selected Bacteria	*Aspergillus parasiticus*	*Aspergillus flavus*	*Penicillium commune*	*Penicillium verrucosum*	*Penicillium griseofulvum*	*Penicillium nordicum*
	Inhibition Zone (cm)					
C1	0.0	0.0	2.4	2.0	3.0	1.0
C4	0.0	0.1	1.5	1.7	3.0	0.9
C5	0.3	0.0	2.5	1.4	3.0	1.0
C6	0.3	0.0	2.7	1.7	3.0	1.1
C11	1.0	0.4	3.0	2.0	3.0	1.4
C12	0.5	0.3	3.0	1.7	2.3	1.3
C13	0.7	0.6	3.0	1.8	2.1	1.0
C14	0.5	0.0	2.0	1.4	3.0	1.1
C15	0.4	0.6	3.0	1.6	3.0	1.2
C16	0.5	0.0	3.0	1.5	0.6	0.7
C17	0.3	0.0	0.6	2.0	3.0	0.8
C19	0.3	0.0	1.6	1.7	3.0	1.0
C20	0.8	0.9	3.0	1.7	1.9	1.3
C27	0.5	0.0	0.9	1.7	1.3	0.5
C28	1.0	0.8	3.0	1.7	1.8	1.3
C29	0.4	0.0	2.9	1.7	1.2	0.9
C30	0.6	0.0	3.0	1.4	1.0	0.9
C31	0.5	0.0	3.0	1.7	1.8	0.9
C35	0.4	0.0	3.0	1.7	1.6	1.1
C36	0.0	0.0	2.7	1.6	3.0	1.0
C37	0.4	0.0	3.0	2.5	1.0	0.8
C38	0.2	0.0	2.4	1.5	3.0	1.1
C39	0.4	0.0	0.0	1.8	1.6	0.4
C44	2.0	0.0	1.4	1.7	3.0	1.0
C45	0.4	0.0	1.1	1.6	3.0	0.8
C46	0.2	0.0	0.5	2.0	3.0	0.7
C47	0.3	0.0	3.0	1.6	2.5	1.1
C48	0.5	0.0	3.0	1.7	1.5	1.1
C50	0.9	0.0	2.5	1.5	1.8	0.8
C56	1.5	1.2	3.0	2.1	1.6	1.1
C57	1.0	0.0	2.8	1.8	0.2	0.6
C58	1.5	1.4	3.0	1.5	3.0	1.5
C59	0.5	0.0	0.6	2.0	1.5	0.5
C60	1.2	1.7	1.6	1.5	2.5	1.2
C66	1.0	1.1	1.0	2.1	0.8	0.6
C68	1.0	0.0	3.0	2.2	2.5	1.1
C69	0.4	0.5	3.0	2.5	3.0	1.1
C70	0.0	0.0	3.0	1.3	3.0	1.2
C71	1.5	1.2	2.6	1.5	3.0	1.3
C72	1.0	1.1	3.0	1.5	1.5	0.9
C79	0.4	0.3	2.7	1.7	3.0	1.1
C80	0.0	0.0	2.7	1.7	3.0	1.0

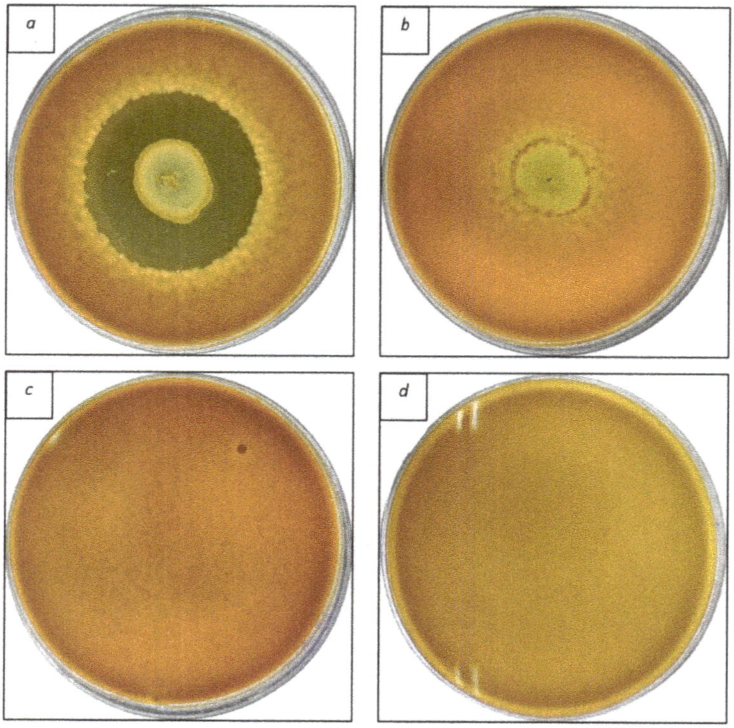

Figure 1. Screening of lactic acid bacteria (LAB) through overlay method. Strong inhibitory action (**a**); slightly inhibitory effect (**b**); no inhibitory action (**c**); control (**d**).

The 14 gram-positive bacteria with antifungal activity against all fungal strains were identified by MALDI-TOF-MS and classified regarding the MTB 7854 and MBT 7311_RUO databases (Bruker Daltonics). The classification to species level was the following: *Pediococcus pentosaceus* C11, *P. pentosaceus* C12, *P. pentosaceus* C13, *P. pentosaceus* C15, *Lactiplantibacillus plantarum* C20, *P. pentosaceus* C28, *P. pentosaceus* C56, *P. pentosaceus* C58, *L. plantarum* C60, *P. pentosaceus* C66, *P. pentosaceus* C69, *P. pentosaceus* C71, *P. pentosaceus* C72, and *P. pentosaceus* C79. Thus, the bacteria identified with potential inhibitory properties were studied to develop an antifungal ingredient against dry-cured meat spoilage agents.

The LAB with antifungal activity are poorly documented, while more attention has been exploited its antibacterial activity. The LAB that produces bacteriocins are isolated from a severe category of food. For example, Delcarlo et al. [34] isolated 22 LAB with antimicrobial action from mussels on the Argentina coast. Parlindungan et al. [35] identified novel probiotic candidates from fermented meats and characterized them by bacteriocin production. Furthermore, Ivanovic et al. [36] isolated, identified, and characterized antibacterial LAB from traditional cheese.

Only a few studies from scientific literature cite a broad spectrum of antifungal activity for LAB. Most of them showed higher strain-specific antifungal activity against one or two mold species [37]. For this reason, one of the goals of this study is to isolate a large-spectrum strain. In this work, we observed that antifungal activity was strain-dependent, as well as fungal species and methodologies analyzed. Moreover, it seems that dry-cured sausages can be used as a potential source of antifungal LAB because they have significant antagonistic properties against the microorganisms tested.

3.2. *In Vitro Antifungal Study of the MB Formulated*

Two methods were performed to determine the fungal sensitivity to bacterial isolates: agar diffusion test and microdilution of BFS in 96-well plates.

The MB was formulated by changing the composition of the MRSb by subtracting the yeast extract and meat extract and adding freeze-dried pork loin as a protein source at 2.00, 4.00, 8.00, and 10.00 g/L (MB2, MB4, MB8, and MB10, respectively). These new formulations were fermented with the 14 bacteria strains identified in Section 3.1, and a first qualitative evaluation was performed using the agar-diffusion test to assess the antifungal properties. Moreover, the fermented MRSb was also prepared with the same bacteria strains to compare the effectiveness of the newly formulated ingredient.

After fermentation, only the MB10 exhibited antifungal activity similar to the fermented MRSb used as a control reference. As presented in Table 3, the inhibition halos obtained varied according to the fungal strain tested and the bacteria used as inoculum. In particular, the most sensitive strains to the MB10 were *P. commune* CECT 20767, *P. nordicum* CECT 2320, and *P. verrucosum* VTT–D01847. These strains were susceptible to all the formulated MB, with inhibition halos observed on the PDA plates ranging from 0.2 cm (+) to inhibition halos greater than 0.4 cm (+++). The fungal strains with the most significant resistance were those of the *Aspergillus* genera, in which only *Pediococcus pentosaceus* C12, *Pediococcus pentosaceus* C15, and *Lactiplantibacillus plantarum* C60 showed inhibitory activity. This finding agrees with the previous characterization of the LAB isolates by overlay method since *Aspergillus* strains showed higher resistance to the LAB than *Penicillium* strains. For this reason, the formulation MB10, *P. pentosaceus* C12, *P. pentosaceus* C15, and *L. plantarum* C60 isolates were selected, and the preliminary results of the antifungal activity were confirmed through the MIC50, MIC100, and MFC determination in vitro. The MB10 and MRSb that were fermented for 48 h obtained the highest inhibition halo zone. The antifungal effect of MB10 fermented for 24, 48, and 72 h is plotted in Table S2.

Table 3. Antifungal activity of bacterial free supernatant (BFS) extract in agar diffusion method. LAB fermented MRS broth (MRSb) and a meat broth (MB10) for 48 h at 37 °C. Then, the supernatant was freeze-dried and resuspended at a concentration of 250 g/L. These bacteria were previously selected due to their antifungal effect in the overlay method.

Selected Bacteria	Antifungal Activity of the BFS of the Fermented MRS Broth and the Meat Broth (MB10) Fungal Strain											
	A. flavus		*A. parasiticus*		*P. commune*		*P. griseofulvum*		*P. nordicum*		*P. verrucosum*	
	MRSb	**MB10**	**MRSb**	**MB10**	**MRSb**	**MB10**	**MRSb**	**MB10**	**MRSb**	**MB10**	**MRSb**	**MB10**
C11	+	-	-	-	+	+	+	+	+	+	+	+
C12	+	+	++	+	++	++	++	++	+	+	++	++
C13	-	-	+	-	+	+	+	+	+++	+++	+++	+++
C15	++	++	++	++	+++	+++	+++	+++	+++	+++	+++	+++
C20	-	-	+	-	+	+	+	+	++	++	+	+
C28	+	-	-	-	+	+	+	+	+++	+++	+++	+++
C56	-	-	+	-	+	+	+	+	+	+	+	+
C58	-	-	-	-	+	+	++	++	+++	+++	+++	+++
C60	+	+	++	++	++	++	++	++	+++	+++	++	++
C66	+	-	-	-	++	++	+++	+++	++	++	+++	+++
C69	-	-	+	-	+	+	-	-	++	++	+	+
C71	-	-	+	-	+	+	-	-	+	+	+	+
C72	+	-	-	-	++	+	++	++	++	++	+++	+++
C79	+	-	-	-	++	+	++	++	++	++	+++	+++

(-) Represents no halo inhibition; (+) Represents a growth inhibition halo of 0.2 cm; (++) represents a growth inhibition halo of between 0.2 to 0.4 cm; (+++) represents a growth inhibition halo greater than 0.4 cm.

Similar findings in the literature indicated that a range of theories might explain LAB's antifungal activities; Schnürer and Magnusson [38] found that the suppression of mold growth on an agar plate is the consequence of a complex interaction of multiple chemicals and metabolites, all of which contribute to the overall antifungal activity.

The microdilution assay demonstrated that the BFS of the fermented MRSb was the most active extract against all fungi tested, regardless of the bacteria isolated (Table 4). The MIC50 obtained ranged from 5–21 g/L; the MIC100 ranged from 21–85 g/L; and the MFC ranged from 21–170 g/L. However, for MB10 formulation, the MIC50, MIC100, and MFC parameters ranged from 11–43 g/L, 21–85 g/L, and 43–170 g/L, respectively. It is important to underline that the BFS of the MRSb and the MB10 fermented by *P. pentosaceus* C15 showed lower concentrations required to achieve fungal growth against all the toxigenic strains tested compared to *P. pentosaceus* C12 and *L. plantarum* C60 fermented BFS. Although the BFS of the MRSb evidenced a higher antifungal effect, a poor difference was evidenced when compared to MB10. In some cases, there were no significant differences ($p \leq 0.05$) between the treatments, e.g., MB10 fermented by *P. pentosaceus* C15 had no statistical difference to MRSb comparing the average values of MIC100 and MFC (Table 4). In other words, *P. pentosaceus* C15 was able to ferment MB10 and produce antifungal compounds, which inhibited fungal growth at similar conditions to MRSb. In addition, MRSb fermented by *P. pentosaceus* C15 obtained MIC50, MIC100, and MFC values lower than *L. plantarum* C60 and *P. pentosaceus* C12, regardless of the fungal strain essayed.

Table 4. Minimal inhibitory concentration (MIC) and minimal fungicidal concentration (MFC) of bacterial-free supernatant (BFS). MRS broth (MRSb) and meat broth (MB10) were fermented by LAB for 48 h at 37 °C and centrifuged to obtain the BFS.

Fungi	**BFS Concentration (g/L) of the Fermented MRSb**								
	***Lactiplantibacillus plantarum* C60**			***Pediococcus pentosaceus* C12**			***Pediococcus pentosaceus* C15**		
	MIC [1]	**MIC [2]**	**MFC**	**MIC [1]**	**MIC [2]**	**MFC**	**MIC [1]**	**MIC [2]**	**MFC**
A. flavus	11	85	170	21	85	170	11	21	85
A. parasiticus	11	43	170	11	43	170	11	21	85
P. commune	21	43	85	21	43	85	11	21	85
P. griseofulvum	21	43	170	21	43	170	5	21	170
P. nordicum	11	43	85	11	43	85	5	21	21
P. verrucosum	21	43	85	21	43	85	11	43	43
Mean	16	50	128	18	50	128	9	25	82
Fungi	BFS concentration (g/L) of the MB10								
	MIC [1]	MIC [2]	MFC	MIC [1]	MIC [2]	MFC	MIC [1]	MIC [2]	MFC
A. flavus	21	85	170	43	85	170	21	43	85
A. parasiticus	21	43	170	21	43	170	21	43	85
P. commune	43	85	85	43	85	85	21	21	85
P. griseofulvum	43	85	170	43	85	170	11	21	170
P. nordicum	21	43	85	43	85	85	11	21	43
P. verrucosum	43	85	85	43	85	85	11	43	43
Mean	32 *	71	128	39 *	78 *	128	16 *	35	85

[1]: 50% of visual inhibition growth; [2]: 100% of visual inhibition growth. *: results were statistically analyzed by the *t*-test, comparing the Bacterial free supernatant (BFS) of MRSb with the BFS of MB10. Means are significantly different if $p \leq 0.05$.

The MRSb was applied in both experiments as a control group since previous studies have demonstrated its antifungal activity while being fermented by LAB and producing antimicrobial substances. For instance, Taroub et al. [39] identified *P. pentosaceus* and *L. plantarum* strains that showed good antifungal activity against *A. niger* and *A. carbonarius*. In addition, the strains showed a high capacity for the degradation of OTA, one of the main toxins produced by these fungi. However, the authors could not identify the compounds responsible for the antifungal and antimicotoxigenic activities of the strains. In other studies, the application of MRS fermented by *L. plantarum* inhibited the *A. flavus and F.*

verticillioides growth and reduced the production of aflatoxin and fumonisin in cereals such as maize [26].

Therefore, MRSb has proven to be an interesting medium for fermenting bacteria as antimicrobial metabolites are produced. However, the use of MRSb has some disadvantages, such as its high cost. In addition, the complexity of some of their ingredients that convert MRSb might not be allowed as an ingredient in meat products. Therefore, one of the objectives of the work was to develop a meat broth with similar antifungal characteristics but, at the same time, could be incorporated as an antifungal ingredient.

Overall, the *P. pentosaceus* C15 showed higher antifungal potential in the agar diffusion test and microdilution of BFS; hence, these results led us to select this strain for the fermentation of MB10 and elaborate a postbiotic antifungal ingredient.

3.3. Chemical Characterization of the Postbiotic Antifungal Ingredient

3.3.1. Organic and Phenolic Acid Production

The metabolites generated after fermentation of the antifungal strains selected were investigated in the BFS of the MRSb and MB10. As plotted in Figure 2, two organic acids, lactic acid and acetic acid, were identified in both formulations. For lactic acid, the mean values of this metabolite ranged from 2.49 to 3.00 g/L, whereas for acetic acid, the concentration detected was lower, with mean values ranging from 0.24 to 0.38 g/L. The highest amount of lactic acid was detected in the fermented MRSb (2.973 g/L) and MB10 (3.00 g/L) fermented by the *P. pentosaceus* C15 strain. There were no statistical differences in lactic acid production between these two formulations ($p \leq 0.05$). Regarding acetic acid, the higher concentration ($p \leq 0.05$) was also quantified in the fermented MB10 with *P. pentosaceus* C15 (0.38 g/L), followed by the MRSb of *P. pentosaceus* C15 (0.32 g/L) and the fermented MRSb with *L. plantarum* C60 (0.31 g/L).

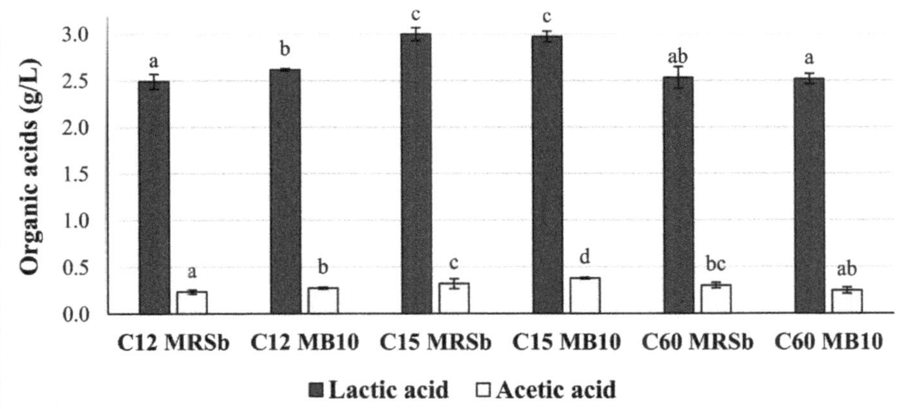

Figure 2. The concentration of organic acids produced by LAB in meat broth (MB10) and MRS broth (MRSb) after incubation for 48 h at 37 °C. The meat broth was prepared with 10% freeze-dried loin pork and fermented by *P. pentosaceus* C12, *P. pentosaceus* C15, and *L. plantarum* C60. Different letters represent statistical differences in the same group of molecules between treatments ($p \leq 0.05$).

Concerning the phenolic acids, a total of 12 different metabolites, described in the literature as antifungals, were identified and quantified through the UHPLC Q-TOF MS technique. Among them, it is noteworthy that the concentration of benzoic acid, DL–3–phenyllactic acid, syringic acid, and vanillic acid stood out regardless of the fermented broth and bacteria tested (Table 5). In general, it could be noted that MRSb produced higher concentrations of phenolic acids compared to its analog MB10. However, in most cases the difference was not significant ($p \leq 0.05$). In view of these results, we suggest further studies to evaluate the synergistic effect of these four compounds in order to evaluate their antifungal activity and to increase the understanding about the mechanism of antimicrobial action of LAB.

Table 5. The concentration of phenolic compounds in Man, Rogosa and Sharpe broth (MRSb) and meat broth (MB10) fermented by antifungal LAB. The broths were fermented for 48 h at 37 °C.

Phenolic Compound	The Concentration of Phenolic CompoundsMean ± SD (μg/L)					
	Pediococcus pentosaceus C12		*Pediococcus pentosaceus* C15		*Lactiplantibacillus plantarum* C60	
	MRSb	MB10	MRSb	MB10	MRSb	MB10
1,2–Dihydroxybenzene	9.39 ± 2.02 ^{a}	2.51 ± 0.41 ^{b}	5.86 ± 2.74 ^{cd}	5.64 ± 2.22 ^{cd}	7.12 ± 1.3 ^{d}	4.64 ± 1.39 ^{c}
Propionic acid	7.14 ± 1.79 ^{a}	7.08 ± 1.34 ^{a}	9.66 ± 2.64 ^{a}	9.36 ± 0.62 ^{a}	6.19 ± 2.85 ^{a}	13.44 ± 5.58 ^{b}
Benzoic acid	31.65 ± 4.42 ^{a}	18.09 ± 5.35 ^{b}	24.69 ± 0.72 ^{c}	22.12 ± 7.32 ^{bc}	19.38 ± 5.91 ^{bc}	23.18 ± 0.35 ^{bc}
Caffeic acid	4.07 ± 1.46 ^{a}	3.99 ± 2.46 ^{a}	8.37 ± 1.26 ^{b}	7.72 ± 1.87 ^{bc}	3.40 ± 0.97 ^{a}	6.27 ± 1.55 ^{c}
DL–3–Phenyllactic acid	18.17 ± 3.52 ^{a}	18.30 ± 1.54 ^{ab}	31.90 ± 7.49 ^{c}	22.68 ± 3.25 ^{b}	16.98 ± 3.66 ^{a}	15.85 ± 0.61 ^{a}
Ferulic acid	n.d	n.d	4.79 ± 1.81 ^{ab}	5.85 ± 4.02 ^{ab}	3.70 ± 0.72 ^{a}	6.18 ± 1.26 ^{b}
Gallic acid	n.d	n.d	8.04 ± 2.83 ^{a}	6.06 ± 1.88 ^{a}	6.58 ± 2.05 ^{a}	7.44 ± 1.25 ^{a}
p–Coumaric acid	22.59 ± 9.91 ^{a}	9.37 ± 1.72 ^{b}	6.28 ± 0.07 ^{b}	7.56 ± 0.44 ^{b}	5.29 ± 1.47 ^{b}	4.99 ± 2.03 ^{b}
Sinapic acid	4.87 ± 1.04 ^{ab}	6.43 ± 0.44 ^{ac}	5.85 ± 2.38 ^{ad}	7.86 ± 3.66 ^{cd}	3.54 ± 0.86 ^{b}	7.22 ± 1.32 ^{cd}
Syringic acid	26.78 ± 3.52 ^{a}	19.47 ± 5.28 ^{b}	17.73 ± 4.12 ^{b}	21.12 ± 0.39 ^{b}	17.42 ± 1.40 ^{b}	12.83 ± 3.40 ^{c}
Vanillic acid	20.84 ± 4.56 ^{ac}	18.81 ± 2.82 ^{a}	30.83 ± 6.19 ^{b}	25.18 ± 1.13 ^{d}	27.60 ± 2.94 ^{bd}	24.00 ± 2.12 ^{cd}
Vanillin	11.68 ± 1.26 ^{a}	13.76 ± 1.08 ^{a}	16.85 ± 5.47 ^{a}	13.73 ± 1.72 ^{a}	7.57 ± 1.65 ^{b}	8.46 ± 1.51 ^{b}

Different letters mean a significant statistical difference in the concentration of phenolic compounds among fermented broths ($p \leq 0.05$). n.d means not detected.

The selected microorganisms grew on the MB10 and produced a broad variety of metabolites previously described in the literature as antimicrobial compounds. The phenolic acid most produced by *P. pentosaceus* C12 in MB10 was syringic acid, showing values of 19.47 μg/L: these values are significantly higher than bacteria C60. Aziz et al. [40] demonstrated that syringic acid at 300 mg/L inhibited *A. flavus and A. parasiticus* growth and aflatoxin production. In addition, Ren et al. [41] demonstrated that lower doses of syringic acid (100 mg/L) avoided the growth of *A. niger*. Therefore, it seems that a higher concentration of phenolic acids is required to achieve an antifungal effect when they are applied in isolation. These findings reinforce the hypothesis that the LAB antifungal effect is obtained via a synergistic effect.

Regarding *P. pentosaceus* C15 and *L. plantarum* C60, vanillic acid was the most produced phenolic acid, reaching values of 25.18 and 24.00 μg/L in MB10 and 30.83 and 27.60 μg/L in MRSb, respectively. An oxidized derivative of vanillin, vanillic acid, is a monohydroxybenzoic acid composed of a 4–hydroxybenzoic acid with a methoxy group at position 3 [41]. It has been noted for its antioxidant properties, but its antifungal activity is poorly reported, and more research is required. In contrast, its precursor, vanillin, has been the subject of extensive research, and its antifungal activity varies depending on the microorganism tested [42,43].

These findings also suggest that the antifungal ingredient obtained from MB10 could be incorporated into meat food products to provide a significant source of phenolic and organic acids.

LAB can synthesize a wide variety of antifungal compounds such as organic acids, phenolic acids, antimicrobial peptides, diacetyl, and reuterin [44]. During fermentation, carbon metabolism produces organic acids such as lactic acid, acetic acid, and propionic acid [45]. Among these metabolites, the most studied are organic acids. It is important to mention that these metabolites can constitute synergistic activity. However, this synergistic activity's exact mechanism is unknown [46].

The results also suggest that the combination of organic acids and phenolic acids produced by LAB could be responsible for the inhibitory activity of the spoilage fungi in vitro. However, it is essential to underline that the antifungal properties of the MB10 are probably not exclusively produced by organic and phenolic acids; for instance, several volatile substances can act synergistically and potentialize the antifungal properties [16].

Moreover, the decrease in pH collaborates to a more efficient antifungal activity [47]. In this context, Peyer et al. [48] found that organic acids and phenolic acids generated by some LAB strains as antifungal metabolites have synergistic actions against *F. culmorum*. Likewise, in bakery products, organic acids and antifungal peptides produced by LAB showed a significant synergistic effect which allowed biopreservation and enhanced the shelf life of quinoa and rice bread [49].

In addition to the synergistic compound effect, the combination of different microorganisms can also promote a positive effect. For instance, Ruggirello et al. [50] associated yeast with LAB and obtained a great antifungal effect against strains of *Aspergillus* and *Penicillium* genera, fermenting cocoa beans. The authors reported that the antifungal potential could be a result of association between organic acids of LAB with proteinaceous substances of yeasts. According to Christ-Ribeiro et al. [51], phenolic chemicals avoid fungal development by inhibiting the production of cell wall components including glucan, chitin, and mannoproteins, as well as cell membrane components like ergosterol. This process happens by damaging the cell wall and cell membrane, affecting nutrient influx regulation. As a result, phenolic chemicals impede fungal cell production of proteins, amino acids, and sphingolipids. Moreover, they obstruct electron transit and the preservation of cellular integrity.

3.3.2. VOC Analysis in MB10 Formulation

The fermented meat broth VOCs (MB10) were analyzed through HS-SPME coupled to the GC-MS technique. A total of 24 compounds were identified through MS comparison with the NIST library and calculation of the LRI and belonged to several chemical classes: acid (2), alcohol (8), aldehyde (6), ketone (5), and pyrazine (2) (Table S3). Moreover, the percentage peak area (%PA) of each identified compound was calculated, and the results obtained are reported in Table 6.

Table 6. Percentage area (%) of the volatile organic compounds (VOCs) identified in the formulated meat broth (MB10) fermented with different lactic acid bacteria strains (*P. pentosaceus* C12, *P. pentosaceus* C15, and *L. plantarum* C60). Results are expressed as mean ± standard deviation.

N°	Rt	Compound	Control	C15	C60	C12
			Acid			
1	3.11	Acetic acid	n.d	13.74 ± 1.12 [a]	11.11 ± 0.62 [ab]	8.95 ± 2.63 [b]
2	15.49	Nonanoic acid	n.d	5.45 ± 1.69 [a]	6.94 ± 1.00 [a]	3.70 ± 1.13 [a]
			Alcohol			
3	8.97	2–Octanol	8.12 ± 2.16 [a]	1.77 ± 0.16 [b]	0.39 ± 0.05 [b]	0.87 ± 0.28 [b]
4	11.19	1–Octanol	2.16 ± 0.07 [a]	2.70 ± 0.10 [ab]	3.03 ± 0.25 [ab]	4.16 ± 1.11 [b]
5	11.62	2–Nonanol	n.d	0.58 ± 0.08 [a]	0.53 ± 0.14 [a]	0.66 ± 0.06 [a]
6	13.43	Phenylethyl alcohol	1.24 ± 0.13 [a]	6.57 ± 0.45 [b]	1.60 ± 0.20 [a]	2.44 ± 0.88 [a]
7	13.6	1–Nonanol	1.26 ± 0.17 [a]	3.65 ± 0.05 [b]	1.95 ± 0.18 [ab]	3.80 ± 1.49 [b]
8	15.88	1–Decanol	n.d	1.22 ± 0.12 [a]	2.18 ± 0.70 [a]	1.86 ± 0.35 [a]
9	16.27	2–Undecanol	0.86 ± 0.03 [a]	1.15 ± 0.16 [a]	1.09 ± 0.03 [a]	1.76 ± 0.42 [b]
			Aldehyde			
10	6.25	Heptanal	3.51 ± 0.43 [ab]	1.04 ± 0.21 [b]	2.13 ± 0.35 [b]	5.71 ± 2.66 [a]
11	8.89	Octanal	13.45 ± 2.80 [a]	3.81 ± 0.33 [b]	3.75 ± 0.63 [b]	4.98 ± 0.06 [b]
12	11.02	Benzeneacetaldehyde	20.31 ± 2.76 [a]	19.27 ± 2.39 [a]	19.82 ± 3.34 [a]	18.36 ± 3.74 [a]
13	11.42	Nonanal	1.24 ± 0.13 [a]	2.91 ± 0.53 [a]	2.43 ± 0.16 [a]	2.94 ± 1.44 [a]
14	15.82	2–Decenal	0.97 ± 0.18 [a]	0.53 ± 0.07 [a]	1.44 ± 0.13 [a]	1.47 ± 0.72 [a]
15	18.5	Dodecanal	1.28 ± 0.07 [a]	0.27 ± 0.05 [b]	0.81 ± 0.19 [ab]	0.94 ± 0.56 [ab]

Table 6. *Cont.*

N°	Rt	Compound	Control	C15	C60	C12
			Alkane			
16	10.77	Decane, 2–methyl	n.d	0.85 ± 0.05^{a}	2.42 ± 0.04^{b}	1.81 ± 0.63^{b}
17	11.69	Undecane	1.66 ± 0.41^{ab}	1.47 ± 0.31^{ab}	0.78 ± 0.22^{a}	2.21 ± 0.66^{b}
			Ketone			
18	5.78	2–Heptanone	n.d	1.73 ± 0.40^{a}	2.80 ± 0.53^{a}	2.82 ± 0.72^{a}
19	8.71	2–Octanone	2.52 ± 0.45^{a}	2.89 ± 0.19^{a}	1.92 ± 0.15^{a}	2.32 ± 0.95^{a}
20	11.28	2–Nonanone	1.59 ± 0.03^{ab}	2.82 ± 0.55^{b}	0.88 ± 0.05^{a}	2.13 ± 1.28^{ab}
21	16.11	2–Undecanone	n.d	0.63 ± 0.10^{a}	1.29 ± 0.38^{b}	0.82 ± 0.12^{ab}
22	20.45	2–Tridecanone	n.d	0.60 ± 0.08^{a}	2.13 ± 0.35^{b}	1.13 ± 0.20^{a}
			Pyrazine			
23	4.55	Pyrazine, methyl–	10.24 ± 1.00^{a}	5.30 ± 0.09^{b}	6.91 ± 1.73^{b}	4.34 ± 1.11^{b}
24	6.37	Pyrazine, 2,5–dimethyl	29.58 ± 2.79^{a}	19.05 ± 1.87^{b}	21.68 ± 0.80^{b}	19.79 ± 0.07^{b}

n.d = no-detected. Different letters represent statistical differences in the same molecules between treatments ($p \leq 0.05$).

Aldehydes were the most abundant compounds in the samples analyzed and represented a proportion between 27.8–40.8% of the total VOCs detected (Figure 3). In particular, five linear aldehydes (heptanal, octanal, nonanal, 2–decenal, and dodecanal) were identified in all MB10; nonetheless, the greatest observed concentration was of an aromatic aldehyde, benzeneacetaldehyde, which %PA ranged between 18.4 and 20.3% depending on the formulation analyzed. Saturated aldehydes are lipid-derived volatiles produced mainly by the oxidation of oleic acid, a characteristic fatty acid of raw pork meat [52–54]. Regarding benzeneacetaldehyde, this compound could be synthesized using phenylalanine as a precursor through the Maillard reaction during the sterilization step of the MB10 since it is mainly detected in cooked meat [55,56]. It was noted that aldehyde content in the fermented MB10 was statistically lower ($p \leq 0.05$) in comparison with the control group, which corroborated the findings of Kwaw et al. [57], which described a decline of aldehydes in mulberry juice when LAB fermentation was applied.

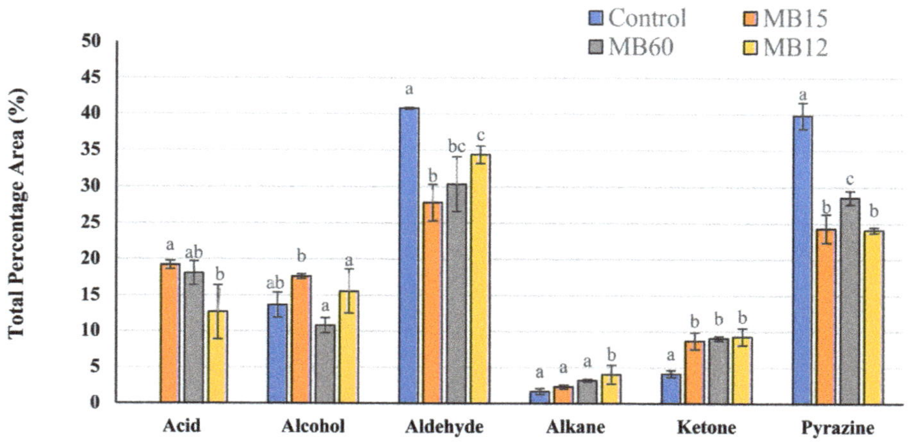

Figure 3. Total percentage area (%) of the chemical classes identified in the volatile fraction of the Meat Broth formulated with 10% of lyophilized pork loin and fermented by *P. pentosaceus* C12 (MB12), *P. pentosaceus* C15 (MB15), and *L. plantarum* C60 (MB60). Different letters represent statistical differences in the same group of molecules between treatments ($p \leq 0.05$).

Pyrazines were the second abundant group detected in MB10 formulations (%PA ranging from 24.1 to 39.8). It was observed that the %PA statistically decreased in the fermented MB10 formulations ($p \leq 0.05$) compared to control formulations. This phenomenon was also described by Kurt et al. [58], which evidenced that the pyrazine content in spirulina water solutions (4% w/v) was reduced after LAB fermentation.

The greatest variety of chemical compounds found in the MB10 were alcohols, representing a mean %PA value ranging from 10.8 to 17.6%. Among the alcohols identified, phenylethyl alcohol (PEA) was detected in a higher proportion (6.6 ± 0.5 %PA) in the MB10 fermented with C15 (MB10-C15) in comparison with other formulations ($p \leq 0.05$). This active compound has been studied for its antifungal potential and could explain the lower MIC and MFC values detected in the MB10-C15, as previously reported in Section 3.2. For instance, Gong et al. [59] determined that the antifungal properties of *Enterobacter absuriae* Vt–7 were mainly due to the volatile antifungal PEA, and it effectively controlled the development of the toxigenic fungi *Aspergillus flavus* in peanuts. Similar results were obtained by Wonglom et al. [60], who associated the antifungal potential of *Trichoderma* sp. T76–12/2 against *Sclerotium* fruit rot due to the synthesis of PEA and other VOCs.

Regarding ketones, LAB not only significantly increased ($p \leq 0.05$) the ketone levels in the formulated MB10, but also introduced three new ketones that were not present in the control group, such as 2–heptanone, 2–undecanone, and 2–tridecanone (Table 6). Some ketones may be synthesized through microbial oxidation of fatty acids, and this could explain its higher proportion in the MB10 formulations when compared to the control formulation (MB10 without fermentation) [61].

Only two acids, acetic acid and nonanoic acid, were identified in the fermented MB10. Nonanoic acid is mainly produced from the degradation of unsaturated fatty acids such as oleic acid, whereas acetic acid is produced because of the heterofermentative metabolism of LAB [62]. Previous studies conducted observed an increase in these chemical compounds when LAB are employed in different food matrices such as pumpkin and watermelon juices [63,64]. Regarding its biological properties, it is important to emphasize that nonanoic acid has evidenced antimicrobial properties against several pathogens such as *Alternaria alternata*, *Botrytis cinerea* [65], *Candida albicans* [66], *Salmonella enterica* [67], and *Escherichia coli* O157:H7 [68]. Thus, volatile acids combined with the other biological compounds found in the MB10 (such as PEA, organic acids, and phenolic acids) could contribute to the antifungal properties of the formulated ingredient since LAB antifungal potential is related to the synergistic action of the different metabolites synthesized [38].

3.3.3. Principal Component Analysis (PCA) of the MB10 with LAB

In order to understand the differences between bioactive metabolite production (organic acids, phenolic acids, and VOCs) in the fermented MB formulations, a PCA was realized, and the results obtained are outlined in Figure 4. The sum of the first two principal components (PC) achieved 90.1% of the total explained variance, in which PC1 explained up to 64.5% of the total variance, while PC2 explained 25.6% (Figure 4a). The PC1 distributed on the positive axis both *P. pentosaceus* strains (C12 and C15), whereas *L. plantarum* C60 strain was positioned on the negative axis. The PC2 allowed the distinction of the C12 and C15 strains according to their bioactive metabolite profile since C12 was positioned on the negative axis and C15 was positioned on the positive axis.

(**a**)

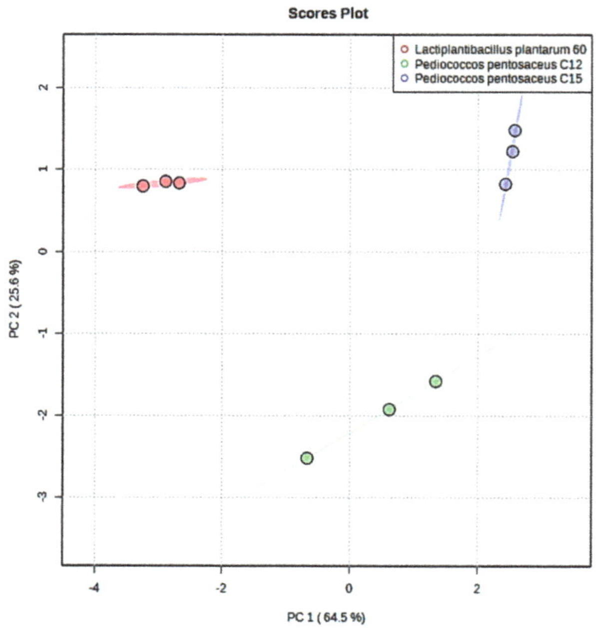

(**b**)

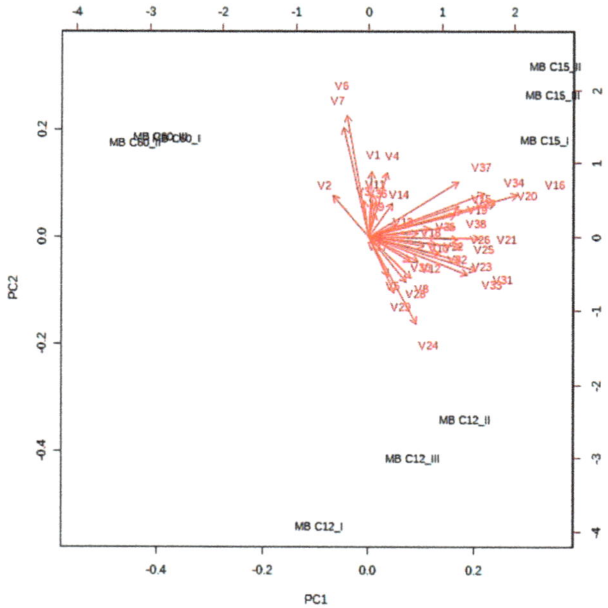

Figure 4. Principal component analysis (PCA) scores plot of the bioactive compounds (organic acids, phenolic acids, and volatile organic compounds) found in the meat broth fermented by *P. pentosaceus* C12 (MB C12), *P. pentosaceus* C15 (MB C15), and *L. plantarum* C60 (MB C60) (**a**) and relative loadings of the variables employed (**b**).

According to the loading plot that represents the relative importance of the variables analyzed (Figure 4b), MB10-C12 was distinguished from MB10-C15 for its higher volatile aldehyde production; specifically, this was found for specific compounds such as heptanal (V24), 2–decenal (V28), and dodecanal (V29). The MB10 formulated with C15 was mainly characteristic from the other formulations due to the higher production of different antifungal compounds such as nonanoic acid (V16) and phenyl ethyl alcohol (V20).

Regarding the MB10 formulated with *L. plantarum* strain (MB10-C60), the main variable that distinguishes this formulation from those prepared with *P. pentosaceus* strains was the higher production of the phenolic compound 3–(4–hydrodoxy–3–methoxyphenyl) propionic (V2). Furthermore, the ferulic acid (V6) and gallic acid content (V7) positioned on the positive axis according to the second component the MB10-C60 formulation and permitted the distinction between the MBC12 formulation. The variable that was positioned on the negative axis according to the first component of the MB C60 was the lower content in volatile alcohols in comparison with C12 and C15, such as 1–nonanol (V21) and 2–undecanol (V23).

4. Conclusions

In the study, 42 antifungal LAB isolated from dry-cured sausages were evaluated against 6 fungi of the *Aspergillus* and *Penicillium* genera. Firstly, a battery of bacteria was isolated from dry-cured meat products and characterized according to their antifungal properties. Among them, 14 presented antifungal capacity and 3 inhibited the growth of all fungi assayed. The *P. pentosaceus* C15 showed higher antifungal activity in vitro; these bacteria were selected for the preparation of several fermented-meat broth (MB2, MB4, MB8 and MB10). MB were elaborated changing de MRSb formulation. The MB10 and MB formulated with 10% of freeze-dried loin pork demonstrated higher antifungal activity through agar diffusion and microdilution method.

In particular, the MB10 fermented with *P. pentosaceus* C15 was selected to elaborate a postbiotic antifungal ingredient due to their higher antifungal activity. The chemical characterization highlighted that the MB10 formulation was rich in phenolic acids and organic acids such as lactic acid and acetic acid. Moreover, the postbiotic product changed the VOCs composition and increased the concentration and number of antifungal compounds detected.

Discovering the synergistic process amongst antifungal agents might provide understanding in increasing the effect while modifying the implicated bacterial or nutritional compositions, subsequently leading to food applications. PCA analysis indicated that some antifungal agents enabled us to differentiate strains by their production. For instance, *P. pentosaceus* C15 and C12 were distinguished by the production of aldehydes such as heptanal, decenal, and dodecanal. In contrast, *L. plantarum* C60 was differentiated from *P. pentosaceus* strains due to the production of 3–(4–hydrodoxy–3–methoxyphenyl) propionic.

Further research will focus on the application of the postbiotic antifungal ingredient (MB10-C15) in the development of dry-cured meat products, as well as study c inhibition of mycotoxins synthesis that can potentially harm the consumers health.

Supplementary Materials: The following supporting information can be downloaded at: https://www.mdpi.com/article/10.3390/foods12071427/s1, Table S1: Elaboration of meat broths for fermentation for lactic acid bacteria; Table S2: Antifungal activity of formulated meat broths (MB) and MRS broth fermented by Pediococcus pentosaceus C15 during 24, 48, and 72 h at 37 °C. The bacterial-free supernatant (BFS) was freeze-dried, resuspended at a concentration of 500 g/L, and tested against six toxigenic fungi; Table S3: Identification of Volatile Organic Compounds (VOCs) of the fermented Meat Broth 10, with retention time, chemical class, calculated LRI (LRI exp.), and references [69–86].

Author Contributions: Conceptualization, J.M. and G.M.; methodology, T.d.M.N.; formal analysis, T.d.M.N., J.C. and C.L.; investigation, T.d.M.N.; resources, J.M. and G.M.; data curation, T.d.M.N., J.C.

and C.L.; writing—original draft preparation, T.d.M.N., J.C., C.L., J.M. and G.M.; visualization, J.C. and C.L.; writing—review and editing, T.d.M.N., J.C., C.L., J.M. and G.M.; supervision, G.M.; project administration, J.M. and G.M.; funding acquisition, J.M. and G.M. All authors have read and agreed to the published version of the manuscript.

Funding: This research was funded by the Spanish Ministry of Science and Innovation, grant number PID2019-108070RB-100.

Data Availability Statement: The data presented in this study are available on request from the corresponding author.

Acknowledgments: The authors would like to thank the program of the University of Valencia (Atracció de Talent UV-INV-PREDOC19F1-1006684).

Conflicts of Interest: The authors declare no conflict of interest.

References

1. Manassi, C.F.; de Souza, S.S.; de Souza Hassemer, G.; Sartor, S.; Lima, C.M.G.; Miotto, M.; De Dea Lindner, J.; Rezzadori, K.; Pimentel, T.C.; Ramos, G.L.D.P.A.; et al. Functional meat products: Trends in pro-, pre-, syn-, para- and post-biotic use. *Food Res. Int.* **2022**, *154*, 111035. [CrossRef]
2. Whitnall, T.; Pitts, N. Global trends in meat consumption. *Agric. Commod.* **2019**, *9*, 96–99.
3. Lippolis, V.; Ferrara, M.; Cervellieri, S.; Damascelli, A.; Epifani, F.; Pascale, M.; Perrone, G. Rapid prediction of ochratoxin A-producing strains of Penicillium on dry-cured meat by MOS-based electronic nose. *Int. J. Food Microbiol.* **2016**, *218*, 71–77. [CrossRef] [PubMed]
4. Simoncini, N.; Pinna, A.; Toscani, T.; Virgili, R. Effect of added autochthonous yeasts on the volatile compounds of dry-cured hams. *Int. J. Food Microbiol.* **2015**, *212*, 25–33. [CrossRef]
5. Rodríguez, A.; Rodríguez, M.; Martín, A.; Nuñez, F.; Córdoba, J.J. Evaluation of hazard of aflatoxin B1, ochratoxin A and patulin production in dry-cured ham and early detection of producing moulds by qPCR. *Food Control* **2012**, *27*, 118–126. [CrossRef]
6. De Ruyck, K.; De Boevre, M.; Huybrechts, I.; De Saeger, S. Dietary mycotoxins, co-exposure, and carcinogenesis in humans: Short review. *Mutat. Res. Mutat. Res.* **2015**, *766*, 32–41. [CrossRef]
7. Perrone, G.; Samson, R.A.; Frisvad, J.C.; Susca, A.; Gunde-Cimerman, N.; Epifani, F.; Houbraken, J. Penicillium salamii, a new species occurring during seasoning of dry-cured meat. *Int. J. Food Microbiol.* **2015**, *193*, 91–98. [CrossRef]
8. Sánchez-Montero, L.; Córdoba, J.J.; Peromingo, B.; Álvarez, M.; Núñez, F. Effects of environmental conditions and substrate on growth and ochratoxin A production by Penicillium verrucosum and Penicillium nordicum: Relative risk assessment of OTA in dry-cured meat products. *Food Res. Int.* **2019**, *121*, 604–611. [CrossRef] [PubMed]
9. Ferrara, M.; Perrone, G.; Gallo, A.; Epifani, F.; Visconti, A.; Susca, A. Development of loop-mediated isothermal amplification (LAMP) assay for the rapid detection of Penicillium nordicum in dry-cured meat products. *Int. J. Food Microbiol.* **2015**, *202*, 42–47. [CrossRef] [PubMed]
10. Sarmast, E.; Fallah, A.A.; Jafari, T.; Mousavi Khaneghah, A. Occurrence and fate of mycotoxins in cereals and cereal-based products: A narrative review of systematic reviews and meta-analyses studies. *Curr. Opin. Food Sci.* **2021**, *39*, 68–75. [CrossRef]
11. Dalié, D.K.D.; Deschamps, A.M.; Richard-Forget, F. Lactic acid bacteria—Potential for control of mould growth and mycotoxins: A review. *Food Control* **2010**, *21*, 370–380. [CrossRef]
12. Thøgersen, R.; Bertram, H.C. Reformulation of processed meat to attenuate potential harmful effects in the gastrointestinal tract—A review of current knowledge and evidence of health prospects. *Trends Food Sci. Technol.* **2021**, *108*, 111–118. [CrossRef]
13. Raman, J.; Kim, J.-S.; Choi, K.R.; Eun, H.; Yang, D.; Ko, Y.-J.; Kim, S.-J. Application of Lactic Acid Bacteria (LAB) in Sustainable Agriculture: Advantages and Limitations. *Int. J. Mol. Sci.* **2022**, *23*, 7784. [CrossRef]
14. Li, H.; Liu, L.; Zhang, S.; Uluko, H.; Cui, W.; Lv, J. Potential use of Lactobacillus casei AST18 as a bioprotective culture in yogurt. *Food Control* **2013**, *34*, 675–680. [CrossRef]
15. Olaniyan, O.T.; Adetunji, C.O. *Biochemical Role of Beneficial Microorganisms: An Overview on Recent Development in Environmental and Agro Science BT—Microbial Rejuvenation of Polluted Environment: Volume 3*; Adetunji, C.O., Panpatte, D.G., Jhala, Y.K., Eds.; Springer: Singapore, 2021; pp. 21–33. ISBN 978-981-15-7459-7.
16. Garnier, L.; Penland, M.; Thierry, A.; Maillard, M.-B.; Jardin, J.; Coton, M.; Leyva Salas, M.; Coton, E.; Valence, F.; Mounier, J. Antifungal activity of fermented dairy ingredients: Identification of antifungal compounds. *Int. J. Food Microbiol.* **2020**, *322*, 108574. [CrossRef] [PubMed]
17. Schmidt, M.; Lynch, K.M.; Zannini, E.; Arendt, E.K. Fundamental study on the improvement of the antifungal activity of *Lactobacillus reuteri* R29 through increased production of phenyllactic acid and reuterin. *Food Control* **2018**, *88*, 139–148. [CrossRef]
18. Crowley, S.; Mahony, J.; van Sinderen, D. Current perspectives on antifungal lactic acid bacteria as natural bio-preservatives. *Trends Food Sci. Technol.* **2013**, *33*, 93–109. [CrossRef]
19. Omedi, J.O.; Huang, W.; Zheng, J. Effect of sourdough lactic acid bacteria fermentation on phenolic acid release and antifungal activity in pitaya fruit substrate. *LWT* **2019**, *111*, 309–317. [CrossRef]

20. Pagnussatt, F.A.; Del Ponte, E.M.; Garda-Buffon, J.; Badiale-Furlong, E. Inhibition of *Fusarium graminearum* growth and mycotoxin production by phenolic extract from *Spirulina* sp. *Pestic. Biochem. Physiol.* **2014**, *108*, 21–26. [CrossRef]
21. Da Rocha Neto, A.C.; Maraschin, M.; Di Piero, R.M. Antifungal activity of salicylic acid against *Penicillium expansum* and its possible mechanisms of action. *Int. J. Food Microbiol.* **2015**, *215*, 64–70. [CrossRef]
22. Karwowska, M.; Munekata, P.E.S.; Lorenzo, J.M.; Tomasevic, I. Functional and Clean Label Dry Fermented Meat Products: Phytochemicals, Bioactive Peptides, and Conjugated Linoleic Acid. *Appl. Sci.* **2022**, *12*, 5559. [CrossRef]
23. Ben Belgacem, Z.; Dousset, X.; Prévost, H.; Manai, M. Polyphasic taxonomic studies of lactic acid bacteria associated with Tunisian fermented meat based on the heterogeneity of the 16S–23S rRNA gene intergenic spacer region. *Arch. Microbiol.* **2009**, *191*, 711–720. [CrossRef] [PubMed]
24. Jesper, M.; Johan, S. Lactobacillus coryniformis subsp.coryniformis Strain Si3 Produces a Broad-Spectrum Proteinaceous Antifungal Compound. *Appl. Environ. Microbiol.* **2001**, *67*, 1–5. [CrossRef]
25. Maier, T.; Klepel, S.; Renner, U.; Kostrzewa, M. Fast and reliable MALDI-TOF MS–based microorganism identification. *Nat. Methods* **2006**, *3*, i–ii. [CrossRef]
26. De Melo Nazareth, T.; Luz, C.; Torrijos, R.; Quiles, J.M.; Luciano, F.B.; Mañes, J.; Meca, G. Potential application of lactic acid bacteria to reduce aflatoxin B1 and fumonisin B1 occurrence on corn kernels and corn ears. *Toxins* **2019**, *12*, 21. [CrossRef]
27. Fredua-Agyeman, M.; Gaisford, S. Assessing inhibitory activity of probiotic culture supernatants against *Pseudomonas aeruginosa*: A comparative methodology between agar diffusion, broth culture and microcalorimetry. *World J. Microbiol. Biotechnol.* **2019**, *35*, 49. [CrossRef]
28. CLSI. M38-A2 Reference Method for Broth Dilution Antifungal Susceptibility Testing of Filamentous Fungi. In *Approved Standard*, 2nd ed.; CLSI document M38-A2; Clinical and Laboratory Standards Institute: Wayne, PA, USA, 2008.
29. Özcelik, S.; Kuley, E.; Özogul, F. Formation of lactic, acetic, succinic, propionic, formic and butyric acid by lactic acid bacteria. *LWT—Food Sci. Technol.* **2016**, *73*, 536–542. [CrossRef]
30. Brosnan, B.; Coffey, A.; Arendt, E.K.; Furey, A. The QuEChERS approach in a novel application for the identification of antifungal compounds produced by lactic acid bacteria cultures. *Talanta* **2014**, *129*, 364–373. [CrossRef] [PubMed]
31. Denardi-Souza, T.; Luz, C.; Mañes, J.; Badiale-Furlong, E.; Meca, G. Antifungal effect of phenolic extract of fermented rice bran with *Rhizopus oryzae* and its potential use in loaf bread shelf life extension. *J. Sci. Food Agric.* **2018**, *98*, 5011–5018. [CrossRef]
32. Luz, C.; Quiles, J.M.; Romano, R.; Blaiotta, G.; Rodríguez, L.; Meca, G. Application of whey of Mozzarella di Bufala Campana fermented by lactic acid bacteria as a bread biopreservative agent. *Int. J. Food Sci. Technol.* **2021**, *56*, 4585–4593. [CrossRef]
33. Russo, P.; Arena, M.P.; Fiocco, D.; Capozzi, V.; Drider, D.; Spano, G. *Lactobacillus plantarum* with broad antifungal activity: A promising approach to increase safety and shelf-life of cereal-based products. *Int. J. Food Microbiol.* **2017**, *247*, 48–54. [CrossRef] [PubMed]
34. Delcarlo, S.B.; Parada, R.; Schelegueda, L.I.; Vallejo, M.; Marguet, E.R.; Campos, C.A. From the isolation of bacteriocinogenic LAB strains to the application for fish paste biopreservation. *LWT* **2019**, *110*, 239–246. [CrossRef]
35. Parlindungan, E.; Lugli, G.A.; Ventura, M.; van Sinderen, D.; Mahony, J. Lactic acid bacteria diversity and characterization of probiotic candidates in fermented meats. *Foods* **2021**, *10*, 1519. [CrossRef] [PubMed]
36. Ivanovic, M.; Mirkovic, N.; Mirkovic, M.; Miocinovic, J.; Radulovic, A.; Knudsen, T.S.; Radulovic, Z. *Autochthonous enterococcus durans pfmi*565 and lactococcus lactis subsp. Lactis bgbu1–4 in bio-control of listeria monocytogenes in ultrafiltered cheese. *Foods* **2021**, *10*, 1448. [CrossRef] [PubMed]
37. Hernández, A.; Rodríguez, A.; Córdoba, M.G.; Martín, A.; Ruiz-Moyano, S. Fungal control in foods through biopreservation. *Curr. Opin. Food Sci.* **2022**, *47*, 100904. [CrossRef]
38. Schnürer, J.; Magnusson, J. Antifungal lactic acid bacteria as biopreservatives. *Trends Food Sci. Technol.* **2005**, *16*, 70–78. [CrossRef]
39. Taroub, B.; Salma, L.; Manel, Z.; Ouzari, H.-I.; Hamdi, Z.; Moktar, H. Isolation of lactic acid bacteria from grape fruit: Antifungal activities, probiotic properties, and in vitro detoxification of ochratoxin A. *Ann. Microbiol.* **2019**, *69*, 17–27. [CrossRef]
40. Aziz, N.H.; Farag, S.E.; Mousa, L.A.; Abo-Zaid, M.A. Comparative antibacterial and antifungal effects of some phenolic compounds. *Microbios* **1998**, *93*, 43–54.
41. Ren, B.; Xia, B.; Li, W.; Wu, J.; Zhang, H. Two novel phenolic compounds from *Stenoloma chusanum* and their antifungal activity. *Chem. Nat. Compd.* **2009**, *45*, 182–186. [CrossRef]
42. Pei, X.; Tekliye, M.; Dong, M. Isolation and identification of fungi found in contaminated fermented milk and antifungal activity of vanillin. *Food Sci. Hum. Wellness* **2021**, *10*, 214–220. [CrossRef]
43. Fitzgerald, D.J.; Stratford, M.; Gasson, M.J.; Narbad, A. Structure−Function Analysis of the Vanillin Molecule and Its Antifungal Properties. *J. Agric. Food Chem.* **2005**, *53*, 1769–1775. [CrossRef] [PubMed]
44. Amiri, S.; Aghamirzaei, M.; Mostashari, P.; Sarbazi, M.; Tizchang, S.; Madahi, H. The impact of biotechnology on dairy industry. In *Microbial Biotechnology in Food and Health*; Elsevier: Amsterdam, The Netherlands, 2021; pp. 53–79.
45. Punia Bangar, S.; Suri, S.; Trif, M.; Ozogul, F. Organic acids production from lactic acid bacteria: A preservation approach. *Food Biosci.* **2022**, *46*, 101615. [CrossRef]
46. Chen, O.; Hong, Y.; Ma, J.; Deng, L.; Yi, L.; Zeng, K. Screening lactic acid bacteria from pickle and cured meat as biocontrol agents of *Penicillium digitatum* on citrus fruit. *Biol. Control* **2021**, *158*, 104606. [CrossRef]
47. Cortés-Zavaleta, O.; López-Malo, A.; Hernández-Mendoza, A.; García, H.S. Antifungal activity of lactobacilli and its relationship with 3-phenyllactic acid production. *Int. J. Food Microbiol.* **2014**, *173*, 30–35. [CrossRef] [PubMed]

48. Peyer, L.C.; Axel, C.; Lynch, K.M.; Zannini, E.; Jacob, F.; Arendt, E.K. Inhibition of *Fusarium culmorum* by carboxylic acids released from lactic acid bacteria in a barley malt substrate. *Food Control* **2016**, *69*, 227–236. [CrossRef]
49. Axel, C.; Brosnan, B.; Zannini, E.; Furey, A.; Coffey, A.; Arendt, E.K. Antifungal sourdough lactic acid bacteria as biopreservation tool in quinoa and rice bread. *Int. J. Food Microbiol.* **2016**, *239*, 86–94. [CrossRef]
50. Ruggirello, M.; Nucera, D.; Cannoni, M.; Peraino, A.; Rosso, F.; Fontana, M.; Cocolin, L.; Dolci, P. Antifungal activity of yeasts and lactic acid bacteria isolated from cocoa bean fermentations. *Food Res. Int.* **2019**, *115*, 519–525. [CrossRef]
51. Christ-Ribeiro, A.; Graça, C.S.; Kupski, L.; Badiale-Furlong, E.; de Souza-Soares, L.A. Cytotoxicity, antifungal and anti mycotoxins effects of phenolic compounds from fermented rice bran and *Spirulina* sp. *Process Biochem.* **2019**, *80*, 190–196. [CrossRef]
52. Estévez, M.; Morcuende, D.; Ventanas, S.; Cava, R. Analysis of volatiles in meat from Iberian pigs and lean pigs after refrigeration and cooking by using SPME-GC-MS. *J. Agric. Food Chem.* **2003**, *51*, 3429–3435. [CrossRef]
53. Wang, X.; Zhu, L.; Han, Y.; Xu, L.; Jin, J.; Cai, Y.; Wang, H. Analysis of volatile compounds between raw and cooked beef by HS-SPME–GC–MS. *J. Food Process. Preserv.* **2018**, *42*, e13503. [CrossRef]
54. Domínguez, R.; Purriños, L.; Pérez-Santaescolástica, C.; Pateiro, M.; Barba, F.J.; Tomasevic, I.; Campagnol, P.C.B.; Lorenzo, J.M. Characterization of Volatile Compounds of Dry-Cured Meat Products Using HS-SPME-GC/MS Technique. *Food Anal. Methods* **2019**, *12*, 1263–1284. [CrossRef]
55. Watanabe, A.; Kamada, G.; Imanari, M.; Shiba, N.; Yonai, M.; Muramoto, T. Effect of aging on volatile compounds in cooked beef. *Meat Sci.* **2015**, *107*, 12–19. [CrossRef]
56. Tamura, Y.; Iwatoh, S.; Miyaura, K.; Asikin, Y.; Kusano, M. Metabolomic profiling reveals the relationship between taste-related metabolites and roasted aroma in aged pork. *LWT* **2022**, *155*, 112928. [CrossRef]
57. Kwaw, E.; Ma, Y.; Tchabo, W.; Apaliya, M.T.; Sackey, A.S.; Wu, M.; Xiao, L. Effect of pulsed light treatment on the phytochemical, volatile, and sensorial attributes of lactic-acid-fermented mulberry juice. *Int. J. Food Prop.* **2018**, *21*, 213–228. [CrossRef]
58. Kurt, H.; Isleten Hosoglu, M.; Guneser, O.; Karagul-Yuceer, Y. Influence of Different Bacteria Species in Chemical Composition and Sensory Properties of Fermented *Spirulina*. *Food Chem.* **2023**, *400*, 133994. [CrossRef]
59. Gong, A.D.; Dong, F.Y.; Hu, M.J.; Kong, X.W.; Wei, F.F.; Gong, S.J.; Zhang, Y.M.; Zhang, J.B.; Wu, A.B.; Liao, Y.C. Antifungal activity of volatile emitted from Enterobacter asburiae Vt-7 against *Aspergillus flavus* and aflatoxins in peanuts during storage. *Food Control* **2019**, *106*, 106718. [CrossRef]
60. Wonglom, P.; Daengsuwan, W.; Ito, S.I.; Sunpapao, A. Biological control of Sclerotium fruit rot of snake fruit and stem rot of lettuce by *Trichoderma* sp. T76-12/2 and the mechanisms involved. *Physiol. Mol. Plant Pathol.* **2019**, *107*, 1–7. [CrossRef]
61. Chen, C.; Lu, Y.; Yu, H.; Chen, Z.; Tian, H. Influence of 4 lactic acid bacteria on the flavor profile of fermented apple juice. *Food Biosci.* **2019**, *27*, 30–36. [CrossRef]
62. Pang, X.N.; Chen, C.; Huang, X.N.; Yan, Y.Z.; Chen, J.Y.; Han, B.Z. Influence of indigenous lactic acid bacteria on the volatile flavor profile of light-flavor Baijiu. *LWT* **2021**, *147*, 111540. [CrossRef]
63. Mandha, J.; Shumoy, H.; Devaere, J.; Schouteten, J.J.; Gellynck, X.; de Winne, A.; Matemu, A.O.; Raes, K. Effect of lactic acid fermentation of watermelon juice on its sensory acceptability and volatile compounds. *Food Chem.* **2021**, *358*, 129809. [CrossRef]
64. Sun, X.; Zhang, Y.; Li, F.; Jiao, X.; Ma, D.; Zhang, L.; Yang, B.; Zhao, J.; Han, J.; Li, Q. Effects of lactic acid bacteria fermentation on chemical compounds, antioxidant capacities and hypoglycemic properties of pumpkin juice. *Food Biosci.* **2022**, *50*, 102126. [CrossRef]
65. Chen, J.H.; Xiang, W.; Cao, K.X.; Lu, X.; Yao, S.C.; Hung, D.; Huang, R.S.; Li, L.B. Characterization of Volatile Organic Compounds Emitted from Endophytic *Burkholderia cenocepacia* ETR-B22 by SPME-GC-MS and Their Inhibitory Activity against Various Plant Fungal Pathogens. *Molecules* **2020**, *25*, 3765. [CrossRef] [PubMed]
66. Lee, J.; Kim, Y.; Khadke, S.K.; Lee, J. Antibiofilm and antifungal activities of medium-chain fatty acids against *Candida albicans* via mimicking of the quorum-sensing molecule farnesol. *Microb. Biotechnol.* **2021**, *14*, 1353–1366. [CrossRef] [PubMed]
67. White, E.; Kumar, G.D.; da Silva, A.L.B.R.; Kerr, W.L.; Cimowsky, S.; Widmer, J.A.; Dunn, L.L. Postharvest Reduction of *Salmonella enterica* on Tomatoes Using a Pelargonic Acid Emulsion. *Foods* **2021**, *10*, 178. [CrossRef]
68. Cimowsky, S.; Kumar, G.D.; Biscaia Ribeiro da Silva, A.L.; White, E.; Kerr, W.L.; Rodrigues, C.; Juneja, V.K.; Dunn, L.L. Postharvest control of *Escherichia coli* O157:H7 on romaine lettuce using a novel pelargonic acid sanitizer. *LWT* **2022**, *154*, 112168. [CrossRef]
69. Lu, C.-Y.; Hao, Z.; Payne, R.; Ho, C.-T. Effects of water content on volatile generation and peptide degradation in the Maillard reaction of glycine, diglycine, and triglycine. *J. Agric. Food Chem.* **2005**, *53*, 6443–6447. [CrossRef]
70. Pino, J.A.; Marbot, R.; Bello, A. Volatile compounds of *Psidium salutare* (H.B.K.) Berg. fruit. *J. Agric. Food Chem.* **2002**, *50*, 5146–5148. [CrossRef]
71. Ziegenbein, F.C.; Hanssen, H.-P.; König, W.A. Secondary metabolites from *Ganoderma lucidum* and *Spongiporus leucomallellus*. *Phytochemistry* **2006**, *67*, 202–211. [CrossRef]
72. Pino, J.; Almora, K.; Marbot, R. Volatile components of papaya (*Carica papaya* L., maradol variety) fruit. *Flavour Fragr. J.* **2003**, *18*, 492–496. [CrossRef]
73. Sampaio, T.S.; Nogueira, P.C.L. Volatile components of mangaba fruit (*Hancornia speciosa* Gomes) at three stages of maturity. *Food Chem.* **2006**, *95*, 606–610. [CrossRef]
74. Xian, Q.; Chen, H.; Zou, H.; Yin, D. Chemical composition of essential oils of two submerged macrophytes, *Ceratophyllum demersum* L. and *Vallisneria spiralis* L. *Flavour Fragr. J.* **2006**, *21*, 524–526. [CrossRef]

75. Nickavar, B.; Kamalinejad, M.; Mohandesi, S. Comparison of the components of the essential oils from leaves and fruits of *Grammosciadium platycarpum*. *Chem. Nat. Compd.* **2006**, *42*, 686–688. [CrossRef]
76. Pavlovic, M.; Tzakou, O.; Petrakis, P.V.; Couladis, M. The essential oil of *Hypericum perforatum* L., *Hypericum tetrapterum* Fries and *Hypericum olympicum* L. growing in Greece. *Flavour Fragr. J.* **2006**, *21*, 84–87. [CrossRef]
77. Bredie, W.L.P.; Mottram, D.S.; Guy, R.C.E. Effect of temperature and pH on the generation of flavor volatiles in extrusion cooking of wheat flour. *J. Agric. Food Chem.* **2002**, *50*, 1118–1125. [CrossRef]
78. Bylaite, E.; Meyer, A.S. Characterisation of volatile aroma compounds of orange juices by three dynamic and static headspace gas chromatography techniques. *Eur. Food Res. Technol.* **2006**, *222*, 176–184. [CrossRef]
79. Cardeal, Z.L.; da Silva, M.D.R.G.; Marriott, P.J. Comprehensive two-dimensional gas chromatography/mass spectrometric analysis of pepper volatiles. *Rapid Commun Mass Spectrom.* **2006**, *20*, 2823–2836. [CrossRef]
80. Demetzos, C.; Angelopoulou, D.; Perdetzoglou, D. A comparative study of the essential oils of *Cistus salviifolius* in several populations of Crete (Greece). *Biochem. Syst. Ecol.* **2002**, *30*, 651–665. [CrossRef]
81. Gocmen, D.; Gurbuz, O.; Rouseff, R.L.; Smoot, J.M.; Dagdelen, A.F. Gas chromatographic-olfactometric characterization of aroma active compounds in sun-dried and vacuum-dried tarhana. *Eur. Food Res. Technol.* **2004**, *218*, 573–578. [CrossRef]
82. Zhao, J.Y.; Liu, J.M.; Zhang, X.Y.; Liu, Z.J.; Tsering, T.; Zhong, Y.; Nan, P. Chemical composition of the volatiles of three wild *Bergenia* species from western China. *Flavour Fragr. J.* **2006**, *21*, 431–434. [CrossRef]
83. Mahattanatawee, K.; Goodner, K.L.; Baldwin, E.A. Volatile constituents and character impact compounds of selected Florida's tropical fruit. *Proc. Fla. State Hort. Soc.* **2005**, *118*, 414–418.
84. Viana, F.A.; Andrade-Neto, M.; Pouliquen, Y.B.M.; Lucien, V.G. Chemical composition of the essential oil from roots of *Philodendron acutatum* Schott. *J. Essent. Oil Res.* **2002**, *14*, 172–174. [CrossRef]
85. Kukic, J.; Petrovic, S.; Pavlovic, M.; Couladis, M.; Tzakou, O.; Niketic, M. Composition of essential oil of *Stachys alpina* L. ssp. *dinarica* Murb. *Flavour Fragr. J.* **2006**, *21*, 539–542. [CrossRef]
86. Benkaci-Ali, F.; Baaliouamer, A.; Meklati, B.Y.; Chemat, F. Chemical composition of seed essential oils from Algerian *Nigella sativa* extracted by microwave and hydrodistillation. *Flavour Fragr. J.* **2007**, *22*, 148–153. [CrossRef]

foods MDPI

Article

Design of *Lactococcus lactis* Strains Producing Garvicin A and/or Garvicin Q, Either Alone or Together with Nisin A or Nisin Z and High Antimicrobial Activity against *Lactococcus garvieae*

Javier Feito [1], Carlos Araújo [1], Sara Arbulu [1], Diogo Contente [1], Beatriz Gómez-Sala [2,3], Lara Díaz-Formoso [1], Estefanía Muñoz-Atienza [1], Juan Borrero [1], Luis M. Cintas [1] and Pablo E. Hernández [1,*]

1 Grupo de Seguridad y Calidad de los Alimentos por Bacterias Lácticas, Bacteriocinas y Probióticos (Grupo SEGABALBP), Sección Departamental de Nutrición y Ciencia de los Alimentos (Higiene y Seguridad Alimentaria), Facultad de Veterinaria, Universidad Complutense de Madrid, Avda. Puerta de Hierro, s/n., 28040 Madrid, Spain
2 APC Microbiome Ireland, University College Cork, T12 K8AF Cork, Ireland
3 Teagasc Food Research Centre, Moorepark, P61 C996 Cork, Ireland
* Correspondence: ehernan@ucm.es

Abstract: *Lactococcus garvieae* is a main ichthyopathogen in rainbow trout (*Oncorhynchus mykiss*, Walbaum) farming, although bacteriocinogenic *L. garvieae* with antimicrobial activity against virulent strains of this species have also been identified. Some of the bacteriocins characterized, such as garvicin A (GarA) and garvicin Q (GarQ), may show potential for the control of the virulent *L. garvieae* in food, feed and other biotechnological applications. In this study, we report on the design of *Lactococcus lactis* strains that produce the bacteriocins GarA and/or GarQ, either alone or together with nisin A (NisA) or nisin Z (NisZ). Synthetic genes encoding the signal peptide of the lactococcal protein Usp45 (SP_{usp45}), fused to mature GarA (*lgnA*) and/or mature GarQ (*garQ*) and their associated immunity genes (*lgnI* and *garI*, respectively), were cloned into the protein expression vectors pMG36c, which contains the P_{32} constitutive promoter, and pNZ8048c, which contains the inducible P_{nisA} promoter. The transformation of recombinant vectors into lactococcal cells allowed for the production of GarA and/or GarQ by *L. lactis* subsp. *cremoris* NZ9000 and their co-production with NisA by *Lactococcus lactis* subsp. *lactis* DPC5598 and *L. lactis* subsp. *lactis* BB24. The strains *L. lactis* subsp. *cremoris* WA2-67 (pJFQI), a producer of GarQ and NisZ, and *L. lactis* subsp. *cremoris* WA2-67 (pJFQIAI), a producer of GarA, GarQ and NisZ, demonstrated the highest antimicrobial activity (5.1- to 10.7-fold and 17.3- to 68.2-fold, respectively) against virulent *L. garvieae* strains.

Keywords: bacteriocins; garvicin; nisin; *Lactococcus garvieae*; lactic acid bacteria (LAB); heterologous production; *Lactococcus lactis*; probiotics; paraprobiotics; postbiotics

1. Introduction

Bacteriocins produced by lactic acid bacteria (LAB) have been largely valued as potential food preservatives, and the LAB producers of bacteriocins (bacteriocinogenic strains) have been valued as potential starter, protective, probiotic, paraprobiotic and postbiotic cultures [1–3]. Moreover, concerns regarding the increase in antimicrobial resistances (AMRs) confer bacteriocins and the bacteriocinogenic LAB unlimited possibilities for applications in the food industry, human and veterinary medicine and the animal production field [4–6]. Microbial-derived biotics, including bacteriocins, are recognized as functional components of natural and bioengineered probiotic, paraprobiotic and postbiotic cultures [2,7,8]. Bacteriocins, including nisin A (NisA) and nisin Z (NisZ), drive the apoptosis of cancer cells and show low toxicity toward normal cells, making them promising anticancer candidates to replace or be combined with conventional therapeutic agents [9–11]. Bacteriocinogenic LAB and the bacteriocins they produce may also have an impact in the modulation of the microbiota and immune system of their host [12,13].

Most LAB bacteriocins are synthesized as biologically inactive precursors that contain an N-terminal extension, whereas the mature peptides are often cationic, amphiphilic, membrane-permeabilizing molecules. The N-terminal extensions of most bacteriocins are of the so-called double-glycine type (leader sequence) and are cleaved off concomitantly with export across the cytoplasmic membrane by dedicated, ATP-binding cassette transporters (ABC-transporters) and their accessory proteins [14]. However, a few bacteriocins contain N-terminal extensions of the Sec-type (signal peptide), which are proteolytically cleaved concomitantly with peptide externalization by the general secretory pathway (GSP) or the Sec-dependent pathway [15]. Several bacteriocins have been shown to be synthesized without N-terminal sequences and represent bacteriocins with a novel secretion mechanism [16].

The mature bacteriocins are often classified into three main classes: class I, or lantibiotics, which have post-translationally modified amino acid residues, class II bacteriocins, which have unmodified amino acid residues, and class III large, heat-labile bacteriocins [17,18]. However, the class I bacteriocins are currently included in the group of ribosomally synthesized and post-translationally modified peptides (RiPPS) [18–20]. The class II bacteriocins can be further divided into the subclasses pediocin-like (class IIa), two-peptide (class IIb), leaderless (class IIc) and the non-pediocin single (class IId) bacteriocins [13,21]. In most cases, bacteriocin production appears to be regulated and involves a quorum-sensing mode of regulation [22,23]. The production of nisin is autoregulated by NisK, a sensor kinase protein, and NisR, a response regulator, which regulate transcription via signal transduction by a two-component regulatory system. The induction of NisA or NisZ is likely to be dependent on their interaction with NisK [23]. The NisI and the ATP-binding cassette transporter NisEFG support the immunity of the producer cells to nisin [24].

The presence of LAB in many ecological niches, including fish and their rearing environment, is well characterized, and the antimicrobial activity and the probiotic potential of the isolated strains is emphasized [25–29]. However, despite the beneficial role of most LAB, the species *Lactococcus garvieae* has been identified as the aetiological agent of lactococcosis, a major, re-emerging and widely distributed ichthyopathology that causes a hyperacute haemorrhagic septicaemia and relevant economic losses in rainbow trout (*Oncorhynchus mykiss*, Walbaum) farming [25,30]. The reported human cases of pathologies associated with this bacterium are increasing, and its zoonotic potential has been accepted [31–33].

Bacteriocinogenic *L. garvieae* that demonstrate an antagonistic activity against virulent *L. garvieae* strains have been isolated. Their bacteriocins have been characterized, such as garvicin A [34], garvicin AG1 and garvicin AG2 [35], garvicin KS [36], garvicin L1-5 [37], garvicin ML [38] and garvieacin Q [39]. GarA, produced by *L. garvieae* 21881, is encoded in the plasmid pGL5, which holds the structural gene of GarA (*lgnA*), its putative immunity protein (*lgnI*) and the ABC-transporter and its accessory proteins (*lgnC* and *lgnD*). GarA is synthesized as a 63 amino acid precursor with a typical double-glycine leader peptide. Once processed, this peptide results in a 43-aa mature peptide with a theoretical molecular mass of 4645.5 Da. GarA has a narrow activity spectrum and is limited only to other *L. garvieae* strains, suggesting that its mechanism of action is based on the inhibition of cell division, most likely by inhibiting septum formation in target cells [34]. GarQ is produced by *L. garvieae* BCC 43588 and is encoded as the GarQ structural gene (*garQ*), its putative immunity protein (*garI*) and the ABC-transporter (*garT*). GarQ shows a wide spectrum of antimicrobial activity against LAB and other potentially foodborne pathogenic strains [39]. GarQ is synthesized as a 70-aa precursor with a double-glycine cleavage site, resulting in mature GarQ of 50-aa and a theoretical molecular mass of 5340 Da. GarQ utilizes the IID and IIC subunits of the mannose phosphotransferase system (Man-PTS) as a receptor. It kills target bacteria by disrupting the membrane integrity, mainly locking the Man-PTS into a conformation that leads to the formation of a constitutively open pore [40,41]. NisZ, a 34 amino acid pentacyclic peptide naturally produced by *L. lactis*, exhibits antimicrobial activity against several Gram-positive and Gram-negative bacteria [42], and the bacteriocin exerts its antimicrobial activity by both pore formation and the inhibition of cell wall synthesis through specific binding to lipid II, which is an essential precursor of the bacterial cell wall [43].

The cloning and heterologous expression of bacteriocins by LAB, particularly *Lactococcus lactis*, has proven to be a promising approach for obtaining microbial cell factories with a potent antimicrobial activity [44–48]. Moreover, the simultaneous production of bacteriocins of different classes and/or subclasses and distinct modes of action may not only improve their antimicrobial activity and spectrum in a synergistic fashion but may also reduce the presence of bacteria that are resistant to their antagonistic activity [49–51]. In this study, we have proceeded to the design and expression in different *L. lactis* strains of up to three different bacteriocins with antimicrobial activity against virulent *L. garvieae*. These bacteriocins, namely, GarA, GarQ and NisA/NisZ, show different modes of action and other well-described beneficial effects, such as the anticarcinogenic effect of NisA/NisZ and its ability to modulate the microbiota and regulate the immune system of its host. Thus, synthetic genes that encode the signal peptide of the lactococcal secreted protein Usp45 (SP_{usp45}), fused to either mature GarA (*lgnA*) with its putative immunity gene (*lgnI*) and/or to mature GarQ (*garQ*) with its immunity gene (*garI*), were cloned into the protein expression vectors pMG36c which encodes the P_{32} constitutive promoter and in plasmid pNZ8048c, under control of the inducible P_{nisA} promoter. Recombinant *L. lactis* strains were then obtained, and their antimicrobial activity against virulent *L. garvieae* was determined.

2. Materials and Methods

2.1. Bacterial Strains, Plasmids and Growth Conditions

The bacterial strains and plasmids used in this study are listed in Table 1. The *L. lactis* strains were grown at 30 °C in M17 broth (Oxoid Ltd., Basingstoke, UK) supplemented with 0.5% (*w*/*v*) glucose (GM17). *Pediococcus damnosus* CECT4797 was grown in MRS broth (Oxoid Ltd.) at 30 °C. *Escherichia coli* JM109 (Promega, Madison, WI, USA) was grown in Luria–Bertani (LB) broth (Oxoid Ltd.) at 30 °C with shaking. Chloramphenicol (Sigma-Aldrich, St. Louis, MO, USA) was added at 20 µg/mL to select growth of *E. coli* and at 5 µg/mL for the selection of the recombinant lactococcal strains. The cell dry weights of the late exponential phase cultures were determined gravimetrically. Agar plates were made by the addition of 1.5% (*w*/*v*) agar (Oxoid) to the liquid media.

Table 1. Bacterial strains and plasmids used in this study.

Strain or Plasmid	Description [a]	Source and/or Reference [b]
Strains		
Lactococcus lactis subsp. *cremoris* NZ9000	Host strain; non-bacteriocin producer; pepN::*nis*RK.	NIZO, [52]
Lactococcus lactis subsp. *cremoris* WA2-67	Host strain; fish origin; nisin Z producer.	SD-NUTRYCIAL, [53]
Lactococcus lactis subsp. *lactis* DPC5598	Host strain; milk origin; nisin A producer. Plasmid-free derivative of DPC4268.	DPC, [54]
Lactococcus lactis subsp. *lactis* BB24	Host strain; meat origin; nisin A producer.	SD-NUTRYCIAL, [55]
Lactococcus garvieae CF00021	SOAT, ADT, MPA indicator	CEFAS-BCC
Lactococcus garvieae CF01144	MPA indicator	CEFAS-BCC
Lactococcus garvieae CLG4	MPA indicator	LFP-UNIZAR
Lactococcus garvieae CLG5	MPA indicator	LFP-UNIZAR
Lactococcus garvieae CLFP28/06	MPA indicator	LFP-UNIZAR
Lactococcus garvieae JIP29/99	MPA indicator	Jouy-en-Josas SC
Pediococcus damnosus CECT4797	SOAT, ADT indicator	CECT

Table 1. *Cont.*

Strain or Plasmid	Description [a]	Source and/or Reference [b]
Plasmids		
pMG36c	Cm^r; pMG36e derivative; constitutive expression vector carrying the P_{32} promoter.	RUG-MG, [56]
pNZ8048c	Cm^r; inducible expression vector carrying the *nisA* promoter.	NIZO, [52]
pJFAI	Cm^r, pMG36c derivative carrying the PCR product A (P_{32} ribosome binding site and the SP_{usp45} fused to mature *lgnA* and *lgnI*).	This work
pJFQI	Cm^r, pMG36c derivative carrying the PCR product B (P_{32} ribosome binding site and the SP_{usp45} fused to mature *garQ* and *garI*).	This work
pJFAIQI	Cm^r, pMG36c derivative carrying the PCR product C (P_{32} ribosome binding site and the SP_{usp45} fused to mature *lgnA* and *lgnI* and mature *garQ* and *garI*).	This work
pJFQIAI	Cm^r, pMG36c derivative carrying the PCR product D (P_{32} ribosome binding site and the SP_{usp45} fused to mature *garQ* and *garI* and mature *lgnA* and *lgnI*).	This work
pNJFAI	Cm^r, pNZ8048c derivative carrying the PCR product E (SP_{usp45} fused to mature *lgnA* and *lgnI*).	This work
pNJFQI	Cm^r, pNZ8048c derivative corresponding to the PCR product J (SP_{usp45} fused to mature *garQ* and *garI*).	This work
pNJFQIAI	Cm^r, pNZ8048c derivative corresponding to the PCR product K (SP_{usp45} fused to mature *garQ* and *garI* and mature *lgnA* and *lgnI*).	This work

[a] SOAT— stab-on-agar test; ADT—agar diffusion test; MPA—microtiter plate assay; Cm^r—chloramphenicol resistance. [b] NIZO—Department of Biophysical Chemistry, NIZO Food Research (Ede, The Netherlands); SD-NUTRYCIAL—Sección Departamental de Nutrición y Ciencia de los Alimentos, Facultad de Veterinaria, Universidad Complutense de Madrid (Madrid, Spain); DPC—Teagasc Dairy Products Research Centre, Moorepark, Fermoy, Co., (Cork, Ireland); CEFAS-BCC— Centre for Environment Fisheries and Aquaculture Science-Bacterial Culture Collection (Suffolk, United Kingdom); LFP-UNIZAR—Laboratory of Fish Pathology, Universidad de Zaragoza (Zaragoza, Spain); Jouy-en-Josas SC—Jouy-en-Josas Strain Collection, INRAE (Jouy-en-Josas, France); CECT—Spanish Type Culture Collection, Universidad de Valencia (Valencia, Spain); RUG-MG—Department of Molecular Genetics, University of Groningen (Haren, The Netherlands).

2.2. Basic Genetic Techniques and Enzymes

Synthetic gene fragments were designed from the described amino acid sequence of the bacteriocin GarA (*lgnA*) and GarQ (*garQ*), as well as those from their putative immunity proteins GarAI (*lgnI*) and GarQI (*garI*), respectively. In addition, the leader peptide of the native bacteriocins was replaced by the signal peptide of the secreted protein Usp45 (SP_{usp45}), a Sec-dependent protein produced by *L. lactis* MG1363 [45,57]. Similarly, additional sequences containing the *Sac*I cleavage site and the P_{32} ribosome binding site (RBS) as well as the *Sac*I/*Hind*III or the *Bsp*HI/*Hind*III enzyme restriction cleavage sites were added at the 5′ and 3′ ends, respectively, of the designed synthetic gene fragments. Their codon usage was adapted for its expression by *L. lactis*. GeneArt® supplied the synthetic genes into the carrier plasmid pMA-T (Life Technologies S.A., Madrid, Spain). The protein expression vectors pMG36c and pNZ8048c were purified from *E. coli* JM109 by using the NucleoSpin Plasmid Kit (Macherey-Nagel, Düren, Germany). DNA restriction enzymes were supplied by New England Biolabs (Beverly, MA, USA). Ligations were performed with the T4 DNA ligase (Invitrogen, Walthman, MA, USA). Electrocompetent *L. lactis* subsp. *cremoris* NZ9000, *L. lactis* subsp. *cremoris* WA2-67, *L. lactis* subsp. *lactis* DPC5598 and *L. lactis* subsp. *lactis* BB24 cells were obtained after successive growth in the SGGM17 medium, which consisted of of M17 (Oxoid Ltd.) supplemented with 0.5 M sucrose, glucose (0.5%; *w*/*v*) and glycine (2%; *w*/*v*). The cultures were centrifuged and resuspended in a cold wash buffer containing glycerol (20%; *v*/*v*) and 0.5 M sucrose. Aliquots of 50 µL were stored at −80 °C until further use.

2.3. Recombinant Plasmids Derived from pMG36c and Transformation into L. lactis Hosts

The primers and PCR products used for the construction of the pMG36c-derived vectors are listed in Table S1. PCR product amplifications were performed in 50 μL reaction mixtures that contained 20 ng of the synthetic gene fragments included in the carrier pMA-T vectors, 70 pmol of each primer, 1 U of Velocity DNA polymerase (Bioline Reagents, Ltd., London, UK), 10 μL of Hi-Fi buffer and 1.5 μL of 30 mM dNTP's mix in a MJ Mini Gradient Thermal Cycler (BioRad Laboratories). PCR cycling conditions were as follows: denaturation at 98 °C (2 min), 35 cycles of denaturation–annealing–extension (98 °C for 30 s, 60 °C for 30 s and 72 °C for 30 s, respectively) and a final extension step at 72 °C (5 min). The PCR-generated fragments were purified using a NucleoSpin® Gel and PCR clean-up kit (Macherey-Nagel) for cloning and nucleotide sequencing. When required, PCR amplifications were sequenced using the ABI PRISM® BigDye® Terminator cycle sequence reaction kit and the automatic DNA sequencer ABI PRISM, model 377 (Applied Biosystems, Foster City, CA, USA) at the Unidad de Genómica (CAI Técnicas Biológicas, UCM, Madrid, Spain). Digestion of the amplified PCR products with *SacI*/*Hind*III permitted the ligation of the resulting restriction fragments into pMG36c, which was digested with the same enzymes. The resulting pMG36c-derived vectors were transformed into competent lactococcal hosts and electrotransformed with a Gene Pulser™ and Pulse Controller apparatus (Bio-Rad Laboratories, Hercules, CA, USA), according to a previously described procedure [58]. Transformed cells containing the pMG36c-derived vectors pJFAI, pJFQI, pJFAIQI and pJFQIAI (Table 1), were selected for their growth with chloramphenicol and evaluated for their bacteriocinogenity. The total bacterial DNA from the transformed lactococcal strains was purified using the InstaGene Matrix (BioRad Laboratories). It was then submitted to PCR using the primers MGPJ-F and MGPJ-R. The sequencing of the generated PCR products was performed at the Unidad de Genómica (CAI Técnicas Biológicas, UCM).

2.4. Recombinant Plasmids Derived from pNZ8048c and Transformation into L. lactis Hosts

The primers and PCR products used for the construction of the pNZ8048c-derived vectors are listed in Table S1. The initial PCR amplification of the designed gene fragments into carrier pMA-T vectors was performed with primers GARF-BSPHI and GARAIM-R, which were designed to provide the restriction cleavage site for *Bsp*HI/*Hind*III. The digestion of the amplified PCR products with *Bsp*HI/*Hind*III permitted the ligation of the resulting restriction fragments into pNZ8048c, which was digested with *Nco*I and *Hind*III. However, it should be highlighted that construction of the pNZ8048c-derived vectors carrying GarQ or GarQ and GarA and their respective immunity proteins was achieved using a novel, PCR-based, restriction-enzyme-free cloning, or ABC cloning, method [59]. Briefly, the procedure implies the PCR amplification of three overlapping fragments, two from the pNZ8048c vector and one from the previously designed synthetic gene fragments, to generate a single, circular, pNZ8048c-derived vector by using a pair of overlapping primers. For the amplification of the appropriate gene fragments, 50 μL PCR reactions containing 100 ng of plasmid pNZ8048c or the synthetic gene fragments included in the carrier pMA-T vectors, 0.5 μmol of each primer and 25 μL of Phusion Hot Start II High-Fidelity PCR Master Mix (Thermo Scientific, Waltham, MA, USA) were used. PCR cycling conditions were as follows: one initial denaturation step at 98 °C (30 s), 30 cycles of denaturation–annealing–extension (98 °C for 10 s, 49.1–65.8 °C for 20 s and 72 °C for 25 s, respectively) and a final extension step at 72 °C for 5–10 min). Overlapping PCRs were carried out using the three corresponding fragments as templates. Specifically, 1.5×10^{10} copies of fragments derived from amplification of pNZ8048c and 3×10^{10} copies of fragments obtained from amplification of the designed synthetic genes were used. Agarose gel electrophoresis, visualization and sequencing of the generated PCR products were performed essentially as described for the construction of the pMG36c-derived vectors. The resulting pNZ8048c-derived vectors were transformed into competent lactococcal hosts, and the transformed cells containing the pNZ8048c-derived vectors pNJFAI, pNJQI and pNJFQIAI (Table 1) were selected for their growth with chloramphenicol and evaluated for their bacteriocinogenicity. Bacterial DNA from the lactococcal transformed cells was submitted to PCR amplification with the primers NZPJ-F and NZPJ-R. The sequencing of the generated PCR products was performed at the Unidad de Genómica (CAI Técnicas Biológicas, UCM).

2.5. *Antimicrobial Activity of the Recombinant L. lactis Strains*

The direct antimicrobial activity of colonies from the recombinant lactococcal strains was examined by a stab-on-agar test (SOAT) as previously described [60]. When appropriate, cultures were induced with nisin A (Sigma-Aldrich) at a final concentration of 10 ng/mL for the production of the cloned bacteriocins. Cell-free culture supernatants (CFS) were obtained by the centrifugation of cultures at 12,000× *g* at 4 °C for 10 min, adjusted to pH 6.2 with 1 M NaOH, filtered through 0.22 µm pore-size syringe filters (Sartorius, Göttingen, Germany) and stored at −20 °C until further use. The antimicrobial activity of the supernatants was determined by an agar diffusion test (ADT). It was further quantified by a microtiter plate assay (MPA) as previously described [60]. For the MPA, the growth inhibition of sensitive cultures was measured spectrophotometrically at 620 nm with a FLUOstar OPTIMA (BMGLabtech, Ortenberg, Germany) plate reader. One bacteriocin unit (BU) was defined as the reciprocal of the highest dilution of the bacteriocin that caused a growth inhibition of 50% (50% of the turbidity of the control culture without bacteriocin).

2.6. *Purification of Bacteriocins*

Bacteriocins were purified as previously described using a multi-chromatographic procedure [60]. Briefly, 1 L supernatants from early stationary cultures of the recombinant lactococci were precipitated with $(NH_4)_2SO_4$ (50%; *w*/*v*), desalted by gel filtration (PD-10 columns) and subjected to a cation-exchange (SP Sepharose Fast Flow), followed by a hydrophobic interaction (Octyl-Sepharose CL-4B) and reverse-phase chromatography in an ÄKTA purifier Reverse Phase Fast Protein Liquid Chromatography system (RP-FPLC), using the PepRPC HR 5/5 column. Fractions exhibiting the highest bacteriocin activity were pooled and re-chromatographed on the same column until chromatographically pure bacteriocin peptides were obtained. All chromatographic columns and equipment were obtained from GE Healthcare Life Sciences (Barcelona, Spain).

2.7. *Mass Spectrometry (MS) and Multiple Reaction Monitoring (MRM) Analysis of Purified Peptide Fractions from Supernatants of the Recombinant L. lactis Strains*

Purified RP-FPLC fractions from the supernatants of the recombinant lactococcal strains were subjected to matrix-assisted laser desorption–ionization time-of-flight mass spectrometry (MALDI-TOF MS) and multiple reaction monitoring liquid chromatography–electrospray ionization tandem mass spectrometry (MRM-LC-ESI-MS/MS) analyses at the Unidad de Proteómica (CAI Técnicas Biológicas, UCM). Briefly, 1 µL of eluted fractions were spotted onto a MALDI target plate and allowed to air-dry at room temperature. Then, 0.8 µL of a sinapic acid matrix (Sigma-Aldrich) in 30% acetonitrile and 0.3% trifluoroacetic acid was added and allowed to air-dry at room temperature. MALDI-TOF MS analyses were performed using a 4800 Plus Proteomics Analyzer MALDI-TOF/TOF mass spectrometer (Applied Biosystems/MDS Sciex, Toronto, Canada).

For the identification of bacteriocins, the MRM method evaluates a complex mixture of tryptic peptides that can be selectively detected by liquid chromatography coupled to electrospray MS. Briefly, the purified RP-FPLC fractions of interest were dried in Speed-vac and resuspended in 20 µL of 8 M urea. The samples were reduced by adding 10 mM of dithiothreitol for 45 min at 37 °C and alkylated with 55 mM of iodacetamide for 30 min in the dark. The urea was then diluted with 25 mM of ammonium bicarbonate to obtain a molarity of less than 2. When the pH was 8.5, digestion was performed by adding recombinant sequencing grade Trypsin (Roche Molecular Biochemicals, Branchburg, NJ, USA) 1:20 (*w*/*w*) and incubating at 37 °C. After 60 min, an aliquot was taken for the partial digestion of the sample. The rest was incubated overnight. The produced peptides were dried in Speed-vac and resuspended in 2% acetonitrile and 0, 1% formic acid. Skyline (64-bit), version 20.1, was used to build and optimize the MRM for the detection of the peptides of interest [61].

All analyses were performed on a LC-MS/MS Eksigent Nanoflow LC system coupled to a hybrid triple quadrupole/ion trap mass spectrometer, 5500 QTRAP (AB Sciex, Foster City, CA, USA), equipped with a nano electrospray interface operating in the positive ion mode. The MS/MS data were analyzed using Protein Pilot 4.5 software (AB Sciex) or MASCOT 2.3 (MatrixScience, London, UK) to identify the peptides against an in-house DataBase with the fasta sequences of the targeted proteins. The searches were performed assuming a digestion with trypsin with a maximum of 2 missed cleavages, a fragment ion mass tolerance of 0.6 Da and a parent ion tolerance of 0.15 Da. Peptide identifications based on the MS/MS data were accepted if they could be established at a CI of greater than 95% ($p < 0.05$). Data were then processed against the MRM-library on Skyline to ensure consistency between the transitions detected and the sequences of the peptides searched.

3. Results

3.1. Genetic Design and Cloning of Synthetic Genes That Drive the Heterologous Production of GarA and/or GarQ by Recombinant L. lactis Cells

In this work, synthetic genes containing the protein SP_{usp45}, fused to mature GarA (*lgnA*) and its putative immunity protein GarAI (*lgnI*) (AI) encoded by *L. garvieae* 21881 [34], synthetic genes containing the protein SP_{usp45} fused to mature GarQ (*garQ*) and its putative immunity protein GarQI (*lgnI*) (QI) encoded by *L. garvieae* BCC43578 [39], and synthetic genes containing the genetic fusions SP_{usp45}::*lgnA+lgnI+garQ+garI* (AIQI) and SP_{usp45}::*garQ+garI+lgnA+lgnI* (QIAI), were designed for cloning into the protein expression vectors pMG36c which carries the P_{32}constitutive promoter, and in pNZ8048c with the inducible P_{nisA} promoter. PCR-based amplifications of the synthesized gene fragments allowed for the generation of the PCR products shown in Table S1. Cloning the PCR products A, B, C and D in pMG36c resulted in the pMG36c-derived vectors pJFAI, pJFQI, pJFAIQI and pJFQIAI, and cloning the PCR product E in pNZ8048c resulted in the pNZ8048c-derived vector pNJFAI (Table 1). Similarly, the use of a novel, PCR-based, restriction-enzyme-free cloning method [59] for cloning fragments in plasmid pNZ8048 allowed for the construction of the pNZ8048c-derived vectors pNJFQI and pNJFQIAI (Table 1). In this study, no efforts were made to omit the presence of hypothetically redundant genes that encode putative immunity proteins in the designed synthetic gene fragments.

3.2. Antimicrobial Activity of the Recombinant L. lactis Strains as Determined by Their Direct Antagonistic Effect (SOAT) and the Antimicrobial Activity (ADT) of Their Cell-Free Supernatants

The transformation of recombinant plasmids into *L. lactis* subsp. *cremoris* NZ9000 showed that while the control NZ9000 (pMG36c) and NZ9000 (pNZ8048c) cells showed no antimicrobial activity against *L. garvieae* CF00021 or *P. damnosus* CECT4797, all the recombinant NZ9000-derived strains exhibited a measurable antagonistic effect against *L. garvieae* CF00021 (Table 2). Interestingly, results from both the SOAT and ADT tests showed that the recombinant strains NZ9000 (pJFAI) and NZ9000 (pNJFAI), which only encode the production of GarA, showed no antimicrobial activity against *P. damnosus* CECT4797.

However, the recombinant strains derived from the native NisZ producer (*L. lactis* subsp. *cremoris* WA2-67) and the native NisA producers (*L. lactis* subsp. *lactis* DPC5598 and *L. lactis* subsp. *lactis* BB24) showed a higher antimicrobial activity against *P. damnosus* CECT4797 than *L. garvieae* CF00021. The strains that only encoded the production of GarA and/or GarQ but not NisZ or NisA (NZ9000 derivatives) showed a weaker antimicrobial activity against *P. damnosus* CECT4797. A closer evaluation of the antimicrobial activity of the NisZ-producing strain *L. lactis* subsp. *cremoris* WA2-67 suggested that the strains WA2-67 (pJFQI) and WA2-67 (pJFQIAI) displayed slightly larger halos of inhibition against both indicator strains than the rest of the recombinant strains evaluated (Table 2). Furthermore, the evaluation of the diluted supernatants of *L. lactis* subsp. *cremoris* WA2-67, which were transformed with pMG36c or the pMG36c-derived vectors, showed that the antimicrobial activity of the recombinant strains WA2-67 (pJFAI), WA2-67 (pJFQI), WA2-67 (pJFAIQI) and WA2-67 (pJFQIAI) was 2-, 8-, 2- and 16-fold higher, respectively, than the antimicrobial activity of *L. lactis* subsp. *cremoris* WA2-67 (pMG36c) (Figure 1).

Table 2. Direct antimicrobial activity and extracellular antimicrobial activity of recombinant *L. lactis* strains, producing garvicin A (GarA) and/or garvicin Q (GarQ), against different bacterial indicators.

Strain	SOAT [a]		ADT [b]	
	L. garvieae CF00021	*P. damnosus* CECT4797	*L. garvieae* CF00021	*P. damnosus* CECT4797
L. lactis subsp. *cremoris*				
NZ9000 (pMG36c)	**(-)**	**(-)**	**(-)**	**(-)**
NZ9000 (pJFAI)	6.7	(-)	10.1	(-)
NZ9000 (pJFQI)	6.4	6.0	10.3	8.6
NZ9000 (pJFAIQI)	6.6	5.8	10.4	8.9
NZ9000 (pJFQIAI)	6.8	5.9	10.2	8.8
NZ9000 (pNZ8048c)	**(-)**	**(-)**	**(-)**	**(-)**
NZ9000 (pNJFAI)	8.6	(-)	11.9	(-)
NZ9000 (pNJFQI)	8.3	7.1	12.2	11.2
NZ9000 (pNJFQIAI)	8.1	6.5	11.7	9.5
WA2-67 (pMG36c)	**12.6**	**20.5**	**13.7**	**21.7**
WA2-67 (pJFAI)	12.8	20.4	14.0	21.8
WA2-67 (pJFQI)	13.6	21.3	15.6	23.2
WA2-67 (pJFAIQI)	12.7	20.1	13.8	21.7
WA2-67 (pJFQIAI)	14.8	23.0	16.9	25.4
WA2-67 (pNZ8048c)	**12.3**	**20.5**	**13.6**	**21.3**
WA2-67 (pNJFAI)	12.7	20.1	13.8	21.7
WA2-67 (pNJFQI)	12.2	20.8	14.0	22.2
WA2-67 (pNJFQIAI)	12.3	20.3	13.8	22.0
L. lactis subsp. *lactis*				
DPC5598 (pMG36c)	**7.2**	**9.0**	**13.7**	**20.0**
DPC5598 (pJFAI)	7.6	8.7	13.5	19.4
DPC5598 (pJFQI)	7.5	8.9	13.6	19.7
DPC5598 (pJFAIQI)	6.8	7.6	12.8	18.0
DPC5598 (pJFQIAI)	6.2	6.5	11.9	16.7
DPC5598 (pNZ8048c)	**7.6**	**9.5**	**14.0**	**20.6**
DPC5598 (pNJFAI)	7.8	9.1	14.5	21.3
DPC5598 (pNJFQI)	7.3	9.4	14.8	22.0
DPC5598 (pNJFQIAI)	6.8	8.6	12.5	19.1
BB24 (pMG36c)	**7.9**	**9.7**	**12.4**	**18.6**
BB24 (pJFAI)	9.9	10.0	14.5	18.3
BB24 (pJFQI)	9.4	13.4	14.6	20.5
BB24 (pJFAIQI)	9.7	13.8	14.2	19.9
BB24 (pJFQIAI)	9.6	13.9	14.3	20.2
BB24 (pNZ8048c)	**8.5**	**10.1**	**13.0**	**18.7**
BB24 (pNJFAI)	10.3	13.7	15.0	19.3
BB24 (pNJFQI)	10.7	14.2	15.1	20.4
BB24 (pNJFQIAI)	10.2	14.6	15.0	20.0

[a] Direct antimicrobial activity as determined by a stab-on-agar test (SOAT) and [b] extracellular antimicrobial activity as determined by an agar diffusion test (ADT). Both are expressed as halos of growth inhibition (diameter in mm). Most of the data are means from two independent determinations in triplicate. Control strains are in bold. (-)—no activity.

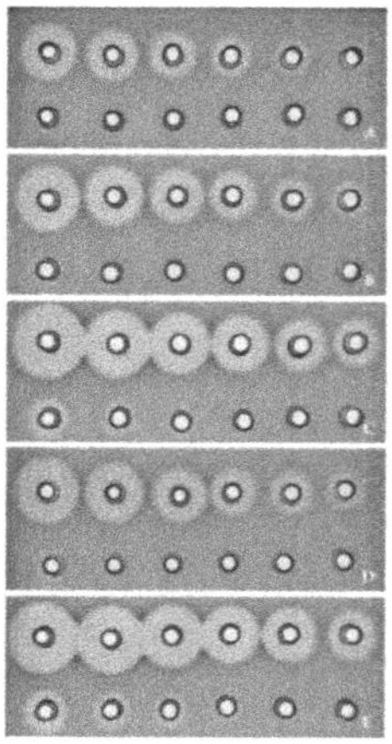

Figure 1. Antimicrobial activity of diluted supernatants of recombinant *L. lactis* strains as determined by an agar diffusion test (ADT) with *L. garvieae* CF00021 as the indicator microorganism. Supernatants from: *L. lactis* subsp. *cremoris* WA2-67 (pMG36c) (**A**), *L. lactis* subsp. *cremoris* WA2-67 (pJFAI) (**B**), *L. lactis* subsp. *cremoris* WA2-67 (pJFQI) (**C**), *L. lactis* subsp. *cremoris* WA2-67 (pJFAIQI) (**D**) and *L. lactis* subsp. *cremoris* WA2-67 (pJFQIAI) (**E**). Supernatants from recombinant strains in the upper lines were diluted 1, 2, 4, 8, 16 and 32 times, and in the bottom lines were diluted 64, 128, 256, 512, 1024 and 2048 times.

The evaluation of the antimicrobial activity of the NisA-producing strain *L. lactis* subsp. *lactis* DPC5598 showed that recombinants encoding the production of either GarAI, GarQI, GarAI+GarQI or GarQI+GarAI displayed similar halos of inhibition to the control strains, DPC5598 (pMG36c) and DPC5598 (pNZ8048c), against both *P. damnosus* CECT4797 and *L. garvieae* CF00021. However, supernatants of the recombinant strains, derived from the NisA producer *L. lactis* subsp. *lactis* BB24, showed slightly larger inhibition halos against *L. garvieae* CF00021 than the control BB24 (pMG36c) and BB24 (pNZ8048) cells (Table 2).

3.3. *Antimicrobial Activity of Recombinant* L. lactis *Strains against Different* L. garvieae *Strains*

The antimicrobial activity of supernatants from all lactococcal strains was quantified against different virulent *L. garvieae* strains by using a more sensitive microtiter plate assay (MPA). Again, *L. lactis* subsp. *cremoris* NZ9000 (pMG36c) showed no antimicrobial activity against any of the *L. garvieae* strains evaluated, while the recombinant NZ9000-derived strains transformed with the constitutive pMG36c-derived vectors showed a weak antimicrobial activity against the virulent *L. garvieae* strains (Table 3). However, the NisZ-producing *L. lactis* subsp. *cremoris* WA2-67 (pMG36c) showed a much higher antimicrobial activity against *L. garvieae*. Remarkably, the antimicrobial activity determined for the recombinant WA2-67-derived strains transformed with the pMG36c-derived vectors to generate the WA2-67 (pJFAI), WA2-67 (pJFQI), WA2-67 (pJFAIQI) and WA2-67 (pJFQIAI) bacteriocin producers was 1.0- to 1.6-fold, 5.1- to 10.7-fold, 0.9- to 1.6-fold, and17.3- to 68.2-fold higher, respectively, than that of the control strain WA2-67 (pMG36c) (Table 3). The antimicrobial activity of the NisA producer *L. lactis* subsp. *lactis* DPC5598 recombinants was 0.5- to 1.2-fold higher than that of the control DPC5598 (pMG36c) strain against *L. garvieae*, while the antagonistic activity of the NisA producer *L. lactis* subsp. *lactis* BB24 recombinants was 1.3- to 3.7-fold higher than that of the control BB24 (pMG36c) strain against the same *L. garvieae* indicator strains (Table 3).

Table 3. Antimicrobial activity [a] of supernatants from recombinant *L. lactis* strains transformed with pMG36c-derived plasmids against virulent *L. garvieae* strains.

Strain	*L. garvieae*					
	CF00021	**CF01144**	**CLG4**	**CLG5**	**CLFP28/06**	**JIP29/99**
L. lactis subsp. *cremoris*						
NZ9000 (pMG36c)	**(-)**	**(-)**	**(-)**	**(-)**	**(-)**	**(-)**
NZ9000 (pJFAI)	88	90	76	69	88	85
NZ9000 (pJFQI)	98	143	112	106	97	100
NZ9000 (pJFAIQI)	121	169	151	153	149	140
NZ9000 (pJFQIAI)	126	145	101	167	107	108
WA2-67 (pMG36c)	**2199**	**1035**	**1148**	**391**	**1087**	**521**
WA2-67 (pJFAI)	2276	1469	1920	475	1566	744
WA2-67 (pJFQI)	11,798	11,082	5935	4201	8864	4821
WA2-67 (pJFAIQI)	2193	1749	1825	468	1637	847
WA2-67 (pJFQIAI)	150,033	19,571	28,362	6780	33,114	9412
L. lactis subsp. *lactis*						
DPC5598 (pMG36c)	**1808**	**1271**	**1865**	**139**	**1042**	**855**
DPC5598 (pJFAI)	1529	1326	2133	198	1318	960
DPC5598 (pJFQI)	1770	1338	1976	129	1288	1066
DPC5598 (pJFAIQI)	1096	1068	1312	92	1064	905
DPC5598 (pJFQIAI)	1033	659	1489	81	1055	815
BB24 (pMG36c)	**1123**	**790**	**443**	**115**	**1479**	**514**
BB24 (pJFAI)	2385	2480	1617	203	2036	1105
BB24 (pJFQI)	2532	2530	1725	202	3115	1243
BB24 (pJFAIQI)	2238	2756	1786	204	2615	1459
BB24 (pJFQIAI)	2535	2699	1641	196	3166	1317

[a] Antimicrobial activity as determined by a microtiter plate assay (MPA) and expressed in bacteriocin units per milligram cell dry weight (BU/mg cdw). Most of the data are means from two independent determinations in triplicate. Control strains are in bold. (-)—no activity.

When the antimicrobial activity of the lactococcal strains that were transformed with the inducible pNZ8048c-derived vectors was determined, *L. lactis* subsp. *cremoris* NZ9000 (pNZ8048c) showed no antimicrobial activity against any of the *L. garvieae* strains, while the recombinant NZ9000 (pNJFAI), NZ9000 (pNJFQI), and NZ9000 (pNJQIAI) strains showed a 12.8- to 18.9-fold, 7.4- to 18.1-fold and 6.9- to 14.4-fold higher antimicrobial activity, respectively, than the cells transformed with the pMG36c-derived vectors (Table 4). However, the antimicrobial activity of the *L. lactis* subsp. *cremoris* WA2-67 (pNZ8048c) cells and their recombinant WA2-67 (pNJFAI), WA2-67 (pNJFQI) and WA2-67 (pNJFQIAI) strains showed a 0.7- to 0.9-fold, 0.5- to 0.8-fold, 0.1- to 0.2-fold and 0.01- to 0.08-fold lower antimicrobial activity, respectively, against the *L. garvieae* strains than the cells transformed with the pMG36c-derived vectors (Table 4). The antimicrobial activity of the NisA-producing *L. lactis* subsp. *lactis* DPC5598 transformed with the pNZ8048c-derived vectors was only slightly (0.9- to 2.0-fold) higher than the antimicrobial activity of the cells transformed with the pMG36c-derived vectors. Similarly, the antagonistic activity of the NisA producer *L. lactis* subsp. *lactis* BB24 recombinants transformed with the pNZ8048c-derived vectors was only slightly (0.6- to 1.9-fold) higher than cells transformed with the pMG36c-derived vectors (Table 4).

Table 4. Antimicrobial activity [a] of supernatants from recombinant *L. lactis* strains transformed with pNZ8048c-derived plasmids against virulent *L. garvieae* strains.

Strain	*L. garvieae*					
	CF00021	CF01144	CLG4	CLG5	CLFP28/06	JIP29/99
L. lactis subsp. *cremoris*						
NZ9000 (pNZ8048c)	**(-)**	**(-)**	**(-)**	**(-)**	**(-)**	**(-)**
NZ9000 (pNJFAI)	1126	997	1440	1295	1600	1263
NZ9000 (pNJFQI)	1214	1069	1395	1194	1440	1220
NZ9000 (pNJFQIAI)	1051	1188	1297	1167	1541	1255
WA2-67 (pNZ8048c)	**1668**	**1002**	**897**	**364**	**987**	**369**
WA2-67 (pNJFAI)	1853	989	1000	412	902	385
WA2-67 (pNJFQI)	2150	1351	1237	565	1163	486
WA2-67 (pNJFQIAI)	2021	1223	1149	595	1245	491
L. lactis subsp. *lactis*						
DPC5598 (pNZ8048c)	**2123**	**1258**	**2317**	**198**	**1097**	**1096**
DPC5598 (pNJFAI)	2760	2699	3443	317	1610	1770
DPC5598 (pNJFQI)	2605	1496	2334	266	1499	1492
DPC5598 (pNJFQIAI)	1150	880	1800	98	1480	920
BB24 (pNZ8048c)	**1203**	**693**	**397**	**158**	**1022**	**685**
BB24 (pNJFAI)	3398	3102	2340	339	2989	1597
BB24 (pNJFQI)	3705	3663	2732	397	4172	1603
BB24 (pNJFQIAI)	3502	3375	2541	299	5016	2001

[a] Antimicrobial activity as determined by a microtiter plate assay (MPA) and expressed in bacteriocin units per milligram cell dry weight (BU/mg cdw). Most of the data are means from two independent determinations in triplicate. Control strains are in bold. (-)—no activity.

3.4. Purification of Bacteriocins Produced by L. lactis subsp. cremoris WA2-67 (pJFQI) and L. lactis subsp. cremoris WA2-67 (pJFQIAI)

The results regarding the purification to homogeneity of the bacteriocins in supernatants of the selected *L. lactis* subsp. *cremoris* recombinants are summarized in Table 5. The evaluation of the most active antimicrobial fractions after the first reversed-phase chromatography step (RP-FPLC) permitted the identification of two active fractions during the purification of the bacteriocins produced by *L. lactis* subsp. *cremoris* WA2-67 (pJFQI) and three active fractions during the purification of the bacteriocins produced by *L. lactis* subsp. *cremoris* WA2-67 (pJFQIAI). Although the antimicrobial activity of the eluted fractions was low, a significant increase in their specific antimicrobial activity was observed.

3.5. Mass Spectrometry (MS) and Multiple Reaction Monitoring (MRM) Analysis of the Purified Bacteriocin Fractions

MALDI-TOF MS analysis of fraction 8 and fraction 7, eluted during the RP-FPLC step of the purification of supernatants from *L. lactis* subsp. *cremoris* WA2-67 (pJFQI) and *L. lactis* subsp. *cremoris* WA2-67 (pJFQIAI), showed major peaks of 3331.4 Da and 3331.3 Da (Figure S1), respectively, matching the molecular mass described for NisZ. In addition, second peaks of 3349.3 Da and 3348.8 Da, respectively, likely correspond to the oxidation of the lanthionine ring of NisZ (Figure S1).

However, MALDI-TOF MS analysis of the eluted fraction 14 from the *L. lactis* subsp. *cremoris* WA2-67 (pJFQI), which encoded GarQ and NisZ, in addition to an analysis of fractions 9 and 12 from the *L. lactis* subsp. *cremoris* WA2-67 (pJFQIAI), which encoded GarA, GarQ and NisZ, could not identify the presence of the bacteriocins GarA and GarQ with predicted molecular masses of 4645.2 Da and 5340 Da, respectively, in the eluted fractions. Since this was a totally unexpected result, the fractions were subjected to MRM-LC-ESI-MS/MS analysis to determine the presence of the expected bacteriocins in the samples. In the MRM method, a series of target tryptic peptides and their associated transitions (fragments m/z) were predicted from the molecular masses and amino acid sequences of the bacteriocins GarA and GarQ. Each targeted peptide has a set of accompanying transitions which are then selectively detected in the second stage of the MS. A summary of the results obtained is shown in Table 6. MRM transitions were established and validated by tandem mass spectrometry (MS/MS). For each bacteriocin, two encrypted peptides were confidently detected in duplicate runs.

Table 5. Purification of NisZ and GarQ produced by *L. lactis* subsp. *cremoris* WA2-67 (pJFQI) and purification of NisZ, GarA and GarQ produced by *L. lactis* subsp. *cremoris* WA2-67 (pJFQIAI).

Purification Stage	Volume (mL)	Total A_{254} [a]	Total Antimicrobial Activity (BU) [b]	Specific Antimicrobial Activity (BU/A_{254}) [c]	Increase in Specific Antimicrobial Activity (Fold) [d]	Recovery Antimicrobial Activity (%)
***L. lactis* subsp. *cremoris* WA2-67 (pJFQI)**						
Culture supernatant	1000	30,500	10.6×10^6	347	1	100
Ammonium sulfate precipitation	100	1930	12.3×10^6	6370	18	116
Gel filtration chromatography	185	872	2.1×10^6	2410	7	20
Cation-exchange chromatography	50	75	11.3×10^6	150,670	434	107
Hydrophobic-interaction chromatography	15	8.4	1.0×10^6	119,050	343	9
Reversed-phase chromatography						
Fraction 8	0.300	0.325	50.0×10^3	153,850	443	0.5
Fraction 14	0.400	0.056	2.3×10^3	41,070	118	0.02
***L. lactis* subsp. *cremoris* WA2-67 (pJFQIAI)**						
Culture supernatant	1000	28,400	1.4×10^8	4930	1	100
Ammonium sulfate precipitation	100	1510	1.5×10^9	993,380	201	1071
Gel filtration chromatography	185	832	2.3×10^8	276,440	56	164
Cation-exchange chromatography	50	119	1.8×10^9	15,130	3	1286
Hydrophobic-interaction chromatography	15	13.8	16.1×10^6	1,166,670	237	11.5
Reversed-phase chromatography						
Fraction 7	0.300	0.125	21,400	171,200	35	0.015
Fraction 9	0.450	0.218	2200	10,100	2	0.0016
Fraction 12	0.500	0.040	200	5000	1	0.00014

[a] Absorbance at 254 nm (A_{254}) multiplied by the volume in milliliters. [b] Antimicrobial activity in bacteriocin units per milliliter (BU/mL), determined by an MPA against *L. garvieae* CF00021 and multiplied by the total volume in milliliters. [c] Specific antimicrobial activity expressed as the total antimicrobial activity (BU) divided by the total A_{254}. [d] Specific antimicrobial activity of a fraction (BU/A_{254}) divided by the specific antimicrobial activity of the initial culture supernatant (BU/A_{254}).

Table 6. Peptide fragments identified by MRM-LC-ESI-MS/MS (QTRAP) in eluted purified fractions from supernatants of *L. lactis* subsp. *cremoris* WA2-67 (pJFQI) and *L. lactis* subsp. *cremoris* WA2-67 (pJFQIAI).

Strain	Fraction	Peptide Sequence	Precursor MW	Detected *m/z*	Retention Time	Detected Transitions	MS/MS Confirmation
L. lactis* subsp. *cremoris							
WA2-67 (pJFQI)	14	GarQ					
		EYHLMNGANGYLTR	1638.76	546.92 (+3)	20.1	4	+
		VNGKYVYR	998.54	499.77 (+2)	8.7	4	+
WA2-67 (pJFQIAI)	9	GarQ					
		EYHLMNGANGYLTR	1638.76	546.92 (+3)	21.3	4	+
		VNGKYVYR	998.54	499.77 (+2)	10.4	4	+
	12	GarA					
		GKINQYRPY	1138.60	569.8 (+3)	14.7	4	+
		INQYRPY	953.48	477.24 (+2)	15.6	5	+

Amino acid sequence of GarQ: EYHLMNGANGYLT**R**VNG**K**YVY**R**VT**K**DPVSAVFGVISNGWGSAGAGFGPQH and GarA: IGGALGNALNGLGTWANMMNGGGFVNQWQVYAN**K**G**K**INQY**R**. PY. Lysine (K) and arginine (R) residues for hydrolysis by trypsin, in bold. +, positive confirmation.

4. Discussion

Virulent *L. garvieae* are the etiological agents of a hyperacute hemorrhagic septicemia in fish, known as lactococcosis. They are also responsible for human pathologies due to their zoonotic character and potential presence in foods [31,33]. Bacteriophages and bacteriocins have potential as complementary strategies for combating *L. garvieae* in foods and fish [26,27], while bacteriocinogenic LAB could be evaluated for their potential use as probiotics, paraprobiotics and postbiotics in food, feed and other biotechnological applications [7,53,62]. The optimization of bacteriocin gene synthesis, expression and production helps the development of LAB as cell factories for the production and delivery of multiple bacteriocins [45,46,63]. The use of synthetic genes that match the codon usage of the producer organisms has a significant impact on gene expression levels and protein folding [47,64].

In this work, the transformation of *L. lactis* subsp. *cremoris* NZ9000 with pMG36c- or pNZ8048c-derived vectors demonstrated that the NZ9000 (pJFAI) and NZ9000 (pNJFAI) recombinant cells, which encode GarAI, showed antimicrobial activity against *L. garvieae* CF00021 but not against *P. damnosus* CECT4797 (Table 2), confirming their production of GarA and previous observations that this bacteriocin was only active against *L. garvieae* [34]. However, differences in the antimicrobial activity of recombinants derived from the NisZ producer *L. lactis* subsp. *cremoris* WA2-67, which was transformed with the pMG36c-derived vectors but not pNZ8048c-derived vectors, were observed against both indicator strains. The obtained results showed that the WA2-67 (pJFQI) cells exhibited larger halos of inhibition than the WA2-67 (pJFAI) cells, and that the WA2-67 (pJFQIAI) cells displayed the largest observed halos (Figure 1). These results suggest that the constitutive expression of GarQI is higher than GarAI or that the specific antimicrobial activity of GarQ is higher than that of GarA. Perhaps the transcription, processing and secretion from genes encoding GarQI+GarAI are more effective than from genes encoding GarAI+GarQI. On the other hand, no remarkable differences were found in the antimicrobial activity of the recombinants derived from the NisA producers transformed with the pMG36c-derived or the pNZ8048c-derived vectors, *L. lactis* subsp. *lactis* DPC5598 and *L. lactis* subsp. *lactis* BB24 (Table 2).

Due to the increasing interest that *L. garvieae* is attracting as not only a relevant bacterial pathogen but also as a zoonotic agent [29,65,66], the antimicrobial activity of the *L. lactis* recombinants was further evaluated and quantified against different virulent *L. garvieae* strains by using a more sensitive microplate assay (MPA). The obtained results showed that the *L. lactis* subsp. *cremoris* NZ9000 cells transformed with the pNZ8048c-derived vectors showed a 6.9- to 18.9-fold higher antimicrobial activity than the recombinant cells bearing the pMG36c-derived vectors (Tables 3 and 4). The enhanced antimicrobial activity in cells with the nisin-inducible constructs may be due to copy number differences between pNZ8048c and pMG36c, but is more likely caused by the promoters used to drive gene expression [67,68]. Plasmid pNZ8048c contains the high-copy number heterogramic replicon of the lactococcal plasmid pSH71 with a unique *NcoI* cleavage site downstream of the *nisA* ribosome binding site (RBS), which is used for translational fusions inducible by NisA [23,69]. To optimize protein production, inducible systems are usually considered superior to constitutive expression systems since the former allow for the achievement of a sufficient biomass before the initiation of target protein expression [70]. The increased antimicrobial activity observed with the NisA-induced cells may also be ascribed to the short induction time for the production of GarA and/or GarQ (3 h), which most likely prevented the secreted bacteriocins from attaching to cell walls to form aggregates and/or to undergo protease degradations.

Moreover and quite remarkably, the antimicrobial activity of *L. lactis* subsp. *cremoris* WA2-67 (pMG36c) and the pMG36c-derived WA2-67 (pJFAI), WA2-67 (pJFQI), WA2-67 (pJFAIQI), and WA2-67 (pJFQIAI) strains showed a 1.0- to 1.6-fold, 5.1- to 10.7-fold, 0.9- to 1.6-fold and 17.3- to 68.2-fold higher antimicrobial activity, respectively, than the control WA2-67 (pMG36c) strain (Table 3). These results also indicate that the expression of QI increases the antimicrobial activity of the producer cells; however, the expression of AI has a much lower effect. Additionally, no increase in antimicrobial activity is observed when QI is expressed as the second of the two modules (AIQI). However, when QI is expressed as the first module (QIAI), a synergistic effect of both modules seems to occur regarding the very high antimicrobial activity of the producer cells. Thus, the AI in the AIQI module appears to prevent the QI from becoming active. However, when QI is expressed first in the QIAI module, the AI appears to synergistically increase the QI activity (Table 3). The pMG36c vector is a shuttle vector. It is based on the low-copy replication origin of pWV01 and is able to replicate in *Escherichia coli*, *Bacillus subtilis* and LAB, whereas the strong P_{32} promoter drives the constitutive transcription of inserted genes into the multicloning site (MCS) of pUC18 [56]. From the results obtained, it may occur that, as previously suggested, the specific antimicrobial activity of GarA against *L. garvieae* is lower and/or its production and stability is less than that of GarQ. Additionally, besides the choice of vector and promoters, other factors such as the activation of quality control networks involving folding factors and housekeeping proteases, the oxidation of methionine to methionine-sulfoxide, bacteriocin self-aggregation and mRNA stability may affect bacteriocin production and activity from the recombinant hosts [46,71]. The coexpression of putative immunity genes may also increase the production of bacteriocins in heterologous hosts. These immunity proteins can act by either affecting bacteriocin pore formation or by perturbing the interaction between the bacteriocin and a membrane-located bacteriocin receptor, thereby preventing producer cells from being killed [16]. The expression in AIQI of LgnI before the expression of GarI could also affect producer protection against GarQ, thereby affecting growth and bacteriocin production by the producer cells. Importantly, *L. lactis* subsp. *cremoris* WA2-67 (pJFQI) and *L. lactis* subsp. *cremoris* WA2-67 (pJFQIAI) showed the highest antimicrobial activity against all virulent strains of *L. garvieae* evaluated (Table 3).

However, the antimicrobial activity of *L. lactis* subsp. *cremoris* WA2-67, which was transformed with the pNZ8048c-derived vectors, produced WA2-67-derivatives that showed a 0.01- to 0.9-fold lower antimicrobial activity, respectively, than the cells transformed with the pMG36c-derived vectors (Table 4). This was an unexpected result since, as previously described, inducible systems are often considered superior to constitutive expression systems for the optimization of protein production. Perhaps NisK and NisR, the two component signal transduction systems for the regulation of NisZ synthesis in *L. lactis* subsp. *cremoris* WA2-67, do not fully activate transcription of the P_{nisA} present in pNZ8048c. It could be also possible that levels of phosphorylated NisR may be not enough to drive the activation of two independent P_{nis} promoters which, in addition, derive from two different *Lactoccocus lactis* subspecies: *cremoris* and *lactis*, respectively. Alternatively, NisZ may not be as efficient as NisA for an interaction with the NisK produced by *L. lactis* subsp. *cremoris* WA2-67, thus constraining the induction of transcription of P_{nisA} in pNZ8048c by blocking NisZ with NisI and NisEFG.

The antimicrobial activity of the NisA-producers *L. lactis* subsp. *lactis* DPC5598 and *L. lactis* subsp. *lactis* BB24 was slightly higher for the cells transformed with the pNZ8048c- than the pMG36c-derived vectors, suggesting that NisA is a better inducer than NisZ for the activation of the transcription of P_{nisA} in pNZ8048c (Tables 3 and 4). However, the antimicrobial activity of the DPC5598-derived recombinants under constitutive or inducible conditions was lower during multi-bacteriocin production. This was probably due to the high energy and metabolic cost linked to plasmid maintenance and replication, to the secretion stress associated with bacteriocin overproduction and/or to the synthesis of proteinases for the elimination of misfolded proteins [72–74]. Differences in the antimicrobial activity of these strains and those from the *L. lactis* subsp. *cremoris* WA2-67 transformed with the pMG36c-derived vectors may be also ascribed to yet-unknown genetic and/or metabolic differences between the strains. *L. lactis* subsp. *lactis* DPC5598 was selected as a potential multi-bacteriocin-producing host because it is a plasmid-free derivative of an industrial strain that is extensively used in fermented dairy products due to its phage insensitivity and fast acid-producing ability [54]. *L. lactis* subsp. *lactis* BB24 is a fermented, meat-derived isolate widely used as an efficient host for the heterologous production of bacteriocins [45,55]. Both multi-bacteriocin producers should be considered for their potential evaluation as probiotics, paraprobiotics and/or postbiotics reducing the increasing presence of virulent and zoonotic *L. garvieae* in selected milk and meat substrates, respectively [31,75].

The MALDI-TOF MS analysis of purified eluted fractions from supernatants of the most active antimicrobial strains, *L. lactis* subsp. *cremoris* WA2-67 (pJFQI) and *L. lactis* subsp. *cremoris* WA2-67 (pJFQIAI) (Table 5), allowed for the detection of NisZ in supernatants of the producer strains (Figure S1), suggesting that this bacteriocin is appropriately processed and transported out of the producer cells. However, the presence of GarA and GarQ could not be detected, suggesting interactions of the bacteriocins with unknown biological compounds, or a low amount and recovery of the bacteriocins during their purification to homogeneity [47,64]. However, bacteriocins in the purified fractions were suitable for MRM evaluation and data analysis, which is an emerging targeted proteomics workflow and a highly selective and sensitive method for detecting peptides in the low ng/mL to sub-ng/mL range concentrations [64,76,77]. When the purified fractions were subjected to MRM-LC-ESI-MS/MS analysis, two encrypted peptides were confidently (99%) detected. The detected peptides were confirmed by MS/MS, and at least four transitions were identified for each (Table 6). The peptide fragments covered 44% of the sequence of GarQ, while the coverage was 21% for GarA. These relatively low-coverage percentages are related to the reduced presence of lysine and arginine (trypsin cleavage sites) in GarA and GarQ, which significatively reduces the number of potentially identifiable target peptides.

Previous studies by our research group identified probiotic features of the native or wild-type NisZ producer *L. lactis* subsp. *cremoris* WA2-67 such as a potent antimicrobial activity against ichthyopathogens, survival in fresh water and the gastrointestinal tract of trout, a resistance to bile and low pH, and an improved colonization ability with respect to the intestinal trout mucosa [26,53,78]. Further in silico analyses of the whole-genome sequence (WGS) of this strain also identified other potential probiotic traits such as the production of vitamins and amino acids, adhesion/aggregation and stress resistance factors and the absence of transferable antibiotic resistance determinants and genes encoding detrimental enzymatic activities or potential virulence factors [79]. Other studies performed by our group demonstrated the effectiveness of *L. lactis* subsp. *cremoris* WA2-67 to protect rainbow trout in vivo against infection of the virulent *L. garvieae* and the relevance of NisZ production as an anti-infective mechanism [53].

The work described in this study constitutes the first report on the design of multi-bacteriocinogenic *L. lactis* subsp. *cremoris* WA2-67 strains with a high antimicrobial activity against virulent *L. garvieae* and a promising role as probiotics, paraprobiotics and/or postbiotics in food, feed and other biotechnological applications. The evaluation of bioengineered strains as probiotics is subjected to approval by regulatory authorities and is performed under strict biological conditions. However, the number of reports on the evaluation of bioengineered bacterial strains as probiotics (live cells), paraprobiotics (dead, non-viable cells) and postbiotics (physical-, chemical- or enzymatic-lysis of probiotic cells) is increasing [7,80,81]. Accordingly, experiments are being planned to evaluate the in vitro effect of *L. lactis* subsp. *cremoris* WA2-67 (pJFQI) and *L. lactis* subsp. *cremoris* WA2-67 (pJFQIAI) on rainbow trout intestinal epithelial cells (RTgutGC) for a transcriptional analysis of several immune, intestinal, barrier-integrity and homeostasis genes and the induction of antimicrobial peptides (AMPs), as well as for their effect on the in vivo modulation of the intestinal microbiota and immune response of rainbow trout (*Oncorhynchus mykiss*, Walbaum) and turbot (*Scophthalmus maximus*).

5. Conclusions

The design of synthetic genes and their cloning into protein expression vectors bearing constitutive or inducible promoters has allowed for the production and functional expression of GarA and/or GarQ by *L. lactis* subsp. *lactis* and *L. lactis* subsp. *cremoris* strains. Most importantly, *L. lactis* subsp. *cremoris* WA2-67, transformed with the pMG36c-derived vectors, allowed for the obtention of *L. lactis* subsp. *cremoris* WA2-67 (pJFQI), a producer of GarQ and NisZ, and *L. lactis* subsp. *cremoris* WA2-67 (pJFQIAI), a producer of GarA, GarQ and NisZ, with a much higher antimicrobial activity (5.1- to 10.7-fold and 17.3- to 68.2-fold, respectively) against virulent *L. garvieae* than the rest of the *L. lactis* strains evaluated.

The concerted use of a sensitive microtiter plate assay (MPA) for the quantification of the antimicrobial activity of supernatants, the use of a multi-chromatographic procedure for the purification of bacteriocins to homogeneity, and the use of a MALDI-TOF MS multiple reaction monitoring (MRM-LC-ESI-MS/MS) analysis of the purified bacteriocins are unavoidable and possibly irreplaceable tools for the identification and characterization of the bacteriocins produced by *L. lactis* subsp. *cremoris* WA2-67 (pJFQI) and *L. lactis* subsp. *cremoris* WA2-67 (pJFQIAI).

Supplementary Materials: The following supporting information can be downloaded at: https://www.mdpi.com/article/10.3390/foods12051063/s1, Table S1: Primers and PCR products used in this study; Figure S1: MALDI-TOF MS analysis of purified NisZ from *Lactococcus lactis* subsp. *cremoris* WA2-67 (pJFQI) (A) and *Lactococcus lactis* subsp. *cremoris* WA2-67 (pJFQIAI (B). Numbers indicate the molecular mass in Daltons of major peptide fragments.

Author Contributions: Conceptualization, P.E.H. and L.M.C.; methodology, J.F., C.A., S.A., E.M.-A. and J.B.; validation, J.F., C.A., S.A., D.C. and B.G.-S.; formal analysis, J.F., L.D.-F., E.M.-A. and J.B.; investigation, J.F., C.A. and S.A.; writing—original draft preparation, J.F.; project administration, P.E.H., L.M.C. and J.F.; funding acquisition, P.E.H. and L.M.C. All authors have read and agreed to the published version of the manuscript.

Funding: This work was supported by project RTI2018-094907-B-I00 from the Ministerio de Ciencia, Innovación y Universidades (MCIU, Madrid, Spain), project FEI16/54 from the Universidad Complutense de Madrid (UCM) and project PR75/18–21627 from Santander/UCM. Javier Feito was supported by a FEI16/54 contract from UCM and held a predoctoral contract from the UCM. Carlos Araújo and Sara Arbulu held a FEI16/54 contract from the UCM.

Institutional Review Board Statement: Not applicable.

Informed Consent Statement: Not applicable.

Data Availability Statement: Data is contained within the article or supplementary material.

Acknowledgments: We thank the Proteomics Unit of the University Complutense of Madrid (UCM), integrated the ProteoRed, PRB3-ISCIII, and support from the Grant PT17/0019 of the PE I + D + I 2013–2016, funded by ISCIII and ERDF, for technical assistance in the proteomic analysis of the samples.

Conflicts of Interest: The authors declare no conflict of interest.

References

1. Chikindas, M.L.; Weeks, R.; Drider, D.; Chistyakov, V.A.; Dicks, L.M. Functions and emerging applications of bacteriocins. *Curr. Opin. Biotechnol.* **2018**, *49*, 23–28. [CrossRef] [PubMed]
2. Aggarwal, S.; Sabharwal, V.; Kaushik, P.; Joshi, A.; Aayushi, A.; Suri, M. Postbiotics: From emerging concept to application. *Front. Sustain. Food Syst.* **2022**, *6*, 887642. [CrossRef]
3. Todorov, S.D.; Popov, I.; Weeks, R.; Chikindas, M.L. Use of bacteriocins and bacteriocinogenic beneficial organisms in food products: Benefits, challenges, concerns. *Foods* **2022**, *11*, 3145. [CrossRef] [PubMed]
4. Vieco-Saiz, N.; Belguesmia, Y.; Raspoet, R.; Auclair, E.; Gancel, F.; Kempf, I.; Drider, D. Benefits and inputs from Lactic Acid Bacteria and their bacteriocins as alternatives to antibiotic growth promoters during food-animal production. *Front. Microbiol.* **2019**, *10*, 57. [CrossRef] [PubMed]
5. O'Connor, P.M.; Kuniyoshi, T.M.; Oliveira, R.P.; Hill, C.; Ross, R.P.; Cotter, P.D. Antimicrobials for food and feed. A bacteriocin perspective. *Curr. Opin. Biotechnol.* **2020**, *61*, 160–167. [CrossRef]
6. Soltani, S.; Hammami, R.; Cotter, P.D.; Rebuffat, S.; Said, L.B.; Gaudreau, H.; Bédard, F.; Biron, E.; Drider, D.; Fliss, I. Bacteriocins as a new generation of antimicrobials: Toxicity aspects and regulations. *FEMS Microbiol. Rev.* **2021**, *45*, fuaa039. [CrossRef]
7. Cuevas-González, P.F.; Liceaga, A.M.; Aguilar-Toalá, J.E. Postbiotics and paraprobiotics: From concepts to applications. *Food Res. Int.* **2020**, *136*, 109502. [CrossRef]
8. Sadeghi, A.; Ebrahimi, M.; Kharazmi, M.S.; Jafari, S.M. Effects of microbial-derived biotics (meta/pharma/post-biotics) on the modulation of gut microbiome and metabolome. General aspects and emerging trends. *Food Chem.* **2023**, *411*, 135478. [CrossRef]
9. Górska, A.; Przystupski, D.; Niemczura, M.J.; Kulbacka, J. Probiotic bacteria: A promising tool in cancer prevention and therapy. *Curr. Microbiol.* **2019**, *76*, 939–949. [CrossRef]
10. Goh, K.S.; Ng, Z.J.; Halim, M.; Oslan, S.N.; Oslan, S.N.H.; Tan, J.S. A comprehensive review on the anticancer potential of bacteriocins: Preclinical and clinical studies. *Int. J. Pept. Res. Ther.* **2022**, *28*, 75. [CrossRef]

11. Molujin, A.M.; Abbasiliasi, S.; Nurdin, A.; Lee, P.-C.; Gansau, J.A.; Jawan, R. Bacteriocins as potential therapeutic approaches in the treatment of various cancers: A review of in vitro studies. *Cancers* **2022**, *14*, 4758. [CrossRef]
12. Umu, Ö.C.O.; Bäuerl, C.; Oostindjer, M.; Pope, P.B.; Hernández, P.E.; Pérez-Martínez, G.; Diep, D.B. The potential of class II bacteriocins to modify gut microbiota to improve host health. *PLoS ONE* **2016**, *11*, e0164036. [CrossRef]
13. Małaczewska, J.; Kaczorek-Łukowska, E. Nisin—A lantibiotic with immunomodulatory properties: A review. *Peptides* **2021**, *137*, 170479. [CrossRef]
14. Havarstein, L.S.; Diep, D.B.; Nes, I.F. A family of bacteriocin ABC transporters carry out proteolytic processing of their substrates concomitant with export. *Mol. Microbiol.* **1995**, *16*, 229–240. [CrossRef]
15. Driessen, A.J.M.; Nouwen, N. Protein translocation across the bacterial cytoplasmic membrane. *Annu. Rev. Biochem.* **2008**, *77*, 643–667. [CrossRef]
16. Pérez-Ramos, A.; Ladjouzi, R.; Benachour, A.; Drider, D. Evidence for the involvement of pleckstrin homology domain-containing proteins in the transport of enterocin DD14 (EntDD14); a leaderless two-peptide bacteriocin. *Int. J. Mol. Sci.* **2021**, *22*, 12877. [CrossRef]
17. Cotter, P.D.; Hill, C.; Ross, R.P. Bacteriocins: Developing innate immunity for food. *Nat. Rev. Microbiol.* **2005**, *3*, 777–788. [CrossRef]
18. Alvarez-Sieiro, P.; Montalbán-López, M.; Mu, D.; Kuipers, O.P. Bacteriocins of Lactic Acid Bacteria: Extending the family. *Appl. Microbiol. Biotechnol.* **2016**, *100*, 2939–2951. [CrossRef]
19. Montalbán-López, M.; Scott, T.A.; Ramesh, S.; Rahman, I.R.; van Heel, A.J.; Viel, J.H.; Bandarian, V.; Dittmann, E.; Genilloud, O.; Goto, Y.; et al. New developments in RiPP discovery, enzymology and engineering. *Nat. Prod. Rep.* **2021**, *38*, 130–239. [CrossRef]
20. Ongpipattanakul, C.; Desormeaux, E.K.; DiCaprio, A.; van der Donk, W.A.; Mitchell, D.A.; Nair, S.K. Mechanism of action of ribosomally synthesized and post-translationally modified peptides. *Chem. Rev.* **2022**, *122*, 14722–14814. [CrossRef]
21. Acedo, J.Z.; Chiorean, S.; Vederas, J.C.; van Belkum, M.J. The expanding structural variety among bacteriocins from Gram-positive bacteria. *FEMS Microbiol. Rev.* **2018**, *42*, 805–828. [CrossRef] [PubMed]
22. Diep, D.B.; Håvarstein, L.S.; Nes, I.F. A bacteriocin-like peptide induces bacteriocin synthesis in *Lactobacillus plantarum* C11: Induction of bacteriocin production. *Mol. Microbiol.* **1995**, *18*, 631–639. [CrossRef] [PubMed]
23. Kuipers, O.P.; Beerthuyzen, M.M.; de Ruyter, P.G.G.A.; Luesink, E.J.; de Vos, W.M. Autoregulation of nisin biosynthesis in *Lactococcus lactis* by signal transduction. *J. Biol. Chem.* **1995**, *270*, 27299–27304. [CrossRef] [PubMed]
24. Pérez-Ramos, A.; Madi-Moussa, D.; Coucheney, F.; Drider, D. Current knowledge of the mode of action and immunity mechanisms of LAB-bacteriocins. *Microorganisms* **2021**, *9*, 2107. [CrossRef] [PubMed]
25. Vendrell, D.; Balcazar, J.; Ruiz-Zarzuela, I.; Deblas, I.; Girones, O.; Muzquiz, J. *Lactococcus garvieae* in fish: A review. *Comp. Immunol. Microbiol. Infect. Dis.* **2006**, *29*, 177–198. [CrossRef]
26. Araújo, C.; Muñoz-Atienza, E.; Nahuelquín, Y.; Poeta, P.; Igrejas, G.; Hernández, P.E.; Herranz, C.; Cintas, L.M. Inhibition of fish pathogens by the microbiota from rainbow trout (*Oncorhynchus mykiss*, Walbaum) and rearing environment. *Anaerobe* **2015**, *32*, 7–14. [CrossRef]
27. Gómez-Sala, B.; Herranz, C.; Díaz-Freitas, B.; Hernández, P.E.; Sala, A.; Cintas, L.M. Strategies to increase the hygienic and economic value of fresh fish: Biopreservation using Lactic Acid Bacteria of marine origin. *Int. J. Food Microbiol.* **2016**, *223*, 41–49. [CrossRef]
28. Gómez-Sala, B.; Muñoz-Atienza, E.; Diep, D.B.; Feito, J.; del Campo, R.; Nes, I.F.; Herranz, C.; Hernández, P.E.; Cintas, L.M. Biotechnological potential and *in vitro* safety assessment of *Lactobacillus curvatus* BCS35, a multibacteriocinogenic strain isolated from dry-salted cod (*Gadus morhua*). *LWT* **2019**, *112*, 108219. [CrossRef]
29. Feito, J.; Araújo, C.; Gómez-Sala, B.; Contente, D.; Campanero, C.; Arbulu, S.; Saralegui, C.; Peña, N.; Muñoz-Atienza, E.; Borrero, J.; et al. Antimicrobial activity, molecular typing and in vitro safety assessment of *Lactococcus garvieae* isolates from healthy cultured rainbow trout (*Oncorhynchus mykiss*, Walbaum) and rearing environment. *LWT* **2022**, *162*, 113496. [CrossRef]
30. Pastorino, P.; Vela Alonso, A.I.; Colussi, S.; Cavazza, G.; Menconi, V.; Mugetti, D.; Righetti, M.; Barbero, R.; Zuccaro, G.; Fernández-Garayzábal, J.F.; et al. A summer mortality outbreak of lactococcosis by *Lactococcus garvieae* in a raceway system affecting farmed rainbow trout (*Oncorhynchus mykiss*) and brook trout (*Salvelinus fontinalis*). *Animals* **2019**, *9*, 43. [CrossRef]
31. Gibello, A.; Galán-Sánchez, F.; Blanco, M.M.; Rodríguez-Iglesias, M.; Domínguez, L.; Fernández-Garayzábal, J.F. The zoonotic potential of *Lactococcus garvieae*: An overview on microbiology, epidemiology, virulence factors and relationship with its presence in foods. *Res. Vet. Sci.* **2016**, *109*, 59–70. [CrossRef]
32. Meyburgh, C.; Bragg, R.; Boucher, C. *Lactococcus garvieae*: An emerging bacterial pathogen of fish. *Dis. Aquat. Org.* **2017**, *123*, 67–79. [CrossRef]
33. Francés-Cuesta, C.; Ansari, I.; Fernández-Garayzábal, J.F.; Gibello, A.; González-Candelas, F. Comparative genomics and evolutionary analysis of *Lactococcus garvieae* isolated from human endocarditis. *Microb. Genom.* **2022**, *8*, 000771. [CrossRef]
34. Maldonado-Barragán, A.; Cardenas, N.; Martinez, B.; Ruiz-Barba, J.L.; Fernandez-Garayzabal, J.F.; Rodriguez, J.M.; Gibello, A. Garvicin A, a novel class IId bacteriocin from *Lactococcus garvieae* that inhibits septum formation in *L. garvieae* strains. *Appl. Environ. Microbiol.* **2013**, *79*, 4336–4346. [CrossRef]
35. Maldonado-Barragán, A.; Alegría-Carrasco, E.; Blanco, M.d.M.; Vela, A.I.; Fernández-Garayzábal, J.F.; Rodríguez, J.M.; Gibello, A. Garvicins AG1 and AG2: Two novel class IId bacteriocins of *Lactococcus garvieae* Lg-Granada. *Int. J. Mol. Sci.* **2022**, *23*, 4685. [CrossRef]

36. Ovchinnikov, K.V.; Chi, H.; Mehmeti, I.; Holo, H.; Nes, I.F.; Diep, D.B. Novel group of leaderless multipeptide bacteriocins from Gram-positive bacteria. *Appl. Environ. Microbiol.* **2016**, *82*, 5216–5224. [CrossRef]
37. Villani, F.; Aponte, M.; Blaiotta, G.; Mauriello, G.; Pepe, O.; Moschetti, G. Detection and characterization of a bacteriocin, garviecin L1-5, produced by *Lactococcus garvieae* isolated from raw cow's milk. *J. Appl. Microbiol.* **2001**, *90*, 430–439. [CrossRef]
38. Borrero, J.; Brede, D.A.; Skaugen, M.; Diep, D.B.; Herranz, C.; Nes, I.F.; Cintas, L.M.; Hernández, P.E. Characterization of garvicin ML, a novel circular bacteriocin produced by *Lactococcus garvieae* DCC43, isolated from mallard ducks (*Anas platyrhynchos*). *Appl. Environ. Microbiol.* **2011**, *77*, 369–373. [CrossRef]
39. Tosukhowong, A.; Zendo, T.; Visessanguan, W.; Roytrakul, S.; Pumpuang, L.; Jaresitthikunchai, J.; Sonomoto, K. Garvieacin Q, a novel class II bacteriocin from *Lactococcus garvieae* BCC 43578. *Appl. Environ. Microbiol.* **2012**, *78*, 1619–1623. [CrossRef]
40. Tymoszewska, A.; Diep, D.B.; Aleksandrzak-Piekarczyk, T. The extracellular loop of man-PTS subunit IID is responsible for the sensitivity of *Lactococcus garvieae* to garvicins A, B and C. *Sci. Rep.* **2018**, *8*, 15790. [CrossRef]
41. Desiderato, C.K.; Hasenauer, K.M.; Reich, S.J.; Goldbeck, O.; Holivololona, L.; Ovchinnikov, K.V.; Reiter, A.; Oldiges, M.; Diep, D.B.; Eikmanns, B.J.; et al. Garvicin Q: Characterization of biosynthesis and mode of action. *Microb. Cell Fact.* **2022**, *21*, 236. [CrossRef]
42. Mulders, J.W.M.; Boerrigter, I.J.; Rollema, H.S.; Siezen, R.J.; Vos, W.M. Identification and characterization of the lantibiotic nisin Z, a natural nisin variant. *Eur. J. Biochem.* **1991**, *201*, 581–584. [CrossRef]
43. Wiedemann, I.; Breukink, E.; van Kraaij, C.; Kuipers, O.P.; Bierbaum, G.; de Kruijff, B.; Sahl, H.-G. Specific binding of nisin to the peptidoglycan precursor lipid II combines pore formation and inhibition of cell wall biosynthesis for potent antibiotic activity. *J. Biol. Chem.* **2001**, *276*, 1772–1779. [CrossRef] [PubMed]
44. Martín, M.; Gutiérrez, J.; Criado, R.; Herranz, C.; Cintas, L.M.; Hernández, P.E. Cloning, production and expression of the bacteriocin enterocin A produced by *Enterococcus faecium* PLBC21 in *Lactococcus lactis*. *Appl. Microbiol. Biotechnol.* **2007**, *76*, 667–675. [CrossRef] [PubMed]
45. Borrero, J.; Jiménez, J.J.; Gútiez, L.; Herranz, C.; Cintas, L.M.; Hernández, P.E. Use of the Usp45 lactococcal secretion signal sequence to drive the secretion and functional expression of enterococcal bacteriocins in *Lactococcus lactis*. *Appl. Microbiol. Biotechnol.* **2011**, *89*, 131–143. [CrossRef]
46. Borrero, J.; Jiménez, J.J.; Gútiez, L.; Herranz, C.; Cintas, L.M.; Hernández, P.E. Protein expression vector and secretion signal peptide optimization to drive the production, secretion, and functional expression of the bacteriocin enterocin A in Lactic Acid Bacteria. *J. Biotechnol.* **2011**, *156*, 76–86. [CrossRef]
47. Jiménez, J.J.; Diep, D.B.; Borrero, J.; Gútiez, L.; Arbulu, S.; Nes, I.F.; Herranz, C.; Cintas, L.M.; Hernández, P.E. Cloning strategies for heterologous expression of the bacteriocin enterocin A by *Lactobacillus sakei* Lb790, *Lb. plantarum* NC8 and *Lb. casei CECT475*. *Microb. Cell. Fact.* **2015**, *14*, 166. [CrossRef]
48. Xu, Y.; Yang, L.; Li, P.; Gu, Q. Heterologous expression of class IIb bacteriocin plantaricin JK in *Lactococcus lactis*. *Protein Expr. Purif.* **2019**, *159*, 10–16. [CrossRef] [PubMed]
49. Bastos, M.d.C.d.F.; Coelho, M.L.V.; Santos, O.C.d.S. Resistance to bacteriocins produced by Gram-positive bacteria. *Microbiology* **2015**, *161*, 683–700. [CrossRef]
50. Mills, S.; Griffin, C.; O'Connor, P.M.; Serrano, L.M.; Meijer, W.C.; Hill, C.; Ross, R.P. A multibacteriocin cheese starter system, comprising nisin and lacticin 3147 in *Lactococcus lactis*, in combination with plantaricin from *Lactobacillus plantarum*. *Appl. Environ. Microbiol.* **2017**, *83*, e00799-17. [CrossRef]
51. Desmond, A.; O'Halloran, F.; Cotter, L.; Hill, C.; Field, D. Bioengineered nisin A derivatives display enhanced activity against clinical neonatal pathogens. *Antibiotics* **2022**, *11*, 1516. [CrossRef]
52. Kuipers, O.P.; de Ruyter, P.G.G.A.; Kleerebezem, M.; de Vos, W.M. Quorum sensing-controlled gene expression in Lactic Acid Bacteria. *J. Biotechnol.* **1998**, *64*, 15–21. [CrossRef]
53. Araújo, C.; Muñoz-Atienza, E.; Pérez-Sánchez, T.; Poeta, P.; Igrejas, G.; Hernández, P.E.; Herranz, C.; Ruiz-Zarzuela, I.; Cintas, L.M. Nisin Z production by *Lactococcus lactis* subsp. *cremoris* WA2-67 of aquatic origin as a defense mechanism to protect rainbow trout (*Oncorhynchus mykiss*, Walbaum) against *Lactococcus garvieae*. *Mar. Biotechnol.* **2015**, *17*, 820–830. [CrossRef]
54. Trotter, M.; Ross, R.P.; Fitzgerald, G.F.; Coffey, A. *Lactococcus lactis* DPC5598, a plasmid-free derivative of a commercial starter, provides a valuable alternative host for culture improvement studies. *J. Appl. Microbiol.* **2002**, *93*, 134–143. [CrossRef]
55. Cintas, L.M.; Casaus, P.; Fernández, M.F.; Hernández, P.E. Comparative antimicrobial activity of enterocin L50, pediocin PA-1, nisin A and lactocin S against spoilage and foodborne pathogenic bacteria. *Food Microbiol.* **1998**, *15*, 289–298. [CrossRef]
56. van de Guchte, M.; van der Vossen, J.M.; Kok, J.; Venema, G. Construction of a lactococcal expression vector: Expression of hen egg white lysozyme in *Lactococcus lactis* subsp. *lactis*. *Appl. Environ. Microbiol.* **1989**, *55*, 224–228. [CrossRef] [PubMed]
57. van Asseldonk, M.; Rutten, G.; Oteman, M.; Siezen, R.J.; de Vos, W.M.; Simons, G. Cloning of Usp45, a gene encoding a secreted protein from *Lactococcus lactis* subsp. *lactis* MG1363. *Gene* **1990**, *95*, 155–160. [CrossRef]
58. Holo, H.; Nes, I.F. High-frequency transformation, by electroporation, of *Lactococcus lactis* subsp. *cremoris* grown with glycine in osmotically stabilized media. *Appl. Environ. Microbiol.* **1989**, *55*, 3119–3123. [CrossRef]
59. El Qaidi, S.; Hardwidge, P.R. ABC cloning: An efficient, simple, and rapid restriction/ligase-free method. *MethodsX* **2019**, *6*, 316–321. [CrossRef] [PubMed]

60. Cintas, L.M.; Casaus, P.; Herranz, C.; Håvarstein, L.S.; Holo, H.; Hernández, P.E.; Nes, I.F. Biochemical and genetic evidence that *Enterococcus faecium* L50 produces enterocins L50A and L50B, the *Sec*-dependent enterocin P, and a novel bacteriocin secreted without an N-terminal extension termed enterocin Q. *J. Bacteriol.* **2000**, *182*, 6806–6814. [CrossRef]
61. MacLean, B.; Tomazela, D.M.; Shulman, N.; Chambers, M.; Finney, G.L.; Frewen, B.; Kern, R.; Tabb, D.L.; Liebler, D.C.; MacCoss, M.J. Skyline: An open source document editor for creating and analyzing targeted proteomics experiments. *Bioinformatics* **2010**, *26*, 966–968. [CrossRef]
62. Pérez-Sánchez, T.; Mora-Sánchez, B.; Balcázar, J.L. Biological approaches for disease control in aquaculture: Advantages, limitations and challenges. *Trends Microbiol.* **2018**, *26*, 896–903. [CrossRef]
63. Gutiérrez, J.; Larsen, R.; Cintas, L.M.; Kok, J.; Hernández, P.E. High-level heterologous production and functional expression of the Sec-dependent enterocin P from *Enterococcus faecium* P13 in *Lactococcus lactis*. *Appl. Microbiol. Biotechnol.* **2006**, *72*, 41–51. [CrossRef]
64. Arbulu, S.; Jiménez, J.J.; Gútiez, L.; Feito, J.; Cintas, L.M.; Herranz, C.; Hernández, P.E. Cloning and expression of synthetic genes encoding native, hybrid- and bacteriocin-derived chimeras from mature class IIa bacteriocins, by *Pichia pastoris* (syn. *Komagataella* spp.). *Food Res. Int.* **2019**, *121*, 888–899. [CrossRef] [PubMed]
65. Ortega, C.; Irgang, R.; Valladares-Carranza, B.; Collarte, C.; Avendaño-Herrera, R. First identification and characterization of *Lactococcus garvieae* isolated from rainbow trout (*Oncorhynchus mykiss*) cultured in Mexico. *Animals* **2020**, *10*, 1609. [CrossRef]
66. Shahin, K.; Veek, T.; Heckman, T.I.; Littman, E.; Mukkatira, K.; Adkison, M.; Welch, T.J.; Imai, D.M.; Pastenkos, G.; Camus, A.; et al. Isolation and characterization of *Lactococcus garvieae* from rainbow trout, *Oncorhynchus mykiss*, from California, USA. *Transbound. Emerg. Dis.* **2022**, *69*, 2326–2343. [CrossRef]
67. Kok, J.; van der Vossen, J.M.; Venema, G. Construction of plasmid cloning vectors for lactic streptococci which also replicate in *Bacillus subtilis* and *Escherichia coli*. *Appl. Environ. Microbiol.* **1984**, *48*, 726–731. [CrossRef]
68. Vos, W.M. Gene cloning and expression in lactic streptococci. *FEMS Microbiol. Rev.* **1987**, *46*, 281–295. [CrossRef]
69. De Ruyter, P.G.; Kuipers, O.P.; de Vos, W.M. Controlled gene expression systems for *Lactococcus lactis* with the food-grade inducer nisin. *Appl. Environ. Microbiol.* **1996**, *62*, 3662–3667. [CrossRef]
70. Kim, J.-H.; Mills, D.A. Improvement of a nisin-inducible expression vector for use in Lactic Acid Bacteria. *Plasmid* **2007**, *58*, 275–283. [CrossRef]
71. Diep, D.; Mathiesen, G.; Eijsink, V.; Nes, I. Use of *Lactobacilli* and their pheromone-based regulatory mechanism in gene expression and drug delivery. *Curr. Pharm. Biotechnol.* **2009**, *10*, 62–73. [CrossRef]
72. Criado, R.; Gutiérrez, J.; Martín, M.; Herranz, C.; Hernández, P.E.; Cintas, L.M. Immunochemical characterization of temperature-regulated production of enterocin L50 (EntL50A and EntL50B), enterocin P, and enterocin Q by *Enterococcus faecium* L50. *Appl. Environ. Microbiol.* **2006**, *72*, 7634–7643. [CrossRef]
73. Serrano-Maldonado, C.E.; Quirasco, M. Enhancement of the antibacterial activity of an *E. faecalis* strain by the heterologous expression of enterocin A. *J. Biotechnol.* **2018**, *283*, 28–36. [CrossRef] [PubMed]
74. Maldonado-Barragán, A.; West, S.A. The cost and benefit of quorum sensing-controlled bacteriocin production in *Lactobacillus plantarum*. *J. Evol. Biol.* **2020**, *33*, 101–111. [CrossRef]
75. Ricci, G.; Ferrario, C.; Borgo, F.; Eraclio, G.; Fortina, M.G. Genome sequences of two *Lactococcus garvieae* strains isolated from meat. *Genome Announc.* **2013**, *1*, e00018-12. [CrossRef] [PubMed]
76. Malmström, J.; Lee, H.; Aebersold, R. Advances in proteomic workflows for systems biology. *Curr. Opin. Biotechnol.* **2007**, *18*, 378–384. [CrossRef] [PubMed]
77. Zhao, Y.; Jensen, O.N. Modification-specific proteomics: Strategies for characterization of post-translational modifications using Enrichment techniques. *Proteomics* **2009**, *9*, 4632–4641. [CrossRef]
78. Araújo, C.; Muñoz-Atienza, E.; Ramírez, M.; Poeta, P.; Igrejas, G.; Hernández, P.E.; Herranz, C.; Cintas, L.M. Safety assessment, genetic relatedness and bacteriocin activity of potential probiotic *Lactococcus lactis* strains from rainbow trout (*Oncorhynchus mykiss*, Walbaum) and rearing environment. *Eur. Food Res. Technol.* **2015**, *241*, 647–662. [CrossRef]
79. Feito, J.; Contente, D.; Ponce-Alonso, M.; Díaz-Formoso, L.; Araújo, C.; Peña, N.; Borrero, J.; Gómez-Sala, B.; del Campo, R.; Muñoz-Atienza, E.; et al. Draft genome sequence of *Lactococcus lactis* subsp. *cremoris* WA2-67: A promising nisin-producing probiotic strain isolated from the rearing environment of a spanish rainbow trout (*Oncorhynchus mykiss*, Walbaum) farm. *Microorganisms* **2022**, *10*, 521. [CrossRef]
80. Dong, H.; Wang, H.; Fu, S.; Zhang, D. CRISPR/Cas tools for enhancing the biopreservation ability of Lactic Acid Bacteria in aquatic products. *Front. Bioeng. Biotechnol.* **2022**, *10*, 1114588. [CrossRef]
81. Mugwanda, K.; Hamese, S.; Van Zyl, W.F.; Prinsloo, E.; Du Plessis, M.; Dicks, L.M.T.; Thimiri Govinda Raj, D.B. Recent advances in genetic tools for engineering probiotic Lactic Acid Bacteria. *Biosci. Rep.* **2023**, *43*, BSR20211299. [CrossRef] [PubMed]

foods

MDPI

Article

Relationship between Volatile Organic Compounds and Microorganisms Isolated from Raw Sheep Milk Cheeses Determined by Sanger Sequencing and GC–IMS

María J. Ruiz [1], José A. Salatti-Dorado [2], María J. Cardador [2], Laureano Frizzo [1], Rafael Jordano [3], Lourdes Arce [2] and Luis M. Medina [3,*]

1 Laboratory of Food Analysis "Rodolfo Oscar Dalla Santina", Institute of Veterinary Science (ICiVet Litoral), National University of the Litoral-National Council of Scientific and Technical Research (UNL/CONICET), Esperanza 3080, Province of Santa Fe, Argentina
2 Analytical Chemistry Department, Institute of Fine Chemistry and Nanochemistry, International Agrifood Campus of Excellence, Marie Curie Annex Building, Campus de Rabanales, University of Cordoba, E-14071 Cordoba, Spain
3 Food Science and Technology Department, International Agrifood Campus of Excellence, Charles Darwin Annex Building, Campus de Rabanales, University of Córdoba, E-14071 Cordoba, Spain
* Correspondence: luismedina@uco.es

Abstract: Recently, the interest of consumers regarding artisan cheeses worldwide has increased. The ability of different autochthonous and characterized lactic acid bacteria (LAB) to produce aromas and the identification of the volatile organic compounds (VOCs) responsible for flavor in cheeses are important aspects to consider when selecting strains with optimal aromatic properties, resulting in the diversification of cheese products. The objective of this work is to determine the relationship between VOCs and microorganisms isolated (*Lacticaseibacillus paracasei*, *Lactiplantibacillus plantarum*, *Leuconostoc mesenteroides* and *Lactococcus lactis* subsp. *hordniae*) from raw sheep milk cheeses (matured and creamy natural) using accuracy and alternative methods. On combining Sanger sequencing for LAB identification with Gas Chromatography coupled to Ion Mobility Spectrometry (GC–IMS) to determinate VOCs, we describe cheeses and differentiate the potential role of each microorganism in their volatilome. The contribution of each LAB can be described according to their different VOC profile. Differences between LAB behavior in each cheese are shown, especially between LAB involved in creamy cheeses. Only L. *lactis* subsp. *hordniae* and L. *mesenteroides* show the same VOC profile in de Man Rogosa and Sharpe (MRS) cultures, but for different cheeses, and show two differences in VOC production in skim milk cultures. The occurrence of *Lactococcus lactis* subsp. *hordniae* from cheese is reported for first time.

Keywords: lactic acid bacteria; GC–IMS; artisan cheese; volatile organic compounds; VOC

1. Introduction

The progressive consumer awareness of healthy, low (or no-)-processed and sustainable products is a fact, with artisanal products (including raw milk cheeses) being preferred in many cases [1]. In recent years, the interest of consumers regarding artisan cheeses worldwide increased [2]. Based on regulations focused on food safety and nutrition and the consumer trend towards products that improve welfare, functional dairy products have become a leading sector in the food industry. Sheep's milk is considered a functional food due to its nutritional quality [3]. Autochthonous microorganisms metabolize the raw materials during fermentation. This process enriches the nutritional value in some fermented foods, imparts health benefits to the consumers and generates gastronomic diversity [4]. In fact, starter and nonstarter lactic acid bacteria (LAB) and coagulating enzymes, used for the production and ripening of cheese, are able to transform protein, fat, lactose and citrate into volatile compounds, enhancing the formation of essential aroma compounds [5]. Raw milk contains natural enzymes, and its microbiota generates and enhances characteristic flavors appreciated by consumers through their metabolic activity. Autochthonous lactic acid bacteria (LAB) are a group of microorganisms including many of those commonly considered as probiotics, as they exert a beneficial effect on the gastrointestinal tract of humans and animals. Furthermore, many LAB genera, such as *Lactobacillus*, *Lactococcus*, *Leuconostoc* and the recently classified *Lactiplantibacillus* and *Lacticaseibacillus* are responsible for the organoleptic characteristics of many traditional foods, improving the nutritional value in some fermented foods [6]. Fermentation processes and food processing and preservation techniques are responsible for the organoleptic characteristics, such as flavor and texture, of this type of food. In this sense, the relationship between volatile organic compounds (VOCs) emitted from raw milk cheeses and the starter and non-starter LAB involved is a matter of recent interest [7].

In the last decade, legislation and market strategies to control and standardize the food supply chain have motivated the development of tools for the authentication of food products [8]. A simple, fast and sensitive analytical technique that has been found in recent years to establish aromatic fingerprints is gas chromatography in conjunction with ion mobility spectrometry (GC–IMS) [9] by extracting VOCs that are generated in the headspace (HS) of the vial. This technique makes it possible to detect VOCs at a part-per-billion (ppb) level in real time in many liquid and solid samples and without any prior treatment. The sensitivity and selectivity of the IMS make it a very suitable tool for use in quality control and food safety, as well as in the characterization, authentication and traceability of food [10]. In fact, GC–IMS can currently be considered an excellent method to separate and identify the volatile contributors to food flavor. It has been already employed to determine the VOC profiles of milk, cheeses and other dairy products [11–13].

Regarding the use of different LAB and their greater or lesser probiotic potential, the conditions are specific for each strain [14]. For this reason, genetic identification is essential for the characterization of this type of bacteria. Molecular analysis based on 16S rRNA, like Sanger sequencing, is one of the methods currently used based on its precision and costs, obtaining bacterial genetic profiles [15].

The objective of this work is to determine the relationship between VOCs and microorganisms isolated from raw sheep milk cheeses, using accuracy and alternative methods. In this way, we identified LAB isolated from artisanal raw sheep milk cheeses using Sanger sequencing and VOCs were detected by GC–IMS.

2. Materials and Methods

2.1. Origin, Isolation and Preparation of Samples and Bacterial Cultures

Six samples of cheese made from raw sheep milk were analyzed. Three samples corresponded to a creamy natural cheese called "Torta del Casar" from Extremadura, Spain. The other three samples were extracted from a matured cheese from Navarra, Spain. In both cheeses, the samples were taken from different parts of the cheese (center, inner and rind areas). A total of 10 g of each sample was weighed, suspended in 90 mL of 0.1% peptone water and homogenized in a Stomacher® Lab Blender (Seward, UK) for 120 s at 230 rpm. Then, 0.1 mL of the 1:10^7 dilution was spread on Man Rogosa and Sharpe agar (MRS, Oxoid, UK) in duplicate. Half of the plates were incubated in anaerobic jars (Thermo Scientific™, Oxoid, UK) at 37 °C for 48 h; the other half were incubated in aerobic conditions at 30 °C for 72 h. Presumptive colonies of LAB were cultured in 3 mL of MRS broth (Oxoid, UK) and incubated in an anaerobic jar (Thermo Scientific™, Oxoid, UK) at 37 °C for 24 h and in aerobiosis at 30 °C for 48 h. Growth colonies were examined for Gram, morphology and catalase determinations prior to bacterial identification by Sanger sequencing.

For the evaluation of LAB spectral fingerprints by GC–IMS, the identified microorganisms were activated in MRS broth (Oxoid, UK) for 24 h at 37 °C, and cultured (pour-plating) on the same agarized medium. Finally, a colony from each culture was collected and suspended in 5 mL of MRS broth (Oxoid, UK) and incubated under the same conditions. Cell concentration was measured by optical density at 540 nm (OD_{540}) in a spectrophotometer (ThermoFisher, Waltham, MA. USA).

2.2. Bacterial Identification by Sanger Sequencing

Sanger sequencing was used as a genotypic characterization method, standardized by the Central Research Support Service of the University of Córdoba, Spain. DNA was extracted from 1.5 mL of culture using the Higher Purity Bacterial Genomic DNA Isolation kit (Canvax, Valladolid, Spain). Good-quality DNA with a ratio of 260/280 between 1.8–2.0 was obtained. A total of 25 ng of DNA was taken to PCR amplify the 16S ribosomal DNA with the universal primers 16SFw: 5′-AGAGTTTGATCCTGGCTCAG-3′ and 16SRv: 5′-GAAAGGAGGTGATCCAGCCG-3′. The reaction was carried out in a 20 μL solution containing 1x reaction buffer, 2.5 mM $MgCl_2$, 100 μM of each dNTP, 0.25 μM of each primer, 1% DMSO and 1 u of BioTaq Taq Polymerase (BioTools). The amplification program includes a first denaturation step at 95 °C for 5 min, followed by 30 cycles of 1 min at 95 °C, 1 min at 55 °C and 1 min at 72 °C, and a final extension step of 7 min at 72 °C. The PCR reactions were purified by precipitation with AcNa/EtOH and suspended in 15 μL of Milli-Q ultrapure water. For each sequencing reaction (16SFw and 16SRv), 6 μL of purified DNA was taken. The sequences obtained from each isolate were aligned to achieve the complete 16S ribosomal sequence (DNAstar megalign program) and analyzed by BLAST in the GeneBank database (https://blast.ncbi.nlm.nih.gov, access on 5 September 2021).

2.3. GC–IMS Analysis Conditions

The analyses of the VOC profiles were made on a headspace gas chromatography-ion mobility spectrometer (HS-GC–IMS) instrument (FlavourSpec®, G.A.S. Gesellschaft für analytische Sensorsysteme mbH. Dortmund, Germany). The accuracy of the GC–IMS method was checked using ketone mix standards. In addition, empty sterile glass vials with magnetic screw caps and septums were analyzed as blanks to determine the presence of compounds retained on the GC column and in the IMS drift tube and run, if necessary, a pre-cleaning method. Working conditions were used according to Gallegos et al. [4] with modifications. Vials were heated at 70 °C for 5 min with a stirring speed of 500 rpm. The injector temperature was 80 °C. A 500 μL aliquot of the headspace was injected into the heated injector and then transferred to a non-polar GC column (SE-54-CB from CS-Chromatographie Service GmbH, Düren, Germany), with a stationary phase formed by 94% methyl–5% phenyl–1% vinyl silicone, with a film thickness of 0.25 μm, 30 m length and 0.25 mm internal diameter. The GC column was operated at a constant temperature (40 °C). The flow rate of N_2 used as sample gas was 5 mL min^{-1}. N_2 was also used as a drift gas at a constant flow of 250 mL min^{-1}. The drift gas enters the device in the opposite direction to the ions, to prevent non-ionized molecules or analytes from entering the ionization chamber. The N_2 of purity grade 5.0 was supplied by Al Air Liquid Spain S.A. (Madrid, Spain). The ions are formed in the tritium source ionization chamber and focused on a louvered grating (BN gate, grating pulse width = 100 μs), and begin to move along the length of the 10 cm long drift tube maintained at 65 °C, under a constant electric field of 400 V cm^{-1}. The separated ions arrive at the detector consisting of a Faraday plate. Data were acquired and recorded during 30 min of analysis. The spectrometer was driven in the positive drift voltage mode. The study of the VOCs of the reference analytes, LAB cultures and cheeses and the analysis of blanks between each sample were carried out. In this way, contamination and non-real signals are avoided.

2.4. Identification of VOCs

Chemical standards were used to unequivocally confirm the presence of volatile compounds in LAB cultures and cheese samples. The identification of volatile metabolites was carried out by analysis of individual standards at 1 or 2 mg L^{-1} by GC–IMS. The commercial software VOCal version 0.2.9 (G.A.S. Gesellschaft für analytische Sensorsysteme mbH. Dortmund, Germany) was used to acquire the data. The coordinate positions (drift time and retention time) of a particular signal were compared to the database to identify the associated compound or metabolite. Identification was performed through comparison of retention and drift time in samples with those of the individual standards and confirmed using a commercial GC–IMS library (Library Plot Module Version 0.1., G.A.S. Gesellschaft für analytische Sensorsysteme mbH, Dortmund, Germany).

The different analytical-grade VOCs (Sigma-Aldrich, St. Louis, MO, USA) were chosen according to studies based on the detection of VOCs in LAB samples and in cheese samples made from raw sheep milk. The compounds studied in this work are summarized in Table 1. Stock solutions at 1000 mg L^{-1} for each compound were prepared by dissolving the appropriate volume of each compound in Milli-Q ultrapure water. Working solutions at 1 mg L^{-1} were prepared by dilution of stock solutions with Milli-Q water. All solutions were stored away from light at 4 °C. The MRS broth (culture medium) was also analyzed to identify the signals corresponding to this solution.

Table 1. Reference compounds of selected cheeses for analysis of LAB samples and raw sheep milk cheese.

Chemical Group	VOC	RI	t_R	t_D	t_D (Database)	Reference *
Ketones	2-Heptanone	892.2	647.6; 647.0	1.3; 1.6	1.3; 1.6	[16–19]
	2-Hexanone	784.2	270.0; 292.0	1.2; 1.5	1.2; 1.5	[16,18]
	2-Pentanone	688.6	140.5; 148.7	1.1; 1.4	1.1; 1.4	[16,18,20]
	2-Butanone	589.4	96.2; 89.3	1.1; 1.3	1.1; 1.2	[7,17,20–22]
	3-Hydroxybutan-2-one	715.12	1384.2	1.8	NA	[7,23]
Esters	Ethyl butanoate	794.9	302.3; 300.6	1.2; 1.6	1.2; 1.6	[16–18,20,22,23]
	Ethyl hexanoate	999.9	1598.8; 1603.4	1.3; 1.8	1.3; 1.8	[7,16–18,22,23]
	Ethyl propanoate	707.7	161.0; 162.6	1.2; 1.5	1.1; 1.4	[16]
	Propyl butanoate	895.9	660.1; 663.7	1.3; 1.7	NA	[18,22]
	Ethyl acetate	668.8; 673.2	138.1; 140.6	1.2; 1.4	1.1; 1.3	[16–18,20]
Aldehydes	2-Methylbutanal	694.6	119.5; 120.9	1.2; 1.4	NA	[17–19,24]
	Benzaldehyde	960.9	1083.9; 1087.7	1.1; 1.5	1.1; 1.5	[7,16,17,19,20,22,23]
	Butanal	602.6; 602.6	85.5; 85.1	1.1; 1.3	1.1; 1.3	[17,20]
	Hexanal	800.2; 800.2	288.9; 291.5	1.3; 1.6	1.2; 1.6	[17–19]
	2-Methylpropanal	551.8; 551.8	75.9; 75.9	1.1; 1.3	1.1; 1.3	[18,20]
	3-Methylthiopropanal	907.7; 907.7	719.0; 721.9	1.1; 1.4	1.1; 1.4	[19]
	Octanal	128.2	1001.9; 1001.9	1.4; 1.8	1.4; 1.8	[4,18,23]
	Acetaldehyde	784.7; 785.5	293.4; 296.1	1.2; 1.4	NA	[18,20,23]
Alcohols	2-Butanol	602.7; 602.7	131.7; 133.0	1.2; 1.4	NA	[16–18,22]
	2-Methyl-1-propanol	629.0; 629.0	102.9; 104.4	1.2; 1.4	NA	[18,20,22,23]
	2-Heptanol	970.5; 970.5	1329.0; 1335.3	1.4; 1.8	NA	[16–18,22,23]
	n-Hexanol	867.4; 867.4	561.8; 566.3	1.3; 1.6	1.9; 1.6	[7,16–18,20,23]
	n-Butanol	660.4; 660.4	129.2; 125.5	1.2; 1.4	1.2; 1.4	[16,17,21,22]
	1-Pentanol	765.9; 765.9	250.9; 252.5	1.2; 1.5	1.2; 1.5	[17,23]
	3-Methyl-1-butanol	724.0; 723.7	133.8; 135.2	1.2; 1.4	NA	[16–19,21,22]
Acids	Acetic acid	621.3684	93.05; -	1.0; -	1.0; 1.5	[16,19–21,23]

RI, retention Index; t_R (s), retention time in seconds; t_D (ms), drift time in milliseconds; t_D (database), drift time of database in milliseconds; NA, data not available in database. * See References section.

2.5. Analysis of Samples

LAB cultures were analyzed with a previous growth of 24 h at 37 °C. Each bacterial culture (1 mL) was poured into a 20 mL glass vial closed with a magnetic screw cap and septum. The vials were individually subjected to GC–IMS analysis using the parameters already mentioned.

2.6. Data Analysis

GC–IMS data analysis was performed with VOCal software version 0.2.9 and VOCal's extended functions: Reporter Plot Version 0.1. and Gallery Plot Module Version 0.1. (G.A.S. Gesellschaft für analytische Sensorsysteme mbH. Dortmund, Germany). A comparative analysis between VOC results was performed using Fisher's exact test.

3. Results and Discussion

3.1. Sequencing of LAB Isolated from Raw Sheep Milk Cheeses

The four LAB strains isolated from matured cheese and creamy natural cheese made with raw sheep milk were identified by Sanger sequencing as *Lacticaseibacillus paracasei*, *Lactiplantibacillus plantarum* (formerly included in *Lactobacillus* spp.), *Lactococcus lactis* subsp. *hordniae* and *Leuconostoc mesenteroides*. *L. paracasei* and *L. lactis* subsp. *hordniae* strains were isolated from matured cheese and *L. plantarum* and *L. mesenteroides* strains were isolated from natural creamy cheese. Genomic identification showed 99% homology with the NCBI database. To the best of the authors' knowledge, the occurrence of *Lactococcus lactis* subsp. *hordniae* has not been reported before in cheeses. It is generally recognized that artisanal cheeses represent one of the ecosystems most used to isolate microorganisms with proper technological potential. These functional characteristics of LAB are strain-dependent [14]. For this reason, genetic identification plays a fundamental role in the selection process of LAB with potential technological properties. Artisanal cheeses derived from raw milk are characterized by different LAB with potential functional properties and certain organoleptic attributes. In recent years, researchers have focused on the isolation of autochthonous LAB from artisanal cheeses made with raw milk without the addition of starter cultures [14]. As in this study, other researchers have isolated strains corresponding to the genera *Lactobacillus* spp., *Lactococcus* spp. and *Leuconostoc* spp. from cheeses made with raw sheep milk [25–27].

Different strains of *L. paracasei* have been described before as good candidates for enhancing flavor in cheeses [28–32] and increasing secondary proteolysis and the production of VOCs, such as 2-methylbutanal, diacetyl, 3-methyl-1-butanol, acetic acid and 3-methylbutanoic acid, among others [28]. Probiotic strains of this species (*L. paracasei* KC39) showed an enhanced microstructure (a crumbly structure that increases adhesiveness and springiness) and favorable sensory characteristics of soft cheeses (especially color and general acceptance) [32].

According to Fernández et al. [33] and Cavanagh et al. [34], *L. lactis* subsp. *hordniae* has never been found in dairy products. *L. lactis* subsp. *hordniae* is considerably under-represented in both biological and genomic studies compared to their dairy-associated lactococci counterparts [35]. Usually, only *L. lactis* subsp. *lactis* and *L. lactis* subsp. *cremoris* are considered to be of industrial interest [36]. This makes our identification remarkable. In this work, this microorganism shows its capacity to produce VOCs from other components. In fact, non-dairy *Lactococcus* strains can be observed to produce relatively high abundances of a broad range of important volatile flavor compounds associated with positive attributes in dairy products [37]. This ability to produce a more varied and diverse volatile profile could be beneficial in dairy applications, leading to flavor diversification, generating unique flavor profiles, or indeed to mask off-flavors created by dairy strains [37].

L. plantarum was tested to evaluate Manchego cheeses made with different starter cultures, as an adjunct culture to starter culture based on two strains of *L. lactis*. Among the compounds produced are some of those detected in our work, such as benzaldehyde, 1-hexanol, 2-heptanol, 2-pentanone and 2-heptanone. Gómez-Ruiz et al. [38] do not detect, in general, a special contribution of using the adjunct culture of *L. plantarum* together with lactococci. However, more recently, Duan et al. [39] consider *L. plantarum* as an interesting adjunct culture due to its proteolytic activity. Likewise, Jia et al. [40] consider *L. plantarum* among the constituents of a starter with a high incidence in the production of volatiles in goat cheeses.

Leuconostocs were major producers of alcohols and esters. The variations observed between them must be studied for each strain [41]. In fact, *L. mesenteroides* is the microorganism more often isolated from cheeses such as Queijo de Azeitao (Portugal), occurring in 37% of samples of different producers. This prevalence is not surprising since the presence of this species has previously been described in other raw ewe milk and raw cow milk cheeses [42]. Generally, leuconostocs are known to produce aromatic compounds, thus contributing to cheese flavor definition [43].

3.2. Identification of VOCs by GC–IMS

VOCs identified in our work are reported in Tables 2 and 3. Twenty-six compounds were detected in cheese samples (all of them in creamy cheese and twenty-one in matured cheese), fourteen were detected in LAB MRS cultures and twenty-three in skim milk cultures (88.5% of total). Only two compounds (3-methylthiopropanal and 1-butanol) were detected in all the samples analyzed.

Table 2. VOCs detected in LAB cultures and creamy raw sheep milk cheese.

Chemical Group	VOC	MRS Media			Milk at 37 °C			Milk at 12 °C			Creamy Cheese		
		CT	Lpp	Lms	CT	Lpp	Lms	CT	Lpp	Lms	Rind	Inner	Center
Ketones	2-Heptanone	-	-	-	+	+	+	+	+	+	+	+	+
	2-Hexanone	+	+	+	-	+	-	-	-	-	+	+	+
	2-Pentanone	-	-	+	+	+	+	+	+	+	+	+	+
	2-Butanone	-	+	+	+	+	+	+	+	+	+	+	+
	3-Hydroxybutan-2-one	-	-	+	+	+	+	+	+	+	+	+	+
Esters	Ethyl butanoate	+	+	+	-	+	-	-	-	-	+	+	+
	Ethyl hexanoate	-	-	-	-	+	-	-	-	-	+	+	+
	Ethyl propanoate	-	-	-	-	+	+	-	-	-	+	+	+
	Propyl butanoate	-	-	-	-	+	-	-	-	-	+	+	+
	Ethyl acetate	-	-	-	-	-	-	-	-	-	+	+	+
Aldehydes	2-Methylbutanal	-	-	-	-	+	+	-	+	+	+	+	+
	Benzaldehyde	+	+	+	-	-	-	-	-	-	+	+	+
	Butanal	-	-	-	+	+	-	-	+	-	-	+	+
	Hexanal	-	-	-	-	-	-	-	-	+	+	+	+
	2-Methylpropanal	-	-	-	-	+	+	-	+	+	+	+	+
	3-Methylthiopropanal	+	+	+	+	+	+	+	+	+	+	+	+
	Octanal	-	-	-	-	-	-	-	-	-	-	-	+
	Acetaldehyde	-	-	-	-	+	+	-	+	+	+	+	+
Alcohols	2-Butanol	-	-	+	-	-	+	-	-	+	+	+	+
	2-Methyl-1-propanol	-	-	-	+	+	+	+	+	+	+	+	+
	2-Heptanol	-	-	+	-	-	+	-	-	+	+	+	+
	n-Hexanol	–	+	+	-	+	-	+	+	+	+	+	+
	1-Butanol	+	+	+	+	+	+	+	+	+	+	+	+
	Pentan-1-ol	-	-	-	-	+	+	-	+	-	+	+	+
	3-Methyl-1-butanol	-	-	+	+	+	+	+	+	+	+	+	+
Acids	Acetic acid	-	-	-	+	+	+	+	+	+	+	+	+

Lpp: Lactiplantibacillus plantarum; Lms: Leuconostoc mesenteroides; + VOC detected; - VOC not detected.

In LAB MRS cultures, seven of the eleven compounds detected were reported in at least one LAB MRS culture and in the MRS control medium. However, the other seven compounds were detected exclusively in the samples from at least one LAB MRS culture (Figure 1). These data indicate that at least these compounds (2-heptanone, 2-pentanone, 2-butanone, 3-hydroxybutan-2-one, 2-butanol, 2-heptanol and 3-methyl-1-butanol) could be generated by LAB activity.

The seven compounds detected exclusively in MRS cultures of LAB were recorded in at least one cheese sample (Figures 1–3). These compounds were investigated to determine whether they occur in different types of artisanal cheeses (creamy and matured). The presence of these compounds was also evaluated in different parts of the cheese (rind, middle (inner) and last heart (center) of the sample). Figure 2 shows the topographical maps obtained for the analysis of a natural creamy cheese and a matured cheese in their three parts. Six of the seven compounds previously identified in the LAB MRS cultures were found in these cheese samples. As can be seen in Figure 2, 3-methyl-1-butanol was not detected in matured cheese samples.

Table 3. VOCs detected in LAB cultures and raw sheep milk matured cheese.

Chemical Group	VOC	MRS Media			Milk at 37 °C			Milk at 12 °C			Matured Cheese		
		CT	Lps	Lcl	CT	Lps	Lcl	CT	Lps	Lcl	Rind	Inner	Center
Ketones	2-Heptanone	-	+	-	+	+	+	+	+	+	+	+	+
	2-Hexanone	+	+	+	-	-	-	-	-	-	+	+	+
	2-Pentanone	-	+	+	+	+	+	+	+	+	+	+	-
	2-Butanone	-	+	+	+	+	+	+	+	+	+	+	+
	3-Hydroxybutan-2-one	-	-	+	+	+	+	+	+	+	+	+	+
Esters	Ethyl butanoate	+	+	+	-	-	-	-	-	-	+	+	+
	Ethyl hexanoate	-	-	-	-	-	-	-	-	-	+	+	+
	Ethyl propanoate	-	-	-	-	+	+	-	-	-	-	-	-
	Propyl butanoate	-	-	-	-	-	-	-	-	-	-	-	-
	Ethyl acetate	-	-	-	-	-	-	-	-	-	-	-	-
Aldehydes	2-Methylbutanal	-	-	-	-	+	+	-	+	+	+	+	+
	Benzaldehyde	+	+	+	-	-	-	-	-	-	+	+	+
	Butanal	-	-	-	+	+	-	-	+	+	+	+	+
	Hexanal	-	-	-	-	-	+	-	-	-	+	+	+
	2-Methylpropanal	-	-	-	-	+	+	-	+	+	+	+	+
	3-Methylthiopropanal	+	+	+	+	+	+	+	+	+	+	+	+
	Octanal	-	-	-	-	-	-	-	-	-	+	+	+
	Acetaldehyde	-	-	-	-	+	+	-	+	+	+	+	+
Alcohols	2-Butanol	-	+	+	-	+	+	-	+	+	+	+	+
	2-Methyl-1-propanol	-	-	-	+	+	+	+	+	+	+	+	+
	2-Heptanol	-	-	+	-	-	+	-	-	+	+	+	+
	n-Hexanol	+	-	+	-	+	-	+	+	+	+	+	+
	1-Butanol	+	+	+	+	+	+	+	+	+	+	+	+
	Pentan-1-ol	-	-	-	-	+	-	-	-	+	-	-	-
	3-Methyl-1-butanol	-	+	+	+	+	+	+	+	+	-	-	-
Acids	Acetic acid	-	-	-	+	+	+	+	+	+	-	-	-

Lps: Lacticaseibacillus paracasei; Lcl: Lactococcus lactis subsp. hordniae; + VOC detected; - VOC not detected.

With respect to the bacterial cultures in skim milk, ten of the twenty-three VOCs detected were identified in both the skim milk control and at least one skim milk culture. The other thirteen compounds were detected exclusively in the samples from at least one LAB skim milk culture (Tables 2 and 3). When the effect of bacterial metabolism of milk at different temperatures (12 and 37 °C) was studied, it was observed that at least these compounds (2-hexanone, esters except ethyl acetate, 2-methylbutanal (at 12 °C), hexanal, 2-methylpropanal, acetaldehyde, 2-butanol, 2-heptanol, n-hexanol (at 37 °C) and pentan-1-ol) were identified by GC–IMS, except for 2-butanol and 2-heptanol (also identified in MRS cultures and appeared to be related to the LAB strain). All compounds detected exclusively in LAB skim milk cultures were found in at least one cheese sample (Figures 2 and 3). The importance of the milk source for LAB activity is demonstrated by the detection of eleven VOCs in skim milk cultures undetected in LAB MRS cultures: ethyl hexanoate, ethyl propanoate and propyl butanoate (only in *L. plantarum* at 37 °C), 2-methylbutanal, butanal, hexanal (only in *L. lactis* subsp. *hordniae*), 2-methylpropanal, acetaldehyde, 2-methyl-1-propanol, pentan-1-ol and acetic acid (the only organic acid we have checked).

Considering VOCs detected in creamy cheeses, twenty-three VOCs (88.5% of total) were detected in skim milk cultures (100% of VOCs detected in this kind of cheese). *L. plantarum* is the most important producer of esters in these last cases. This activity is clearly differentiated from that attributed to Leuconostoc. Furthermore, there were four VOC compounds detected in cheeses only detected in skim milk cultures of *L. plantarum*: 2-hexanone, ethyl butanoate, ethyl hexanoate and propyl butanoate. The activity of both microorganisms in the production of esters and aldehydes from a dairy source (MRS culture) was not detected, except for the cases where even control samples include these compounds (only ethyl butanoate, benzaldehyde and 3-methylthiopropanal). Results of both microorganisms could be statistically differentiated ($p < 0.05$). VOC production of both microorganisms agrees completely with the detection of these compounds in creamy cheese.

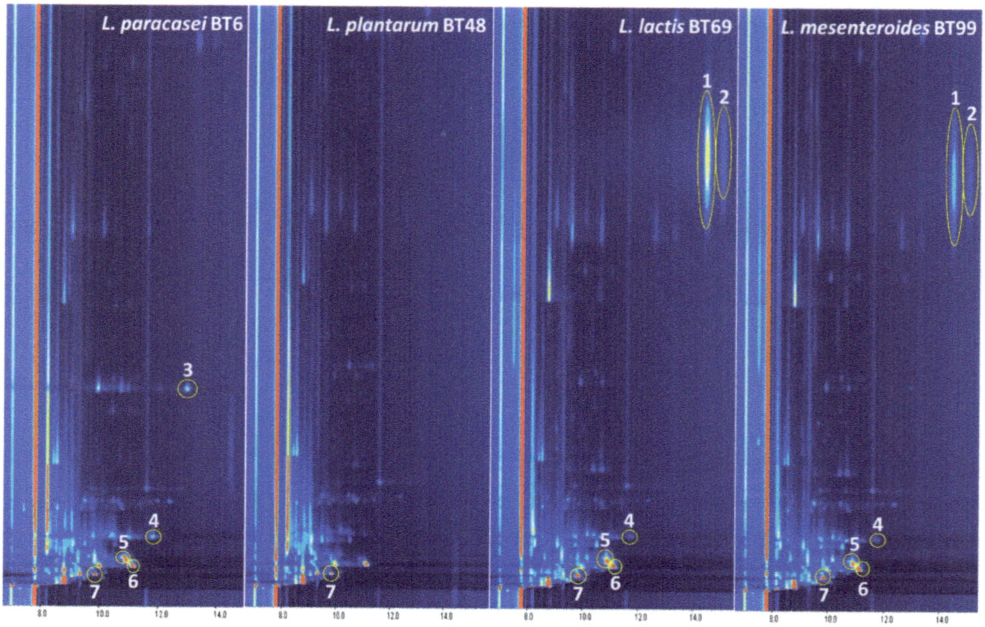

Figure 1. VOCs detected in LAB MRS cultures by GC–IMS. VOCs highlighted: (1) 2-heptanol; (2) 3-hydroxybutan-2-one; (3) 2-heptanone; (4) 3-methyl-1-butanol; (5) 2-butanol; (6) 2-pentanone; (7) 2-butanone.

Regarding matured cheese samples, nineteen VOCs (73% of total) were detected in skim milk cultures. Results are not correlated for 2-hexanone, ethyl butanoate, ethyl hexanoate, benzaldehyde, 3-methyl-1-butanol and acetic acid. These differences could be related to time required to maturation as well as interactions during this process. All the compounds detected by the activity of both microorganisms are present in the matured cheese, with only two exceptions: ethyl propanoate and 3-methyl-1-butanol. A total of 80% of the VOCs detected in this cheese are found from *L. paracasei* and 90% from *L. lactis* subsp. *Hordniae*. Differences between both LAB profiles are 22%. Differences between behavior of LAB at 12 °C or 37 °C are found, but we cannot demonstrate statistical differences.

3.3. Discussion Based on Groups of VOCs

In general, in our work, the production of ketones could be not necessarily dependent on LAB activities because of their detection in skim milk control samples. In fact, ketones are usual contributors to the typical aroma of dairy products [21], although it is generally difficult to determine their origin in cheeses. Some of these compounds appear to derive from animal feed (by cutting the grass or upon drying) or proteolysis (due, for instance, to coagulants used) [5], while others such as methyl ketones seem to be formed during cheese ripening from the metabolic activity of the dominant microbiota [22]. In fact, the detection of four ketones in MRS cultures without occurring in control samples shows the potential activity of all the LAB studied in this sense. Fruity and floral notes are associated with ketones, so the presence of these volatile compounds can be considered a positive for cheese flavor. 2-Butanone, with a butterscotch odor, was identified as the main odorant in cheddar cheese and 2-heptanone, with an herbaceous odor, is an important flavor compound of Emmental and natural and creamy Gorgonzola cheeses [24]. All VOCs corresponding to this group detected in our cheeses, such as 2-butanone, 2-pentanone, 2-hexanone, 2-heptanone and 3-hydroxybutan-2-one, were also found in the MRS LAB cultures and in the skim milk cultures (Tables 2 and 3). 2-Pentanone was previously described as a secondary metabolite by LAB [20].

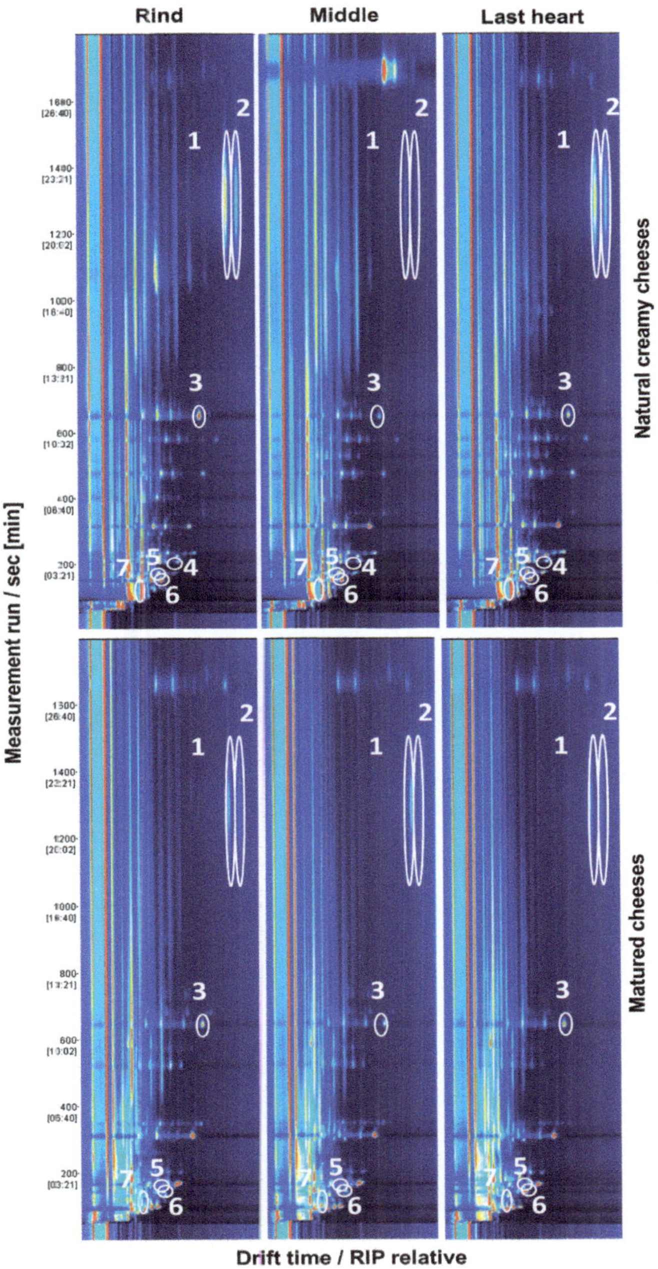

Figure 2. VOCs detected in artisanal raw sheep milk (natural creamy or matured) cheeses by GC–IMS. VOCs highlighted: (1) 2-heptanol; (2) 3-hydroxybutan-2-one; (3) 2-heptanone; (4) 3-methyl-1-butanol; (5) 2-butanol; (6) 2-pentanone; (7) 2-butanone.

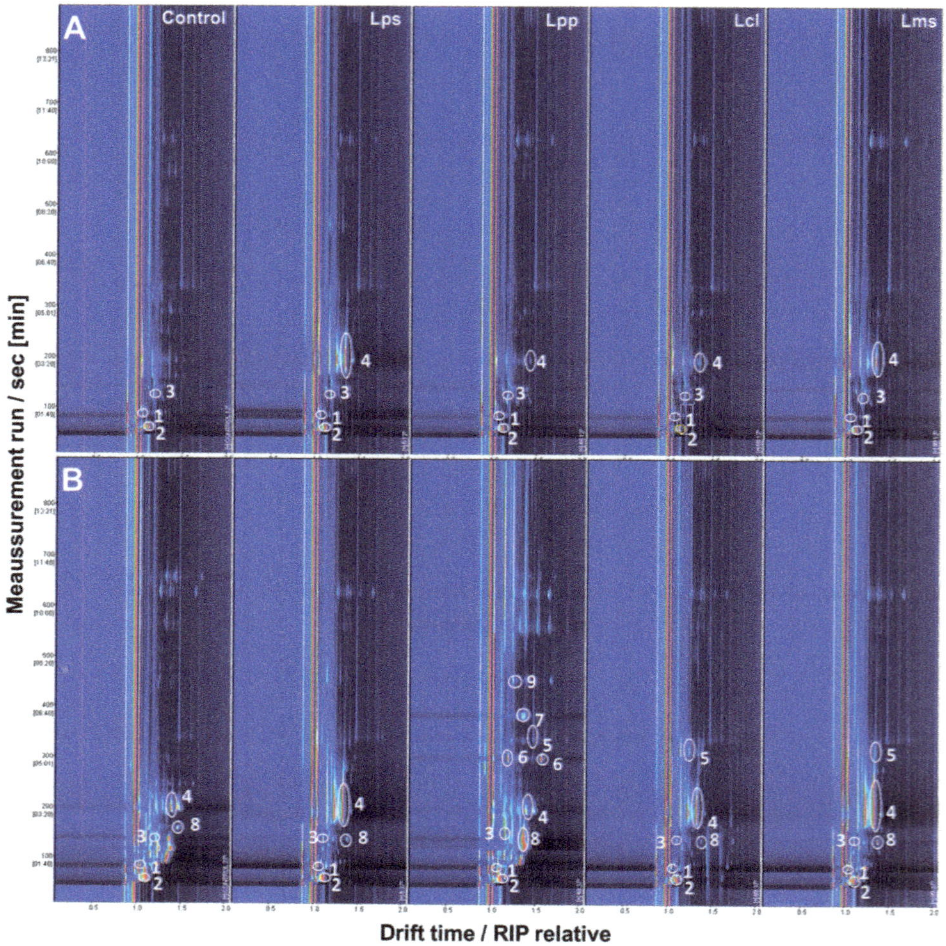

Figure 3. VOCs detected in skim milk LAB cultures after incubation 48 h at 12 °C and 37 °C by GC–IMS. (**A**) At 12 °C; (**B**) At 37 °C; VOCs highlighted: (1) 2-methylbutanal; (2) 2-methylpropanal; (3) Acetaldehyde; (4) Pentan-1-ol; (5) Hexanal; (6) 2-hexanone; (7) Ethyl butanoate; (8) Ethyl propanoate; (9) Propyl butanoate.

One of the main classes of volatile components in cheeses are esters [5]. Esters in fermented dairy products are formed between short-to-medium-chain fatty acids and available primary and secondary alcohols derived from lactose fermentation or from amino acid catabolism [16]. These reactions are performed either by microorganisms or by chemical reactions [44]. The non-detection of esters from the MRS LAB cultures studied in our work is consistent with the absence of lactose in MRS broth. In our work, and as we have described, *L. plantarum* is the most important producer of esters in creamy cheese samples. None of the skim milk cultures at 12 °C produced esters. However, at 37 °C, *L. plantarum* (with the exception of ethyl acetate) and the rest of LAB studied (in the case of ethyl propanoate) show activity, which is also expressed in cheeses, especially creamy cheese. The occurrence of ethyl acetate in these creamy cheeses could be attributed to chemical reactions [44]. No derived secondary alcohols are present. On the other hand, their occurrence in the cheeses studied is expected. In fact, ethyl propanoate and ethyl butanoate predominated in natural creamy cheese and ethyl butanoate and ethyl hexanoate were predominant in both types of cheeses analyzed. Esters, which are produced by esterification of an acid and an alcohol, were the most identified group in cheeses, mainly in creamy natural cheese. Researchers considered them one of the main groups of compounds isolated from the headspace of "Torta del Casar" cheeses [21]. According to these authors, some LAB are involved in this production, as it is our case. These compounds confer, in a synergistic way, floral and fruity notes that probably contribute to the balance of cheese flavor by minimizing the sharpness imparted by the acids [22].

Aldehydes present transitory nature and do not accumulate in cheese since they tend to reduce into the corresponding alcohols or, alternatively, oxidize into the respective acids [45]. However, a diversity of aldehydes in cheese samples was found. In natural creamy cheese, hexanal and 2-methylpropanal were detected. In both matured and creamy cheeses, we detected 2-methylbutanal, butanal, octanal and acetaldehyde (Tables 2 and 3). These compounds are main products of the autoxidation of unsaturated fatty acids, playing a key role in the development of cheese flavor. For instance, octanal could be produced by the autooxidation of oleic acid, and it is characterized by a green grass-like and herbaceous aroma [11]. On the other hand, 2-methylbutanal, with its typically aldehydic odor, may derive from the catabolism of amino acids such as isoleucine [21]. In our study, the absence of fatty acids in MRS broth prevents the occurrence of aldehydes in MRS LAB cultures separately, excluding benzaldehyde and 3-methylthiopropanal, whose fatty acids are probably produced from peptone components. In skim milk cultures, aldehydes are produced in most of the samples, except for benzaldehyde and octanal. The activity of LAB studied in milk produce acetaldehyde, an important organoleptic compound.

In general, raw milk cheeses contain greater amounts of alcohols than pasteurized milk cheeses and this is due to their higher microbial diversity [17]. The formation of primary and secondary alcohols in cheeses could be due to lactose fermentation or generated by dehydrogenation of aldehydes and ketones due to the strong reducing conditions of cheese matrices [18]. Alcohols can produce acids, keto acids and carbonyls by catabolism [22]. In our study, n-hexanol, pentan-1-ol and 3-methyl-1-butanol were detected in natural creamy cheeses and 2-butanol, 2-methyl-1-propanol, 2-heptanol and 1-butanol were detected in both cheeses (Tables 2 and 3). Specifically, various alcohols were detected in MRS LAB and skim milk cultures, as well as in cheese samples. Secondary alcohols such as 2-heptanol and 2-butanol were detected from *L. lactis* subsp. *Hordniae*, *L. mesenteroides* and both cheeses, and from *L. paracasei*, *L. lactis* subsp. *Hordniae*, *L. mesenteroides* and both cheeses, respectively. 2-Heptanol has been identified as a key odorant of Gorgonzola and Grana Padano cheeses and it was the alcohol detected in the highest concentrations in semi-hard Spanish goat cheeses. 2-Butanol was the highest secondary alcohol isolated in the artisanal Manchego cheese [24]. In our study, 3-methyl-1-butanol was detected from *L. paracasei*, *L. lactis* subsp. *Hordniae*, *L. mesenteroides* (MRS cultures), in all the skim milk cultures and in creamy cheese. The presence of branched-chain primary alcohols, such as 3-methyl-1-butanol, indicates the reduction in the aldehyde produced from leucine. This compound has been detected in raw goat's milk cheeses and confers a pleasant fresh cheese flavor [24]. According to Gallegos et al. [11], 1-hexanol, a primary alcohol, is mainly produced by the reduction in its corresponding aldehyde and methyl ketone. In all cheeses, these alcohols give a particular flavor, which makes them distinctive and recognizable by consumers.

Generally, in cheeses, carboxylic acids are abundant volatile compounds. They give acidic, fatty, pungent, balsamic and vinegary odors [46]. Linear carboxylic acids, such as acetic, butanoic, hexanoic and octanoic, decanoic, butyric and acetic acids have often been detected in cheeses. However, the method used in our study has limited the detection of these types of compounds. Acetic acid was the only compound studied that could be identified in the samples of creamy natural cheese, as well as the skim milk cultures (Tables 2 and 3). This is not strange, because this compound derives mainly from the early fermentation of lactate [47].

4. Conclusions

Predicting the ability of different characterized LAB to produce aromas and the identification of the VOCs responsible for flavor in cheeses are important aspects to consider when selecting strains with optimal aromatic properties, resulting in the diversification of cheese products. In our work, the contribution of each LAB studied to the aroma profile of the cheeses can be described according to their different VOC profiles. Differences between LAB behavior in each cheese are shown, especially between LAB involved in creamy cheeses. Only *L. lactis* subsp. *Hordniae* and *L. mesenteroides* show the same VOC profile in MRS cultures, but they participate in different cheeses, and also show two differences in VOC production in skim milk cultures (especially pentan-1-ol).

The combination of Sanger sequencing in identifying LAB from raw milk cheeses with GC–IMS to determinate the VOCs produced helped us to describe cheeses and to differentiate the potential role of each microorganism in their volatilome. Further studies will be useful to determinate the production of other VOCs, especially in searching for different acids. Finally, as far as we know, the occurrence of *Lactococcus lactis* subsp. *Hordniae* in cheese is reported for first time.

Author Contributions: Conceptualization: L.M.M. and L.A.; Methodology: M.J.R., J.A.S.-D., M.J.C., R.J., L.A., L.F. and L.M.M.; Formal analysis: M.J.R., J.A.S.-D. and M.J.C.; Direction: L.M.M., L.F. and L.A.; Funding acquisition: L.M.M., R.J. and L.A.; Writing—original draft preparation: M.J.R., L.M.M.; Writing—review and editing: L.M.M., J.A.S.-D., M.J.C., L.A. and M.J.R. All authors have read and agreed to the published version of the manuscript.

Funding: This research was funded by the Researching Programme of the University of Cordoba, Spain.

Institutional Review Board Statement: Not applicable.

Informed Consent Statement: Not applicable.

Data Availability Statement: Data is contained within the article.

Acknowledgments: The authors are grateful to support of the Researching Programme of the University of Cordoba of the project in which this work is involved, as well as the AgriFood International Excellence Campus (ceiA3), where the University of Cordoba works on the AgriFood Sector. The authors want also to thank Fernando Pérez-Rodriguez for their advice.

Conflicts of Interest: The authors declare no conflict of interest.

References

1. Colonna, A.; Durham, C.; Meunier-Goddik, L. Factors affecting consumers' preferences for and purchasing decisions regarding pasteurized and raw milk specialty cheeses. *J. Dairy Sci.* **2011**, *94*, 5217–5226. [CrossRef]
2. AMRC (Ag Marketing Resource Center). Cheese Industry Profile. USDA Rural Development–Iowa State University. 2022. Available online: https://www.agmrc.org/commodities-products/livestock/dairy/cheese-industry-profile (accessed on 20 October 2022).
3. Mohapatra, A.; Shinde, A.K.; Singh, R. Sheep milk: A pertinent functional food. *Small Rumin. Res.* **2019**, *181*, 6–11. [CrossRef]
4. Gallegos, J.; Arce, C.; Jordano, R.; Arce, L.; Medina, L.M. Target identification of volatile metabolites to allow the differentiation of lactic acid bacteria by gas chromatography-ion mobility spectrometry. *Food Chem.* **2017**, *220*, 362–370. [CrossRef] [PubMed]
5. Soltani, M.; Sahingil, D.; Gokce, Y.; Hayaloglu, A.A. Changes in volatile composition and sensory properties of Iranian ultrafiltered white cheese as affected by blends of *Rhizomucor miehei* protease or camel chymosin. *J. Dairy Sci.* **2016**, *9*, 7744–7754. [CrossRef]
6. Zheng, J.; Wittouck, S.; Salvetti, E.; Franz, C.M.; Harris, H.M.; Mattarelli, P.; O'Toole, P.W.; Pot, B.; Vandamme, P.; Walter, J.; et al. A taxonomic note on the genus *Lactobacillus*: Description of 23 novel genera, emended description of the genus *Lactobacillus* Beijerinck 1901, and union of Lactobacillaceae and Leuconostocaceae. *Int. J. Syst. Evol. Microbiol.* **2020**, *70*, 2782–2858. [CrossRef]
7. Gaglio, R.; Franciosi, E.; Todaro, A.; Guarcello, R.; Alfeo, V.; Randazzo, C.L.; Settanni, L.; Todaro, M. Addition of selected starter/non-starter lactic acid bacterial inoculums to stabilise PDO Pecorino Siciliano cheese production. *Food Res. Int.* **2020**, *136*, 109335. [CrossRef]
8. Galimberti, A.; Casiraghi, M.; Bruni, I.; Guzzetti, L.; Cortis, P.; Berterame, N.M.; Labra, M. From DNA barcoding to personalized nutrition: The evolution of food traceability. *Curr. Res. Food Sci.* **2019**, *28*, 41–48. [CrossRef]
9. Contreras, M.M.; Jurado-Campos, N.; Arce, L.; Arroyo-Manzanares, N. A robustness study of calibration models for olive oil classification: Targeted and non-targeted fingerprint approaches based on GC-IMS. *Food Chem.* **2019**, *288*, 315–324. [CrossRef]

10. Arce, L.; Gallegos, J.; Garrido-Delgado, R.; Medina, L.M.; Sielemann, S.; Wortelmann, T. Ion mobility spectrometry a versatile analytical tool for metabolomics applications in food science. *Curr. Metab.* **2014**, *2*, 264–271. [CrossRef]
11. Gallegos, J.; Garrido-Delgado, R.; Arce, L.; Medina, L.M. Volatile metabolites of goat cheeses determined by Ion Mobility Spectrometry. Potential applications in quality control. *Food Anal. Methods* **2015**, *8*, 1699–1709. [CrossRef]
12. Bertuzzi, A.S.; McSweeney, P.L.H.; Rea, M.C.; Kilcawley, K.N. Detection of volatile compounds of cheese and their contribution to the flavor profile of surface-ripened cheese. *Compr Rev. Food Sci. Food Saf.* **2018**, *17*, 371–390. [CrossRef] [PubMed]
13. Kilcawley, K.N.; Faulkner, H.; Clarke, H.J.; O'Sullivan, M.G.; Kerry, J.P. Factors influencing the flavour of bovine milk and cheese from grass based versus non-grass based milk production systems. *Foods* **2018**, *7*, 37. [CrossRef]
14. Ruiz-Moyano, S.; Goncalves dos Santos, M.T.; Galván, A.I.; Merchán, A.V.; González, E.; Córdoba, M.G.; Benito, M.J. Screening of autochthonous lactic acid bacteria strains from artisanal soft cheese: Probiotic characteristics and prebiotic metabolism. *LWT.* **2019**, *114*, 108388. [CrossRef]
15. Beck, T.F.; Mullikin, J.C.; Biesecker, L.G.; Progra, N.C.S. Systematic evaluation of Sanger validation of next-generation sequencing variants. *Clin. Chem.* **2016**, *62*, 647–654. [CrossRef] [PubMed]
16. Cozzolino, R.; Martignetti, A.; De Giulio, B.; Malorni, L.; Addeo, F.; Picariello, G. SPME GC-MS monitoring of volatile organic compounds to assess typicity of Pecorino di Carmasciano ewe-milk cheese. *Int. J. Dairy Technol.* **2021**, *74*, 383–392. [CrossRef]
17. Guarrasi, V.; Sannino, C.; Moschetti, M.; Bonanno, A.; Di Grigoli, A.; Settanni, L. The individual contribution of starter and non-starter lactic acid bacteria to the volatile organic compound composition of Caciocavallo Palermitano cheese. *Int. J. Food Microbiol.* **2017**, *259*, 35–42. [CrossRef]
18. Marilley, L.; Casey, M.G. Flavours of cheese products: Metabolic pathways, analytical tools and identification of producing strains. *Int. J. Food Microbiol.* **2004**, *90*, 139–159. [CrossRef]
19. Luz, C.; D'Opazo, V.; Quiles, J.M.; Romano, R.; Mañes, J.; Meca, G. Biopreservation of tomatoes using fermented media by lactic acid bacteria. *LWT.* **2020**, *130*, 109618. [CrossRef]
20. Štefániková, J.; Árvay, J.; Miškeje, M.; Kačániová, M. Determination of volatile organic compounds in Slovak Brynda cheese by the electronic nose and the headspace solid-phase microextraction gas chromatography-mass spectrometry. *Slovak J. Food Sci.* **2020**, *14*, 767–773.
21. Delgado-Martínez, F.; Carrapiso, A.I.; Contador, R.; Rosario Ramírez, M. Volatile compounds and sensory changes after high pressure processing of mature "Torta del Casar" (raw ewe's milk cheese) during refrigerated storage. *Innov. Food Sci. Emerg. Technol.* **2019**, *52*, 34–41. [CrossRef]
22. Bozoudi, D.; Kondyli, E.; Claps, S.; Hatzikamari, M.; Michaelidou, A.; Biliaderis, C.G.; Litopoulou-Tzanetaki, E. Compositional characteristics and volatile organic compounds of traditional PDO Feta cheese made in two different mountainous areas of Greece. *Int. J. Dairy Technol.* **2018**, *71*, 673–682. [CrossRef]
23. Reale, A.; Di Renzo, T.; Boscaino, F.; Nazzaro, F.; Fratianni, F.; Aponte, M. Lactic acid bacteria biota and aroma profile of Italian traditional sourdoughs from the irpinian area in Italy. *Front. Microbiol.* **2019**, *10*, 1621. [CrossRef] [PubMed]
24. Delgado, F.; Rodríguez-Pinilla, J.; Márquez, G.; Roa, I.; Ramírez, R. Physicochemical, proteolysis and texture changes during the storage of a mature soft cheese treated by high-pressure hydrostatic. *Eur. Food Res. Technol.* **2015**, *240*, 1167–1176. [CrossRef]
25. Kirmaci, H.A.; Ozer, B.H.; Akcelik, M.; Akcelik, N. Identification and characterisation of lactic acid bacteria isolated from traditional Urfa cheese. *Int. J. Dairy Technol.* **2016**, *69*, 301–307. [CrossRef]
26. Ramírez-López, C.; Vélez-Ruiz, J.F. Isolation, characterization and selection of autochthonous lactic acid bacteria from goat milk and fresh artisanal-goat cheese. *Inf. Tecnol.* **2016**, *27*, 6. [CrossRef]
27. Zoumpopoulou, G.; Papadimitriou, K.; Alexandraki, V.; Mavrogonatou, E.; Alexopoulou, K.; Anastasiou, R.; Georgalaki, M.; Kletsas, D.; Tsakalidou, E.; Giaouris, E. The microbiota of Kalathaki and Melichloro Greek artisanal cheeses comprises functional lactic acid bacteria. *LWT* **2020**, *130*, 109570. [CrossRef]
28. Peralta, G.H.; Wolf, I.V.; Bergamini, C.V.; Perotti, M.C.; Hynes, E.R. Evaluation of volatile compounds produced by *Lactobacillus paracasei* I90 in a hard-cooked cheese model using solid-phase microextraction. *Dairy Sci. Technol.* **2014**, *94*, 73–81. [CrossRef]
29. Poveda, J.M.; Nieto-Arribas, P.; Seseña, S.; Chicón, R.; Castro, L.; Palop, L.; Cabezas, L. Volatile composition and improvement of the aroma of industrial Manchego cheese by using *Lactobacillus paracasei* subsp. *paracasei* as adjunct and other autochthonous strains as starters. *Eur. Food Res. Technol.* **2014**, *238*, 485–494.
30. Stefanovic, E.; Kilcawley, K.N.; Roces, C.; Rea, M.C.; O'Sullivan, M.; Sheehan, J.J.; McAuliffe, O. Evaluation of the potential of *Lactobacillus paracasei* adjuncts for flavor compounds development and diversification in short-aged Cheddar cheese. *Front. Microbiol.* **2018**, *9*, 1506. [CrossRef]
31. Terpou, A.; Bosnea, L.; Kanellaki, M.; Plessas, S.; Bekatorou, A.; Bezirtzoglou, E.; Koutinas, S.A. Growth capacity of a novel potential probiotic *Lactobacillus paracasei* K5 strain incorporated in industrial white brined cheese as an adjunct culture. *J. Food Sci.* **2018**, *83*, 723–731. [CrossRef]
32. Shehata, M.G.; Abd El-Aziz, N.M.; Darwish, A.G.; El-Sohaimy, S.A. *Lacticaseibacillus paracasei* KC39 immobilized on prebiotic wheat bran to manufacture functional soft white cheese. *Fermentation* **2022**, *8*, 496. [CrossRef]
33. Fernández, E.; Alegría, A.; Delgado, S.; Martín, M.C.; Mayo, B. Comparative phenotypic and molecular genetic profiling of wild *Lactococcus lactis* subsp. *lactis* strains of the *L. lactis* subsp. *lactis* and *L. lactis* subsp. *cremoris* genotypes, isolated from starter-free cheeses made of raw milk. *Appl. Environ. Microbiol.* **2011**, *77*, 5324–5335.

34. Cavanagh, D.; Casey, A.; Altermann, E.; Cotter, P.D.; Fitzgerald, G.F.; McAuliffe, O. Evaluation of *Lactococcus lactis* Isolates from Nondairy Sources with Potential Dairy Applications Reveals Extensive Phenotype-Genotype Disparity and Implications for a Revised Species. *Appl. Environ. Microbiol.* **2015**, *81*, 3961–3972. [CrossRef] [PubMed]
35. Kelleher, P.; Bottacini, F.; Mahony, J.; Kilcawley, K.N.; Van Sinderen, D. Comparative and functional genomics of the *Lactococcus lactis* taxon; insights into evolution and niche adaptation. *BMC Genom.* **2017**, *18*, 267. [CrossRef] [PubMed]
36. Laroute, V.; Tormo, H.; Couderc, C.; Mercier-Bonin, M.; Le Bourgeois, P.; Cocaign-Bousquet, M.; Daveran-Mingot, M.L. From genome to phenotype: An integrative approach to evaluate the biodiversity of *Lactococcus lactis*. *Microorganisms* **2017**, *5*, 27. [CrossRef]
37. McAuliffe, O. Symposium review: Lactococcus lactis from non dairy sources: Their genetic and metabolic diversity and potential applications in cheese. *J. Dairy Sci.* **2018**, *101*, 3597–3610. [CrossRef]
38. Gómez-Ruiz, J.A.; Cabezas, L.; Martínez-Castro, I.; González-Viñas, M.A.; Poveda, J.M. Influence of a defined-strain starter and *Lactobacillus plantarum* as adjunct culture on volatile compounds and sensory characteristics of Manchego cheese. *Eur. Food Res. Technol.* **2008**, *227*, 181–190. [CrossRef]
39. Duan, C.; Li, S.; Zhao, Z.; Wang, C.; Zhao, Y.; Yang, G.; Niu, C.; Gao, L.; Liu, X.; Zhao, L. Proteolytic activity of *Lactobacillus plantarum* strains in Cheddar cheese as adjunct cultures. *J. Food Prot.* **2019**, *82*, 2108–2118. [CrossRef]
40. Jia, R.; Zhang, F.; Song, Y.; Lou, Y.; Zhao, A.; Liu, Y.; Peng, H.; Hui, Y.; Ren, R.; Wang, B. Physicochemical and textural characteristics and volatile compounds of semihard goat cheese as affected by starter cultures. *J. Dairy Sci.* **2021**, *104*, 270–280. [CrossRef]
41. Pogačić, T.; Maillard, M.B.; Leclerc, A.; Hervé, C.; Chuat, V.; Valence, F.; Thierry, A. *Lactobacillus* and *Leuconostoc* volatilomes in cheese conditions. *Appl. Microbiol. Biotechnol.* **2016**, *100*, 2335–2346. [CrossRef]
42. Cardinali, F.; Ferrocino, I.; Milanović, V.; Belleggia, L.; Rita Corvaglia, M.; Garofalo, C.; Foligni, R.; Mannozzi, C.; Mozzon, M.; Cocolin, L.; et al. Microbial communities and volatile profile of Queijo de Azeitão PDO cheese, a typical Mediterranean thistle curdled cheese from Portugal. *Food Res. Int.* **2021**, *147*, 110537. [CrossRef] [PubMed]
43. Van Mastrigt, O.; Egas, R.A.; Abee, T.; Smid, E.J. Aroma formation in retentostat co-cultures of *Lactococcus lactis* and *Leuconostoc mesenteroides*. *Food Microbiol.* **2019**, *82*, 151–159. [CrossRef] [PubMed]
44. Majcher, M.A.; Goderska, K.; Pikul, J.; Jelen, H.H. Changes in volatile, sensory and microbial profiles during preparation of smoked ewe cheese. *J. Sci. Food Agric.* **2011**, *91*, 1416–1423. [CrossRef] [PubMed]
45. Coda, R.; Brechany, E.; De Angelis, M.; De Candia, S.; Di Cagno, R.; Gobbetti, M. Comparison of the compositional, microbiological, biochemical, and volatile profile characteristics of nine Italian ewes' milk cheeses. *J. Dairy Sci.* **2006**, *89*, 4126–4143. [CrossRef]
46. The Good Scents Company Information System. Providing Information for the Flavor, Fragrance, Food and Cosmetic Industries. Available online: http://www.thegoodscentscompany.com/index.html (accessed on 20 June 2021).
47. Curioni, P.M.; Bosset, J.O. Key odorants in various cheese types as determined by gas chromatography-olfactometry. *Int. Dairy J.* **2002**, *12*, 959–984. [CrossRef]

foods

MDPI

Review

The Safety of Probiotics Intended for Use in Pregnant and Lactating Women: From a Desirable to a Required Task

Leónides Fernández [1,2], Belén Orgaz [1,2] and Juan M. Rodríguez [2,3,*]

1 Department of Galenic Pharmacy and Food Technology, Complutense University of Madrid, 28040 Madrid, Spain; leonides@ucm.es (L.F.); belen@ucm.es (B.O.)
2 Instituto Pluridisciplinar, Complutense University of Madrid, 28040 Madrid, Spain
3 Department of Nutrition and Food Science, Complutense University of Madrid, 28040 Madrid, Spain
* Correspondence: jmrodrig@ucm.es; Tel.: +34-913943749

Abstract: During pregnancy, women undergo changes that affect virtually every organ, apparatus, or system, including the host microbiota. Most pregnancies progress smoothly despite the common presence of minor side-effects arising from such adaptations. However, some women may experience more serious complications, including gestational diabetes mellitus, preeclampsia, or preterm delivery. Probiotics are one of the products most used to try to prevent or treat any of the minor or severe symptoms or complications that women may experience during pregnancy or lactation; however, most of them have never been tested in such populations and, therefore, their efficacy and safety claims are frequently unsubstantiated. Overall, probiotic trials involving pregnant or lactating women have shown that these products are usually well-tolerated and safe although adverse effects may also exist. Therefore, health professionals attending pregnant or lactating women should be aware of their use and monitor their efficacy and safety. In conclusion, probiotics recommendations for pregnant or lactating women should be based on scientific evidence, opting exclusively for those products that have been designed for the specific target or condition that a pregnant or lactating woman may be experiencing or at risk of, and which efficacy and safety has already been convincingly tested in such populations.

Keywords: probiotics; safety; pregnancy; lactation; infertility; adverse effects

1. Introduction

Probiotic microbes have been consumed by humans since at least the Neolithic era, unintentionally at first through fermented foods (e.g., fermented milks) and, more recently, in a growing variety of functional foods and pharmaceutical-like forms (capsules, sachets, drops, ovules, etc.). The long history of safety associated with many of the species included in probiotic products—mainly from the lactic acid bacteria group and the genus *Bifidobacterium*—along with data from in vitro assays, animal studies, and clinical trials, are the main reasons why probiotics are widely assumed to be safe for most populations [1]. However, some adverse effects and risks associated with probiotic intake have also been reported or suggested, including gastrointestinal side effects, systemic infections, negative metabolic and immunological activities, and the potential transfer of harmful genes to members of the host microbiome [2]. It is important to note that the number of documented adverse effects related to probiotic use is very low compared to their long and widespread use.

Pregnant and lactating women are no exception; throughout history and across most cultures, they have consumed probiotic bacteria. In fact, probiotics present an attractive approach to preventing or treating certain health issues or complications that may arise during these life stages. The use of these products has generally proven to be safe in these populations, although, as with other groups, some adverse effects have been reported in a small percentage of clinical trials. Pregnant and lactating women are usually considered vulnerable populations, making the safety of the foods, supplements, or medications they receive particularly important. Therefore, thorough safety assessments should be included in intervention studies involving probiotics targeting pregnant and lactating women.

This article does not aim to be a systematic review or meta-analysis of the safety of probiotics used by these populations but rather to provide a personal perspective from a research group that has participated in clinical trials with probiotics involving pregnant and/or lactating women for the last 30 years. Our impression is that while the use of such products has notably increased among these populations, there has been a corresponding rise in skepticism about the actual usefulness or safety of probiotics among practitioners. This skepticism is primarily based on the low number of well-designed, large-cohort clinical studies addressing their efficacy and safety in these specific populations. In this context, the objective of this article is to show that (a) some specific probiotic strains may be beneficial during these life stages; (b) most studies conducted so far have reported either a lack of adverse effects or the presence of only minor side effects when probiotics are administered to pregnant or lactating women, including those with major pregnancy complications such as gestational diabetes mellitus; and (c) medical confidence in these products may be enhanced by improving safety assessments in clinical trials targeting pregnant and lactating women.

2. Adaptations of the Maternal Microbiota During Pregnancy and Lactation

During pregnancy, each women undergoes anatomic and physiologic changes that affect virtually every organ, apparatus, or system. These biological adaptations are required to support and protect the growing embryo and fetus, and to prepare the mother-to-be for the subsequent lactation period. The composition and function of the microbiota also change during this period to potentially contribute to positive or negative pregnancy and lactation outcomes [3,4]. The maternal microbiota has been described as central for reproduction [5] because of the relevant roles that it may play, for example, in the induction of a tolerogenic immune state [6] or in the vertical transmission of the first colonizers of the neonatal gut [7].

In relation to the vaginal microbiota, the high circulating estrogen levels and the lack of menstruation usually lead to a decrease on its variability and diversity by increasing the *Lactobacillus* dominance [8–10]. The stability appears to be higher and linked to healthier outcomes when this microbiota is dominated by *Lactobacillus crispatus*. However, diversity may increase during the final trimester of pregnancy, especially when *Lactobacillus iners* is the dominant *Lactobacillus* species [10]. In addition, the risk of yeasts colonization and infection increases as pregnancy progresses because of some pregnancy-related factors, including increased estrogen levels and glycogen production, and decreased cell-mediated immunity [11]. Overall, the composition of the vaginal microbiome may have a strong influence on pregnancy outcomes, including higher or lower risk of miscarriage or preterm delivery [12–14].

The gut microbiota undergoes a profound remodeling during pregnancy. Fecal samples from the first trimester are metataxonomically like those of healthy nonpregnant women while, in contrast, samples from the third trimester are similar to those of patients with metabolic syndrome, enhancing the efficiency of energy extraction [3]. The transfer of fecal microbiota from women at the third trimester of pregnancy to germ-free recipient mice can induce some metataxonomic, metabolic, and immunological changes that are commonly observed in metabolic syndrome [3]. In fact, increases in food intake and fat deposition are normal features from early pregnancy onwards while, in mid-to-late gestation some metabolic, hormonal, and immunological changes promote insulin resistance and facilitate lipolysis and gluconeogenesis, providing a continuous supply of glucose to the fetus [4]. Although these diabetogenic-like changes would be associated with a highly detrimental metabolic disease in men and nonpregnant women, such temporary adaptations provide maternal benefits since they are relevant contributions to support fetal growth and to prepare the host for the highly energy-demanding lactation [15–17].

Changes in gut microbiota and physiology in late pregnancy (increased intestinal permeability, increased pressure of the gravid womb on gut vessels), highly increase the likelihood of bacterial translocation [4]. This process is coupled with the entero-mammary circulation of immune cells, and seems responsible, at least partly, for generating a specific mammary microbiota during late pregnancy and throughout lactation [18–20]. This human milk microbiota is also very relevant to gut colonization during early life and may impact both mammary and infant health [21–24]. Finally, pregnancy increases susceptibility to a variety of oral diseases, from gingivitis to periodontitis, because of the interrelated hormonal, immunological, and microbiota changes in the oral cavity. In this context, growing evidence suggests that the risk of suffering adverse pregnancy outcomes, from preeclampsia to preterm birth or low birth weight, may be strongly influenced by the composition of the oral microbiome [25–28].

3. Health Problems During Pregnancy and Probiotics

Some of the physiological adaptations from early pregnancy onwards, including those affecting the composition of the microbiome, explain most of the minor problems that may appear during a normal gestation including, among others, gingivitis, nausea, vomiting, bloating, abdominal cramps, constipation, hemorrhoids, fatigue, mood swings, headaches, increased urination, occasional genitourinary infections (urinary tract infections, vaginal candidiasis, and bacterial vaginosis), dizziness, or weight gain [29].

Most pregnancies progress smoothly despite the relatively frequent presence of such usually minor side-effects. However, some women may experience more serious complications that require careful management to ensure a healthy mother and/or fetal outcome. Such complications may arise as an aggravation of a pre-existing disease but, in most cases, they appear for the first time during pregnancy. The most common ones are gestational diabetes mellitus (GDM), preeclampsia, and preterm delivery.

GDM is characterized by hyperglycemia, and it usually resolves after pregnancy. Women with GDM are considered within the risk pregnancy group as this condition is associated with higher rates of preeclampsia and cesarean section, predisposes to macrosomia, and poses an increased risk of metabolic syndrome and type 2 diabetes for the mother–infant dyad later in life. Preeclampsia is a multisystemic complication that is usually diagnosed from the week 20 of gestation and is associated with hypertension and proteinuria. Preeclampsia can affect the placenta, predisposing intrauterine growth restriction or preterm delivery. Premature labor is referred to a delivery that occurs before 37 weeks of gestation, and the risk of short term (sepsis, necrotizing enterocolitis, respiratory or gastrointestinal disorders...) or long term (neurodevelopment) consequences increases as gestational age decreases.

The causes that may lead to pregnancy complications depend on each woman but infectious, inflammatory, immunological, or metabolic-based conditions are usually cited among the main predisposing factors. In turn, such conditions are being increasingly regarded as typical consequences of an altered gut microbiota (the so-called dysbiosis). In the last decades, the microbiota (and, particularly, the gut microbiota) has emerged as one of the main health drivers and, therefore, it has become a good example of an in-fashion field (for better or worse) in biomedicine. Consequently, gut dysbiosis has been associated with a plethora of diseases and disorders, virtually encompassing all medical specialties. Paradoxically, a high percentage of the PubMed articles about gut microbiota are reviews or hypothesis about how to manipulate the gut microbiota to restore an eubiosis state, while only a small number of them deal with original scientific works. This fact raises false expectations among the population, especially in relation to disorders and diseases as complex as autism spectrum disorders, multiple sclerosis, obesity, or different types of cancer [30].

In this context, probiotics have been widely marketed as a natural and safe way to restore the microbiota and, thus, to prevent or treat almost any kind of disease. They can be used either by themselves or as an adjunct to conventional therapies. Additionally, they can be found as the so-called functional foods or in a wide spectrum of over-the counter (OTC) pharmaceutical-like presentations, creating a rapidly growing market worldwide. The global market for probiotics reached a value of about 58 billion U.S. dollars in 2022 and was forecast to reach over 85 billion dollars by 2027 [31]. People used to think that OTC are safe just because anyone can buy them, but this is a false perception. As an example, OTC vasoconstrictors containing pseudoephedrine have been linked to a higher risk of heart attack. Anyway, this general perception of safety has made probiotics very popular among general consumers and, also, one of the OTC products most used by pregnant women to try to improve any of the minor or severe symptoms or complications that they may experience during pregnancy [32,33] (Figure 1).

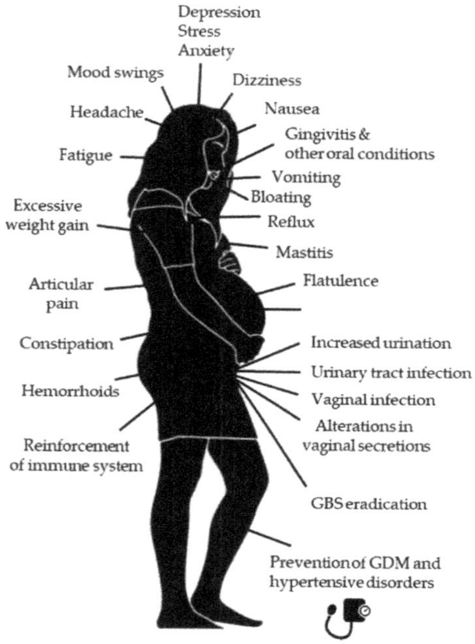

Figure 1. Main minor and severe symptoms or complications of pregnancy and lactation that probiotics claim to prevent or treat, often without any scientific evidence.

Nowadays, there is an overwhelming abundance of messages in commercials and internet and social networks recommending the use of probiotics in pregnancy because they (sic) "help reduce the symptoms of nausea and vomiting (morning sickness)", "control blood sugar levels (gestational diabetes)", "prevent gestational diabetes", "treat blood pressure and inflammation (preeclampsia)", "reduce the risk of preeclampsia", "prevent preterm birth and low birth weight", "support gastrointestinal and immune health in baby and mom", "support healthy bowel movements", "support nutrient absorption", "support vaginal and urinary tract health" "inhibit the growth of harmful bacteria (bacterial vaginosis)", "promote postpartum recovery", "help elevate mood (postpartum depression)", "reduce symptoms of stress, anxiety, and depression", "balance mood and mental well-being", "improve the health status of the mammary gland (mastitis)", "support immune system form mom and baby", "minimize antibiotic-associated risks", "reduce the risk of allergies", "decrease risk of asthma in children", "prevent atopic dermatitis (eczema) and food sensitivity in infants", "decrease fussiness in infants" or provide "digestive harmony" and a "healthy weight gain for mom". Furthermore, these are just only some of "the extraordinary benefits that probiotics offer to expectant mothers, contributing to both maternal and fetal well-being". Such recommendations are based on the "multifaceted benefits" observed in "some" studies [34,35].

Although it may seem incredible for both practitioners and consumers, most of these strong health-related claims are made for strains and products which have never been tested in clinical trials and, therefore, there is an absolute lack of scientific evidence backing their use and claims. In addition, the quality control of a large proportion of these products is, at least, highly questionable. In fact, many studies have found relevant discordances between the information provided in the label and their actual content, both in qualitative (absence of publicized strains) and quantitative (presence at much lower concentrations) terms [36–41] or the presence of health-inadequate formulations [42]. In addition, even for those strains and products that have been tested in clinical trials, a significant proportion suffer from an incorrect design (including very small cohorts) and under-reported safety [43].

The lack of safety data in health-related products recommended to pregnant women (and their fetuses) is particularly concerning due to their vulnerability and may lead to or contribute to unexpected or undesirable outcomes [33]. As stated above, only very few probiotic strains or products have been tested in pregnant women and, among them, only a few contain a thoughtful report of safety and potential adverse effects, providing details such as type, timing, duration, and severity of symptoms [44].

Many healthcare professionals recognize their lack of solid knowledge on what probiotics are, and do not discriminate among the different products available in the market based on scientific evidence of safety and efficacy for a specific target in a specific population [45,46]. So, while some evidence suggests that a few specific probiotics can be helpful for a few specific pregnancy-related issues, there is a strong and urgent need for improving the quality of evidence, safety reporting, public and professional awareness, and a clear regulation of their use and claims [47].

4. Safety of Probiotics During Pregnancy

4.1. Generalities

Probiotic-associated risks during pregnancy have been the specific subject of three reviews and two meta-analyses [33,48,49], where information about the target, strains, and posology is provided. They found no mortality or serious adverse effects (AEs) on mothers, fetuses, or infants associated with the intake of these products. Out of 100 studies dealing with the administration of probiotics during pregnancy, only 28 reported AEs and, among these, only 11 studies linked AEs potentially to the probiotic intervention. Most of the reported AEs were related to maternal gastrointestinal health, and most of them occurred during the third trimester of gestation. They included nausea, abdominal cramping, flatulence, and headache. However, similar side effects have been reported after administration of probiotics to non-pregnant populations [1]. On the other hand, these gastrointestinal symptoms are relatively common during pregnancy even in women not consuming probiotics [50]. It has also been suggested that such minor problems might be attributed to the fact that many of the products included in the reviews and meta-analysis contained not only probiotics but, also, prebiotics, plant extracts, and other ingredients [33]. In addition, one study found an increased risk of changes in stool consistency and of vaginal discharge [51] but such findings were not observed in a subsequent study using the same strains [52].

In parallel, another systematic review and meta-analysis evaluated perinatal outcomes in pregnant women that received probiotic supplements in the frame of randomized controlled trials [53]. Across the 25 trials that were meta-analyzed, the reporting of perinatal outcomes was highly inconsistent, and the only data reported in at least 50% of the studies were gestational age at birth, cesarean section, and birth weight. Overall, probiotic administration of probiotics had neither a positive nor negative impact on these perinatal outcomes. Previously, a comprehensive review of 37 studies involving the use of prenatal probiotics from 1990 to 2011 also found relevant inconsistencies in the reporting of perinatal clinical outcomes. However, no severe AEs were reported among the reviewed studies, suggesting that the different tested strains were safe and well tolerated [54].

Although at much lower extent than oral probiotics, vaginal probiotics may also be used during pregnancy to try to minimize changes in the vaginal secretion or the duration and/or symptoms of vaginal infections [55]. Studies on vaginal probiotics have an even higher frequency of safety reporting inconsistencies or lack of safety data compared to oral probiotics. Nevertheless, some studies reported a good tolerance and no adverse effects derived from the probiotic intervention, even in women at risk of preterm delivery [56]. The European trend to encourage companies to reposition vaginal probiotics as drugs may help to overcome this situation in the future [57].

4.2. Probiotics for the Eradication of Vaginal and Rectal Colonization by Group B Streptococci

Rectal and vaginal colonization by *Streptococcus agalactiae* (Group B Streptococci, GBS) during pregnancy has the potential for vertical mother-to-infant transmission, and, in some cases, can lead to neonatal early-onset sepsis (EOS), a condition with a high morbidity and mortality risk [58–60]. Two main approaches (risk-based and screening-based protocols) are currently followed in Western countries to try to avoid such transmission. Both approaches involve intrapartum antibiotic prophylaxis (IAP) to women that display a risk factor or that are positive after recto-vaginal GBS screening at week 35–38 of pregnancy, respectively. However, IAP-related strategies face some drawbacks and limitations since they do not prevent GBS-related abortions, stillbirths, and preterm births [61], do not guarantee GBS eradication, may increase antibiotic resistance among GBS and other pathogenic bacteria [62–65], and may have a negative impact on the acquisition and development of the infant microbiota [66–70], a fact that may have a lasting impact on health [71,72].

In this frame, the administration of some target-specific probiotic strains during the third trimester of pregnancy has been associated with a variable impact on the GBS recto-vaginal colonization, from no effect to a notable decrease of the colonization rates at 35–38 weeks [73–81]. However, all these probiotic interventions have a common feature: no adverse events associated with probiotic intake were reported. Two of them found a decrease in gastrointestinal symptoms in the probiotic arm compared to the placebo group [73,74]. On the other hand, GBS were eradicated from 70% of GBS-positive pregnant women in one of the studies [78]. Such reduction in the exposure of pregnant women and their infants to GBS and to IAP may be considered, by itself, as a beneficial safety-related outcome.

4.3. Prevention and Management of Gestational Diabetes Mellitus (GDM) and Hypertensive Disorders

As stated above, GDM is one of the most relevant complications during pregnancy since it may imply negative short and long-term outcomes for the mother–infant pair. Although several interventions (physical activity, diet, and/or medication) may be advisable to prevent GDM and restrict gestational weight gain, particularly among overweight and obese pregnant women, their actual efficacy is relatively limited [82]. In fact, the authors of a comparison of different strategies to prevent GDM concluded that "none of the interventions could offer remarkable benefit for GDM prevention" [82].

In this context, it is not surprising that probiotics were considered a promising approach to reduce GDM rates after the publication in 2010 of the results of a randomized, double-blind placebo-controlled trial showing that the GDM frequency was much lower among women receiving probiotics (*Lacticaseibacillus rhamnosus* GG and *Bifidobacterium animalis* ssp. *lactis* Bb12, 1×10^9 colony forming units (CFU) of each strain per day, from week 20 to delivery) and diet counselling (13%) than among those receiving a placebo and diet counselling (36%, $p = 0.003$). In fact, the GDM rate of the latter group was similar to that of control women who did not receive either the probiotic or diet counselling (34%) [83]. In addition, no adverse events, including no differences in the duration of pregnancies and in infant postnatal growth rates, were reported in mothers or children of the probiotic group.

Previous results from the same trial had shown that dietary counselling together with probiotics improved blood glucose control even in normoglycaemic pregnant women, providing a mean for the prevention and management of glucose disorders [84]. A Cochrane systematic review on probiotics for preventing GDM was dedicated exclusively to this trial, reaching the same conclusions [85].

The work of Luoto et al. [83] has been followed by other probiotic interventions targeting GDM rates and/or other metabolic outcomes of pregnancy. The first systematic review on probiotics and GDM that included more than one trial (one Norwegian prospective cohort study [86], one Finnish randomized trial resulting in three publications [83,84,87] and an Iranian randomized trial which results were also published in three different articles [88,89]) concluded that, despite the variety of strains, dosages and outcomes, these products seemed to be safe and reduced maternal fasting glucose, GDM, and pre-eclampsia rates and levels of the C-reactive protein [90]. It also concluded that more randomized controlled trials were urgently required, particularly among overweight and obese pregnant women since they are at high risk of metabolic and hypertensive disorders. Paradoxically, at present, there are more than 110 reviews, systematic reviews, and meta-analyses available in PubMed (29 November 2024) which have repeatedly evaluated the outcomes of a combination of up to only 15 original trials or prospective studies.

Overall, almost all the meta-analysis published up to the date agree that the administration of probiotics during pregnancy is safe and may have the potential to improve glucose metabolism and reduce GDM rates [91–93]. Still, they also highlight the need for further well-designed trials, involving a much higher number of pregnant women and with a proper and clear evaluation of safety and efficacy [94]. It is interesting to note that, among the 15 articles analyzed in the meta-analysis that include the highest number of studies so far [91], 11 studies had been performed in Iran by the same group and, therefore, may not be representative of pregnant women in other geographical, environmental, and socio-economical settings [95]. This problem is shared by most of the meta-analysis since, as another example, eight of these Iranian articles were included among the eleven studies included in other meta-analysis [96].

No adverse effects have been found in a meta-analysis assessing the impact of probiotics on pregnant women already diagnosed with GDM, while this intervention resulted in a significant reduction in insulin resistance [97]. However, again, three out of the four articles included in this meta-analysis were published by the Iranian group. It has been also reported that probiotic supplements in women with GDM may decrease neonatal birth weight [98]. High birth weight is a frequent complication in pregnant women with this condition. In contrast, another meta-analysis found no evidence to conclude that the administration of probiotics to pregnant women with GDM or pregnant women who are overweight may control newborn birth weight [99]. Interestingly, such apparently conflicting conclusions arose from two meta-analysis that shared most of the analyzed studies.

A Cochrane systematic review on the usefulness and safety of probiotics for preventing GDM published in 2021 [100] introduced some controversy in the, until then, general agreement that the use of probiotics during pregnancy was, in the worst of the cases, a safe practice. The conclusion of this review was that while there was not a clear beneficial effect of probiotics on the risk of GDM, the risk of preeclampsia increased with probiotic intake (9.2% in the probiotic group versus 4.9% in the placebo group, $p = 0.09$) and, therefore, the authors urged caution in the use of probiotics during pregnancy. Surprisingly, the conclusion that probiotics increased the risk of preeclampsia was based in only one of the seven studies included in the review, precisely the only one that had been authored by the same authors of the review [101]; in contrast, the remaining six studies supported that probiotics were safe during pregnancy [84,102–106]. This conflicting situation deserves a more detailed comment on the results obtained by the different studies included in the controversial review.

As already stated, the trial of Laitinen et al. [84] found that the probiotic strains used in their intervention (*L. rhamnosus* GG and *B. animalis* ssp. *lactis* Bb12) were safe and useful for the prevention and treatment of glucose disorders during pregnancy. Three additional publications derived from the same trial confirmed safety for mothers and infants and efficacy for reducing the frequency of GDM [83,84,107]. Aaltonen et al. [107] demonstrated that, in comparison with two placebo groups, the simultaneous administration of these two strains did not affect the values of systolic and diastolic blood pressure in any of the three trimesters of gestation, which disagrees with an increased risk of preeclampsia. It is worth mentioning that the strains, posology, and period of treatment of the trial of Callaway et al. [101] were the same that those employed in the Finnish trial [83,84,86,107]. In addition, the trial of Callaway et al. [101] contains some apparent internal contradictions: the authors found that probiotics increased the risk of preeclampsia, but this practice had no negative impact in "gestational hypertension", in "hypertensive disorders of pregnancy", or in the values of systolic and diastolic blood pressure at 28 or 36 weeks of gestation. In the same trial, it was found that the administration of probiotics decreased the rates of both "excessive weight gain" and "small for gestational age" when compared with the placebo arm.

The second trial included in the review of Davidson et al. [100] involved the administration of *Ligilactobacillus salivarius* UCC118 to obese pregnant women and did not report adverse effects related to probiotic intake [102]. In this study, there was no difference in the rate of preeclampsia between the probiotic and the placebo arms. A complementary part of the same trial also failed in finding differences in preeclampsia rates between both the probiotic and placebo groups [108]. The third trial in the review of Davidson et al. [100] reported that probiotic supplementation (*Lactobacillus acidophilus*, *Lacticaseibacillus casei*, *Bifidobacterium bifidum*; one strain of each species) in the first half of pregnancy had beneficial effects on inflammation and insulin and lipid metabolism and, again, no adverse effects were found [103]. The fourth trial [104] showed that the administration of a probiotic strain (*L. rhamnosus* HN001), from week 14 to week 16 of pregnancy, was safe and may be beneficial in reducing GDM prevalence, especially in the case of older women for pregnancy. Finally, the two remaining trials found that the respective probiotics tested in high-risk obese women (*L. rhamnosus* GG and *B. lactis* Bb12 [105]; *L. rhamnosus* HN001 and *B. animalis* ssp. *lactis* 420 [106]) were safe and well tolerated although failed in lowering the risk of GDM [103,104]. The trial of Okesene-Gafa et al. [104] used the same strains and at the same dose that were in the trials of Laitinen et al. [84] and Calloway et al. [101] although the probiotic intervention started later (week 28). In conclusion, it is difficult to blame probiotics of increasing preeclampsia rates based on the low-level evidence of a single trial, particularly when the rest of the trials have failed to find such association. Two more recent systematic reviews and meta-analyses specifically evaluated the impact of probiotics on the incidence of preeclampsia among pregnant women with GDM, and they found no effect on the incidence of preeclampsia in this at-risk population when compared to the placebo groups of the respective trials [109,110].

The fact that most studies have revealed, so far, that ingestion of probiotics during pregnancy and in immunocompromised adults is usually safe [111] does not mean that risk is absent. Safety and efficacy of most of the OTC probiotics ingested by pregnant women have not been previously tested in clinical trials. It has been reported that probiotics may be a cause of sepsis, liver abscess, or endocarditis in vulnerable adults with hypertension and/or diabetes [112–119]. In addition, liver abscess and *Lactobacillus* bacteremia have also been reported for *L. rhamnosus* GG [120–122], one of the strains that was administered to pregnant women in some of the trials cited above [83,84,87,105,107], including the one that raised the alarm of an increased risk of preeclampsia in the probiotic group [101].

5. Safety of Probiotics During Lactation

A variety of microbes inhabits the mammary ecosystem during lactation and, consequently, human milk contains a site-specific microbiota, which is one of the main routes of vertical mother-to-infant transmission of the first gut colonizers [24]. However, milk dysbiosis states, which are characterized by a drastic reduction in bacterial diversity, a notable enrichment in some aerotolerant bacteria, mainly of the genera *Staphylococcus* and

Streptococcus, and a depletion in obligate anaerobes [22,123,124], may occur. These dysbiosis states may eventually lead to mastitis, a condition that represents the major medical reason for an undesired weaning.

Empiric antibiotic therapy is the conventional etiological treatment for mastitis. However, mastitis-causing bacteria can display several resistance mechanisms, including the ability to form thick biofilms inside the lactiferous ducts and the presence of some genes providing either intrinsic or transmissible antibiotic resistance. Therefore, the rates of successful treatment of mastitis with antibiotherapy are relatively low. Additionally, this practice negatively modify the milk microbiota and may impair the acquisition of maternal microbes through breastfeeding [125]. In this frame, strategies for the management of mastitis by the selection and administration of target-specific probiotic strains have been developed [22].

A few specific human milk-derived strains, mainly *L. salivarius* CECT5713, *L. salivarius* PS2 and *Limosilactobacillus fermentum* CECT5716, have been tested in clinical trials to treat or prevent acute and subacute mastitis during late pregnancy and/or the lactation period [126–132]. The results indicate that they were efficient, safe, and well tolerated since only a few cases of flatulence were reported and all of them were associated with the ingestion of *L. fermentum* CECT5716 [127]. In fact, Arroyo et al. [127] showed that the treatment with either *L. salivarius* CECT5713 or *L. fermentum* CECT5716 was not only significantly more efficient than antibiotherapy but, also, safer. While 9% of the women who received antibiotics developed vaginal candidiasis, this negative therapy-associated effect was not observed in any of the women that received one of these two probiotic strains. Overall, the above cited trials led to a significant reduction in the milk concentration of the mastitis-causing bacteria and, also, of the residues of some drugs (mainly anti-inflammatory and analgesic drugs) [133]. Therefore, the administration of these probiotic strains to women increased the safety of their infants. Two recent systematic reviews and meta-analyses, providing information about the strains and posology, support the efficacy of such strategy [134,135]. The conclusion of Alemu et al. [135], after analyzing randomized controlled trials involving the administration of maternal probiotic on breast milk and infant gut microbiome and health, was that "maternal probiotic supplements effectively orchestrate the breast milk and infant gut microbiome with a wide range of clinical benefits and safety".

6. Safety of Probiotics Targeting Infertility

As previously stated, vaginal lactobacilli may play relevant reproductive functions, including a fine hormonal and immunological tuning of the reproductive ecosystem during the fertility window of a menstrual cycle. Recently, it has been shown that the administration of two *L. salivarius* strains to women with reproductive failure (either repetitive abortion or infertility without a known cause, and concomitant with low concentrations of vaginal lactobacilli) led to a significant increase in the rate of successful pregnancies [136,137]. Interestingly, the strains were able to increase the vaginal levels of vascular endothelial growth factor (VEGF) and transforming growth factor β1 and β2 (TGF-β1, TGF-β2) in those women who achieved a term pregnancy. The expression of these growth factors is highly regulated and coordinated during the early implantation stages, a fact that highlights their joint relevance for a successful pregnancy [138]. VEGF is a glycoprotein with a critical role for embryogenesis and embryo implantation because of its function as regulator of the vasculogenesis, angiogenesis, and vascular functions in the human endometrium [139]. TGF-β1 and β2 play well-known roles in the induction of mucosal immune-tolerance, a process that is essential for embryo tolerance [140]. Both trials showed that the strains were safe and well tolerated.

There have been some other oral or intravaginal probiotic trials aimed at increasing fertility rates during assisted-reproduction therapies with variable outcomes. None of them described any negative safety-related issue but it is not clear if it was because of a lack of AEs or because they were not assessed during the trial [141–145].

7. Recommendations

Pregnancy involves several adaptations in a woman's body, which can sometimes lead to some symptoms or complications. Probiotics are often used to address these issues due to their general perception of safety. However, while probiotics may have the potential to interfere or counteract these physiological adaptations, this may derive in an unexpected negative effect for which, most probably, the probiotic will not be blamed.

AEs derived from a probiotic intake exist and expanding their use to vulnerable populations, such as pregnant or lactating women (a practice that may also affect their fetuses and infants), require well-informed recommendations by health professionals [2]. Probiotic trials (or, at least, part of them) specifically designed to evaluate safety are scarce, and safety concerns may be related to the strain(s), the posology, the state of each treated person (intended use), and the quality of the product [146,147].

The safety of probiotics should be based on in silico, in vitro, and in vivo studies. A thoughtful safety assessment should monitor acute and long-term risks. General recommendations about them and to guide the medical and scientific communities about probiotic safety have been already proposed [2,148]. Safety assessment may pay special attention to the target population, particularly if it can be considered vulnerable, including extra monitoring by well-informed practitioners. Whole genome sequencing of the strain(s) included in a probiotic product seems an essential requirement, since this enables the detection of some safety-related traits, including deleterious metabolic activities, antimicrobial resistance, toxin production, or genetic stability [2,149]. In vitro characterization of the strain(s) (e.g., antibiotic resistance phenotype) is also crucial but, again, such an assessment must have in account the intended use since, as an example, probiotic strains may interfere with estrogen metabolism, a process that may be relevant during pregnancy [150]. Finally, the reporting of adverse effects in randomized controlled trials have a paramount importance in safety assessment and should follow well-respected guidelines, such as updated Consolidated Standards of Reporting Trials (CONSORT) statements for the complete and detailed reporting of harms.

8. Conclusions

In conclusion, trials in which probiotic strains have been provided to pregnant or lactating women are scarce although, in most of them, their administration was well tolerated and safe even by obese women with GDM. However, one study has reported an increased risk of preeclampsia when probiotics were administered to overweight and obese women [101]. Although this study showed some contradictory results regarding preeclampsia markers and provided opposite results to those of other trials in which pregnant women received the same strains [83,84,87,105,107], such a risk should be considered in practice. Consequently, the administration of probiotics to obese pregnant women should be monitored to detect any increased risk of preeclampsia. Anyway, health professionals attending pregnant or lactating women should be aware of any probiotic that they may be taking during such stages, to follow up their actual efficacy, to ensure their safety, and to report any potential AE derived from probiotic intake that, otherwise, may remain unnoticed. It is obvious that absolute safety does not exist, neither for probiotics nor for drugs or foods, but the systematic collection of such information may allow for the improvement of our knowledge on the risk-to-benefit ratio of using probiotics and to select the best options in every case when they may be useful on a risk-benefit basis.

Health professionals should recommend probiotics based on their own clinical experience and the guidelines provided by medical associations in each country but, also, on scientific evidence, opting for those products that have been designed for the specific target or condition that a pregnant or lactating woman may be experiencing or at high risk of, and which efficacy and safety has already been convincingly tested in such populations (Figure 2).

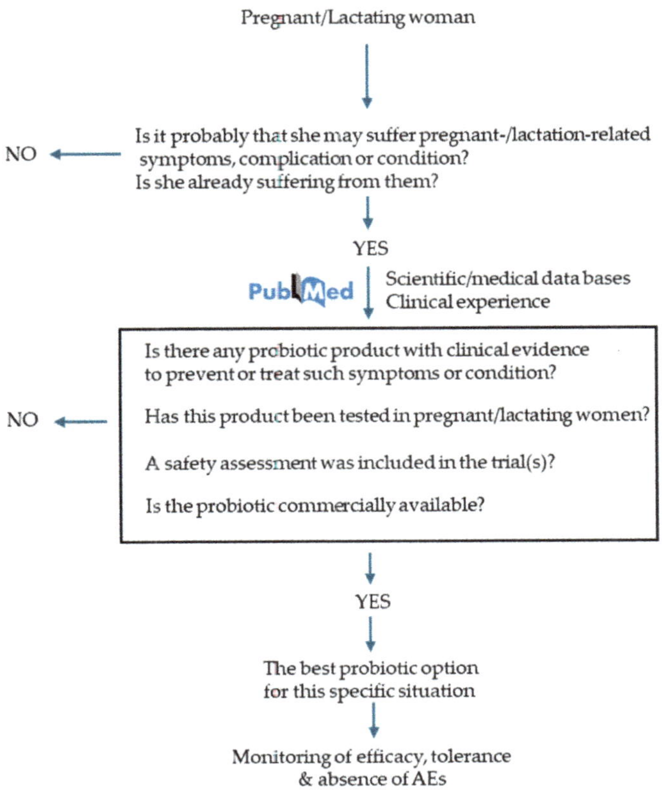

Figure 2. A proposal of a decision-support system for health professionals when deciding if it is worth prescribing a probiotic product to pregnant or lactating women and which product may be better in each case.

Author Contributions: Conceptualization, L.F., B.O. and J.M.R.; resources, J.M.R.; writing—original draft preparation, J.M.R.; writing—review and editing, L.F., B.O. and J.M.R. All authors have read and agreed to the published version of the manuscript.

Funding: The work of the author in this field is funded by grant PID2022-139262OB-I00 (Spanish Ministry of Science and Innovation).

Institutional Review Board Statement: Not applicable.

Informed Consent Statement: Not applicable.

Data Availability Statement: No new data were created or analyzed in this study. Data sharing is not applicable to this article.

Conflicts of Interest: The authors declare no conflicts of interest.

References

1. Doron, S.; Snydman, D.R. Risk and safety of probiotics. *Clin. Infect. Dis.* **2015**, *60* (Suppl. S2), S129–S134. [CrossRef] [PubMed]
2. Merenstein, D.; Pot, B.; Leyer, G.; Ouwehand, A.C.; Preidis, G.A.; Elkins, C.A.; Hill, C.; Lewis, Z.T.; Shane, A.L.; Zmora, N.; et al. Emerging issues in probiotic safety: 2023 perspectives. *Gut Microbes* **2023**, *15*, 2185034. [CrossRef] [PubMed]
3. Koren, O.; Goodrich, J.K.; Cullender, T.C.; Spor, A.; Laitinen, K.; Bäckhed, H.K.; Gonzalez, A.; Werner, J.J.; Angenent, L.T.; Knight, R.; et al. Host remodeling of the gut microbiome and metabolic changes during pregnancy. *Cell* **2012**, *150*, 470–480. [CrossRef]
4. Gosalbes, M.J.; Compte, J.; Moriano-Gutierrez, S.; Vallès, Y.; Jiménez-Hernández, N.; Pons, X.; Artacho, A.; Francino, M.P. Metabolic adaptation in the human gut microbiota during pregnancy and the first year of life. *EBioMedicine* **2019**, *39*, 497–509. [CrossRef]

5. Reid, G.; Brigidi, P.; Burton, J.P.; Contractor, N.; Duncan, S.; Fargier, E.; Hill, C.; Lebeer, S.; Martín, R.; McBain, A.J.; et al. Microbes central to human reproduction. *Am. J. Reprod. Immunol.* **2015**, *73*, 1–11. [CrossRef]
6. Mor, G.; Aldo, P.; Alvero, A.B. The unique immunological and microbial aspects of pregnancy. *Nat. Rev. Immunol.* **2017**, *17*, 469–482. [CrossRef]
7. Milani, C.; Duranti, S.; Bottacini, F.; Casey, E.; Turroni, F.; Mahony, J.; Belzer, C.; Delgado Palacio, S.; Arboleya Montes, S.; Mancabelli, L.; et al. The first microbial colonizers of the human gut: Composition activities, and health implications of the infant gut microbiota. *Microbiol. Mol. Biol. Rev.* **2017**, *81*, e00036-17. [CrossRef]
8. Gajer, P.; Brotman, R.M.; Bai, G.; Sakamoto, J.; Schutte, U.M.E.; Zhong, X.; Koenig, S.S.K.; Fu, L.; Ma, Z.; Zhou, X.; et al. Temporal dynamics of the human vaginal microbiota. *Sci. Transl. Med.* **2012**, *4*, 132ra52. [CrossRef]
9. Walther-Antonio, M.R.S.; Jeraldo, P.; Berg Miller, M.E.; Yeoman, C.J.; Nelson, K.E.; Wilson, B.A.; White, B.A.; Chia, N.; Creedon, D.J. Pregnancy's stronghold on the vaginal microbiome. *PLoS ONE* **2014**, *9*, e98514. [CrossRef]
10. Goltsman, D.S.A.; Sun, C.L.; Proctor, D.M.; DiGiulio, D.B.; Robaczewska, A.; Thomas, B.C.; Shaw, G.M.; Stevenson, D.K.; Holmes, S.P.; Banfield, J.F.; et al. Metagenomic analysis with strain-level resolution reveals fine-scale variation in the human pregnancy microbiome. *Genome Res.* **2018**, *28*, 1467–1480. [CrossRef]
11. Aguin, T.J.; Sobel, J.D. Vulvovaginal candidiasis in pregnancy. *Curr. Infect. Dis. Rep.* **2015**, *17*, 462. [CrossRef] [PubMed]
12. Al-Memar, M.; Bobdiwala, S.; Fourie, H.; Mannino, R.; Lee, Y.S.; Smith, A.; Marchesi, J.R.; Timmerman, D.; Bourne, T.; Bennett, P.R.; et al. The association between vaginal bacterial composition and miscarriage: A nested case-control study. *BJOG* **2020**, *127*, 264–274. [CrossRef] [PubMed]
13. Fettweis, J.M.; Serrano, M.G.; Brooks, J.P.; Edwards, D.J.; Girerd, P.H.; Parikh, H.I.; Huang, B.; Arodz, T.J.; Edupuganti, L.; Glascock, A.L.; et al. The vaginal microbiome and preterm birth. *Nat. Med.* **2019**, *25*, 1012–1021. [CrossRef] [PubMed]
14. Powell, A.M.; Ali Khan, F.Z.; Ravel, J.; Elovitz, M.A. Untangling associations of microbiomes of pregnancy and preterm birth. *Clin. Perinatol.* **2024**, *51*, 425–439. [CrossRef] [PubMed]
15. Di Cianni, G.; Miccoli, R.; Volpe, L.; Lencioni, C.; Del Prato, S. Intermediate metabolism in normal pregnancy and in gestational diabetes. *Diabetes Metab. Res. Rev.* **2003**, *19*, 259–270. [CrossRef]
16. Lain, K.Y.; Catalano, P.M. Metabolic changes in pregnancy. *Clin. Obstet. Gynecol.* **2007**, *50*, 938–948. [CrossRef]
17. Nelson, S.M.; Matthews, P.; Poston, L. Maternal metabolism and obesity: Modifiable determinants of pregnancy outcome. *Hum. Reprod. Update* **2010**, *16*, 255–275. [CrossRef]
18. Fernández, L.; Langa, S.; Martín, V.; Maldonado, A.; Jiménez, E.; Martín, R.; Rodríguez, J.M. The human milk microbiota: Origin and potential roles in health and disease. *Pharmacol. Res.* **2013**, *69*, 1–10. [CrossRef]
19. Rodríguez, J.M. The origin of human milk bacteria: Is there a bacterial entero-mammary pathway during late pregnancy and lactation? *Adv. Nutr.* **2014**, *5*, 779–784. [CrossRef]
20. de Andrés, J.; Jiménez, E.; Chico-Calero, I.; Fresno, M.; Fernández, L.; Rodríguez, J.M. Physiological translocation of lactic acid bacteria during pregnancy contributes to the composition of the milk microbiota in mice. *Nutrients.* **2017**, *10*, 14. [CrossRef]
21. Martín, V.; Maldonado-Barragán, A.; Moles, L.; Rodriguez-Baños, M.; Campo, R.; Fernández, L.; Rodríguez, J.M.; Jiménez, E. Sharing of bacterial strains between breast milk and infant feces. *J. Hum. Lact.* **2012**, *28*, 36–44. [CrossRef] [PubMed]
22. Fernández, L.; Arroyo, R.; Espinosa, I.; Marín, M.; Jiménez, E.; Rodríguez, J.M. Probiotics for human lactational mastitis. *Benef. Microbes.* **2014**, *5*, 169–183. [CrossRef] [PubMed]
23. Bäckhed, F.; Roswall, J.; Peng, Y.; Feng, Q.; Jia, H.; Kovatcheva-Datchary, P.; Li, Y.; Xia, Y.; Xie, H.; Zhong, H.; et al. Dynamics and stabilization of the human gut microbiome during the first year of life. *Cell Host Microbe* **2015**, *17*, 690–703. [CrossRef] [PubMed]
24. Fernández, L.; Pannaraj, P.S.; Rautava, S.; Rodríguez, J.M. The microbiota of the human mammary ecosystem. *Front. Cell. Infect. Microbiol.* **2020**, *10*, 586667. [CrossRef] [PubMed]
25. Balan, P.; Brandt, B.W.; Chong, Y.S.; Crielaard, W.; Wong, M.L.; Lopez, V.; He, H.G.; Seneviratne, C.J. Subgingival microbiota during healthy pregnancy and pregnancy gingivitis. *JDR Clin. Trans. Res.* **2021**, *6*, 343–351. [CrossRef]
26. Meriç, P.; Silbereisen, A.; Emingil, G.; Öztürk, V.Ö.; Bostanci, N. Clinical, oral immunological and microbiological shifts during and after pregnancy. *Clin. Oral Investig.* **2023**, *28*, 60. [CrossRef]
27. Yang, I.; Claussen, H.; Arthur, R.A.; Hertzberg, V.S.; Geurs, N.; Corwin, E.J.; Dunlop, A.L. Subgingival microbiome in pregnancy and a potential relationship to early term birth. *Front. Cell Infect. Microbiol.* **2022**, *12*, 873683. [CrossRef]
28. Saadaoui, M.; Singh, P.; Al Khodor, S. Oral microbiome and pregnancy: A bidirectional relationship. *J. Reprod. Immunol.* **2021**, *145*, 103293. [CrossRef]
29. Nilsen, A.B.V.; Eri, T.S.; Downe, S.; Haukeland Fredriksen, E. Exploring women's self-reported health problems in pregnancy in the UK and Norway. *Health Care Women Int.* **2023**, *44*, 1438–1453. [CrossRef]
30. Lynch, S.V.; Pedersen, O. The human intestinal microbiome in health and disease. *N. Engl. J. Med.* **2016**, *375*, 2369–2379. [CrossRef]
31. Statista. Available online: https://www.statista.com/statistics/821259/global-probioticsl-market-value (accessed on 29 July 2024).
32. Gilmartin, C.E.; Vo-Tran, T.H.; Leung, L. Complementary medicines in pregnancy: Recommendations and information sources of healthcare professionals in Australia. *Int. J. Clin. Pharm.* **2018**, *40*, 421–427. [CrossRef] [PubMed]
33. Sheyholislami, H.; Connor, K.L. Are probiotics and prebiotics safe for use during pregnancy and lactation? A systematic review and meta-analysis. *Nutrients* **2021**, *13*, 2382. [CrossRef] [PubMed]
34. Vagibiom. Available online: https://vagibiom.com/the-remarkable-benefits-of-probiotics-for-women-during-pregnancy/ (accessed on 9 November 2024).

35. Mom Junction. Available online: https://www.momjunction.com/articles/probiotics-during-pregnancy-safety-benefits_001050690/ (accessed on 9 November 2024).
36. Aureli, P.; Fiore, A.; Scalfaro, C.; Casale, M.; Franciosa, G. National survey outcomes on commercial probiotic food supplements in Italy. *Int. J. Food Microbiol.* **2010**, *137*, 265–273. [CrossRef] [PubMed]
37. Drago, L.; Rodighiero, V.; Celeste, T.; Rovetto, L.; De Vecchi, E. Microbiological evaluation of commercial probiotic products available in the USA in 2009. *J. Chemother.* **2010**, *22*, 373–377. [CrossRef]
38. Morovic, W.; Hibberd, A.A.; Zabel, B.; Barrangou, R.; Stahl, B. Genotyping by PCR and high-throughput sequencing of commercial probiotic products reveals composition biases. *Front. Microbiol.* **2016**, *7*, 1747. [CrossRef]
39. Lugli, G.A.; Mangifesta, M.; Mancabelli, L.; Milani, C.; Turroni, F.; Viappiani, A.; van Sinderen, D.; Ventura, M. Compositional assessment of bacterial communities in probiotic supplements by means of metagenomic techniques. *Int. J. Food Microbiol.* **2019**, *294*, 1–9. [CrossRef]
40. Jackson, S.A.; Schoeni, J.L.; Vegge, C.; Pane, M.; Stahl, B.; Bradley, M.; Goldman, V.S.; Burguière, P.; Atwater, J.B.; Sanders, M.E. Improving end-user trust in the quality of commercial probiotic products. *Front. Microbiol.* **2019**, *10*, 739. [CrossRef]
41. Fusco, V.; Fanelli, F.; Chieffi, D. Authenticity of probiotic foods and dietary supplements: A pivotal issue to address. *Crit. Rev. Food. Sci. Nutr.* **2022**, *62*, 6854–6871. [CrossRef]
42. Biagioli, M.; Laghi, L.; Carino, A.; Cipriani, S.; Distrutti, E.; Marchianò, S.; Parolin, C.; Scarpelli, P.; Vitali, B.; Fiorucci, S. Metabolic variability of a multispecies probiotic preparation impacts on the anti-inflammatory activity. *Front. Pharmacol.* **2017**, *8*, 505. [CrossRef]
43. Lerner, A.; Shoenfeld, Y.; Matthias, T. Probiotics: If it does not help it does not do any harm. Really? *Microorganisms* **2019**, *7*, 104. [CrossRef]
44. Oliver, L.; Rasmussen, H.; Gregoire, M.B.; Chen, Y. Health care providers' knowledge; perceptions, and use of probiotics and prebiotics. *Top. Clin. Nutr.* **2014**, *29*, 139–149. [CrossRef]
45. Fijan, S.; Frauwallner, A.; Varga, L.; Langerholc, T.; Rogelj, I.; Lorber, M.; Lewis, P.; Pržan, P.P. Health professionals' knowledge of probiotics: An international survey. *Int. J. Environ. Res. Public Health* **2019**, *16*, 3128. [CrossRef] [PubMed]
46. Wilson, Z.; Whitehead, K. A cross sectional survey to assess healthcare professionals' attitudes to and understanding of probiotics. *Clin. Nutr. ESPEN* **2019**, *34*, 104–109. [CrossRef] [PubMed]
47. Suez, J.; Zmora, N.; Segal, E.; Elinav, E. The pros, cons, and many unknowns of probiotics. *Nat. Med.* **2019**, *25*, 716–729. [CrossRef]
48. Elias, J.; Bozzo, P.; Einarson, A. Are probiotics safe for use during pregnancy and lactation? *Can. Fam. Physician* **2011**, *57*, 299–301.
49. Jarde, A.; Lewis-Mikhael, A.M.; Moayyedi, P.; Stearns, J.C.; Collins, S.M.; Beyene, J.; McDonald, S.D. Pregnancy outcomes in women taking probiotics or prebiotics: A systematic review and meta-analysis. *BMC Pregnancy Childbirth* **2018**, *18*, 14. [CrossRef]
50. Body, C.; Christie, J.A. Gastrointestinal diseases in pregnancy: Nausea, vomiting, hyperemesis gravidarum, gastroesophageal reflux disease, constipation, and diarrhea. *Gastroenterol. Clin. N. Am.* **2016**, *45*, 267–283. [CrossRef]
51. Gille, C.; Böer, B.; Marschal, M.; Urschitz, M.S.; Heinecke, V.; Hund, V.; Speidel, S.; Tarnow, I.; Mylonas, I.; Franz, A.; et al. Effect of probiotics on vaginal health in pregnancy. EFFPRO, a randomized controlled trial. *Am. J. Obstet. Gynecol.* **2016**, *215*, 608.e1–608.e7. [CrossRef]
52. Yang, S.; Reid, G.; Challis, J.R.G.; Gloor, G.B.; Asztalos, E.; Money, D.; Seney, S.; Bocking, A.D. Effect of oral probiotic *Lactobacillus rhamnosus* GR-1 and *Lactobacillus reuteri* RC-14 on the vaginal microbiota, cytokines and chemokines in pregnant women. *Nutrients* **2020**, *12*, 368. [CrossRef]
53. Pérez-Castillo, Í.M.; Fernández-Castillo, R.; Lasserrot-Cuadrado, A.; Gallo-Vallejo, J.L.; Rojas-Carvajal, A.M.; Aguilar-Cordero, M.J. Reporting of perinatal outcomes in probiotic randomized controlled trials. A systematic review and meta-analysis. *Nutrients* **2021**, *13*, 256. [CrossRef]
54. VandeVusse, L.; Hanson, L.; Safdar, N. Perinatal outcomes of prenatal probiotic and prebiotic administration: An integrative review. *J. Perinat. Neonatal Nurs.* **2013**, *27*, 288–301, quiz E1-2. [CrossRef] [PubMed]
55. Rostok, M.; Hütt, P.; Rööp, T.; Smidt, I.; Štšepetova, J.; Salumets, A.; Mändar, R. Potential vaginal probiotics: Safety, tolerability and preliminary effectiveness. *Benef. Microbes* **2019**, *10*, 385–393. [CrossRef] [PubMed]
56. Bayar, E.; MacIntyre, D.A.; Sykes, L.; Mountain, K.; Parks, T.P.; Lee, P.P.; Bennett, P.R. Safety, tolerability, and acceptability of *Lactobacillus crispatus* CTV-05 (LACTIN-V) in pregnant women at high-risk of preterm birth. *Benef. Microbes* **2023**, *14*, 45–56. [CrossRef] [PubMed]
57. CosmosID. Available online: https://www.cosmosid.com/blog/reclassification-of-vaginal-probiotics-in-the-eu-as-pharmaceutical-drugs/ (accessed on 9 November 2024).
58. Seale, A.C.; Bianchi-Jassir, F.; Russell, N.J.; Kohli-Lynch, M.; Tann, C.J.; Hall, J.; Madrid, L.; Blencowe, H.; Cousens, S.; Baker, C.J.; et al. Estimates of the Burden of Group B streptococcal disease worldwide for pregnant women, stillbirths, and children. *Clin. Infect. Dis.* **2017**, *65* (Suppl. S2), S200–S219. [CrossRef]
59. Paul, P.; Gonçalves, B.P.; Le Doare, K.; Lawn, J.E. 20 million pregnant women with group B streptococcus carriage: Consequences, challenges, and opportunities for prevention. *Curr. Opin. Pediatr.* **2023**, *35*, 223–230. [CrossRef]
60. Coggins, S.A.; Puopolo, K.M. Neonatal group B *Streptococzcus* disease. *Pediatr. Rev.* **2024**, *45*, 63–73. [CrossRef]
61. Lawn, J.E.; Bianchi-Jassir, F.; Russell, N.J.; Kohli-Lynch, M.; Tann, C.J.; Hall, J.; Madrid, L.; Baker, C.J.; Bartlett, L.; Cutland, C.; et al. Group B streptococcal disease worldwide for pregnant women, stillbirths, and children: Why, what, and how to undertake estimates? *Clin. Infect. Dis.* **2017**, *65* (Suppl. S2), S89–S99. [CrossRef]

62. Greve, N.B.; Slotved, H.C.; Olsen, J.E.; Thomsen, L.E. Identification of antibiotic induced persister cells in *Streptococcus agalactiae*. *PLoS ONE* **2024**, *19*, e0303271. [CrossRef]
63. Campisi, E.; Rosini, R.; Ji, W.; Guidotti, S.; Rojas-López, M.; Geng, G.; Deng, Q.; Zhong, H.; Wang, W.; Liu, H.; et al. genomic analysis reveals multi-drug resistance clusters in group B *Streptococcus* CC17 hypervirulent isolates causing neonatal invasive disease in Southern mainland China. *Front. Microbiol.* **2016**, *7*, 1265. [CrossRef]
64. Gao, K.; Guan, X.; Zeng, L.; Qian, J.; Zhu, S.; Deng, Q.; Zhong, H.; Pang, S.; Gao, F.; Wang, J.; et al. An increasing trend of neonatal invasive multidrug-resistant group B *Streptococcus* infections in southern China, 2011–2017. *Infect. Drug Resist.* **2018**, *11*, 2561–2569. [CrossRef]
65. Hayes, K.; O'Halloran, F.; Cotter, L. A review of antibiotic resistance in Group B *Streptococcus*: The story so far. *Crit. Rev. Microbiol.* **2020**, *46*, 253–269. [CrossRef] [PubMed]
66. Tanaka, S.; Kobayashi, T.; Songjinda, P.; Tateyama, A.; Tsubouchi, M.; Kiyohara, C.; Shirakawa, T.; Sonomoto, K.; Nakayama, J. Influence of antibiotic exposure in the early postnatal period on the development of intestinal microbiota. *FEMS Immunol. Med. Microbiol.* **2009**, *56*, 80–87. [CrossRef] [PubMed]
67. Arboleya, S.; Sánchez, B.; Solís, G.; Fernández, N.; Suárez, M.; Hernández-Barranco, A.M.; Milani, C.; Margolles, A.; de Los Reyes-Gavilán, C.G.; Ventura, M.; et al. Impact of prematurity and perinatal antibiotics on the developing intestinal microbiota: A functional inference study. *Int. J. Mol. Sci.* **2016**, *17*, 649. [CrossRef] [PubMed]
68. Nogacka, A.; Salazar, N.; Suárez, M.; Milani, C.; Arboleya, S.; Solís, G.; Fernández, N.; Alaez, L.; Hernández-Barranco, A.M.; de los Reyes-Gavilán, C.G.; et al. Impact of intrapartum antimicrobial prophylaxis upon the intestinal microbiota and the prevalence of antibiotic resistance genes in vaginally delivered full-term neonates. *Microbiome* **2017**, *5*, 93. [CrossRef] [PubMed]
69. Dierikx, T.H.; Visser, D.H.; Benninga, M.A.; van Kaam, A.H.L.C.; de Boer, N.K.H.; de Vries, R.; van Limbergen, J.; de Meij, T.G.J. The influence of prenatal and intrapartum antibiotics on intestinal microbiota colonisation in infants: A systematic review. *J. Infect.* **2020**, *81*, 190–204. [CrossRef]
70. Arboleya, S.; Saturio, S.; Gueimonde, M. Impact of intrapartum antibiotics on the developing microbiota: A review. *Microbiome Res. Rep.* **2022**, *1*, 22. [CrossRef]
71. Cox, L.M.; Yamanishi, S.; Sohn, J.; Alekseyenko, A.V.; Leung, J.M.; Cho, I.; Kim, S.G.; Li, H.; Gao, Z.; Mahana, D.; et al. Altering the intestinal microbiota during a critical developmental window has lasting metabolic consequences. *Cell* **2014**, *158*, 705–721. [CrossRef]
72. Lynch, C.M.K.; Cowan, C.S.M.; Bastiaanssen, T.F.S.; Moloney, G.M.; Theune, N.; van de Wouw, M.; Florensa Zanuy, E.; Ventura-Silva, A.P.; Codagnone, M.G.; Villalobos-Manríquez, F.; et al. Critical windows of early-life microbiota disruption on behaviour, neuroimmune function, and neurodevelopment. *Brain Behav. Immun.* **2023**, *108*, 309–327. [CrossRef]
73. Hanson, L.; VandeVusse, L.; Duster, M.; Warrack, S.; Safdar, N. Feasibility of oral prenatal probiotics against maternal group B *Streptococcus* vaginal and rectal colonization. *J. Obstet. Gynecol. Neonatal Nurs.* **2014**, *43*, 294–304. [CrossRef]
74. Ho, M.; Chang, Y.Y.; Chang, W.C.; Lin, H.C.; Wang, M.H.; Lin, W.C.; Chiu, T.H. Oral *Lactobacillus rhamnosus* GR-1 and *Lactobacillus reuteri* RC-14 to reduce Group B *Streptococcus* colonization in pregnant women: A randomized controlled trial. *Taiwan J. Obstet. Gynecol.* **2016**, *55*, 515–518. [CrossRef]
75. Di Pierro, F.; Parolari, A.; Brundu, B.; Nigro, R. Positive clinical outcomes derived from using a proprietary mixture of selected strains during pregnancy. *Acta Biomed.* **2016**, *87*, 259–265. [PubMed]
76. Aziz, N.; Spiegel, A.; Bentley, J.; Yoffe, P.; Klikoff, A.; Ehrlich, K.; El-Sayed, Y.; Norton, M.; Taslimi, M. Evaluation of probiotic oral supplementation effects on Group B *Streptococcus* rectovaginal colonization in pregnant women: A randomized double-blind placebo-controlled trial. *Am. J. Obstet. Gynecol.* **2018**, *218*, S509–S510. [CrossRef]
77. Olsen, P.; Williamson, M.; Traynor, V.; Georgiou, C. The impact of oral probiotics on vaginal Group B Streptococcal colonisation rates in pregnant women: A pilot randomised control study. *Women Birth* **2018**, *31*, 31–37. [CrossRef]
78. Martín, V.; Cárdenas, N.; Ocaña, S.; Marín, M.; Arroyo, R.; Beltrán, D.; Badiola, C.; Fernández, L.; Rodríguez, J.M. Rectal and vaginal eradication of *Streptococcus agalactiae* (GBS) in pregnant women by using *Lactobacillus salivarius* CECT 9145, a target-specific probiotic strain. *Nutrients* **2019**, *11*, 810. [CrossRef]
79. Sharpe, M.; Shah, V.; Freire-Lizama, T.; Cates, E.C.; McGrath, K.; David, I.; Cowan, S.; Letkeman, J.; Stewart-Wilson, E. Effectiveness of oral intake of *Lactobacillus rhamnosus* GR-1 and *Lactobacillus reuteri* RC-14 on Group B *Streptococcus* colonization during pregnancy: A midwifery-led double-blind randomized controlled pilot trial. *J. Matern.-Fetal Neonatal Med.* **2019**, *34*, 1814–1821. [CrossRef]
80. Farr, A.; Sustr, V.; Kiss, H.; Rosicky, I.; Graf, A.; Makristathis, A.; Makristathis, A.; Petricevic, L. Oral probiotics to reduce vaginal group B streptococcal colonization in late pregnancy. *Sci. Rep.* **2020**, *10*, 19745. [CrossRef]
81. Hanson, L.; VandeVusse, L.; Malloy, E.; Garnier-Villarreal, M.; Watson, L.; Fial, A.; Forgie, M.; Nardini, K.; Safdar, N. Probiotic interventions to reduce antepartum Group B *Streptococcus* colonization: A systematic review and meta-analysis. *Midwifery* **2022**, *105*, 103208. [CrossRef]
82. Wu, S.; Jin, J.; Hu, K.L.; Wu, Y.; Zhang, D. Prevention of gestational diabetes mellitus and gestational weight gain restriction in overweight/obese pregnant women: A systematic review and network meta-analysis. *Nutrients* **2022**, *14*, 2383. [CrossRef]
83. Luoto, R.; Laitinen, K.; Nermes, M.; Isolauri, E. Impact of maternal probiotic-supplemented dietary counselling on pregnancy outcome and prenatal and postnatal growth: A double-blind, placebo-controlled study. *Br. J. Nutr.* **2010**, *103*, 1792–1799. [CrossRef]

84. Laitinen, K.; Poussa, T.; Isolauri, E.; The Nutrition, Allergy, Mucosal immunology and Intestinal microbiota Group. Probiotics and dietary counselling contribute to glucose regulation during and after pregnancy: A randomised controlled trial. *Br. J. Nutr.* **2009**, *101*, 1679–1687. [CrossRef]
85. Barrett, H.L.; Dekker Nitert, M.; Conwell, L.S.; Callaway, L.K. Probiotics for preventing gestational diabetes. *Cochrane Database Syst. Rev.* **2014**, *2014*, CD009951. [CrossRef] [PubMed]
86. Brantsaeter, A.L.; Myhre, R.; Haugen, M.; Myking, S.; Sengpiel, V.; Magnus, P.; Jacobsson, B.; Meltzer, H.M. Intake of probiotic food and risk of preeclampsia in primiparous women. The Norwegian Mother and Child cohort study. *Am. J. Epidemiol.* **2011**, *174*, 807–815. [CrossRef] [PubMed]
87. Ilmonen, J.; Isolauri, E.; Poussa, T.; Laitinen, K. Impact of dietary counselling and probiotic intervention on maternal anthropometric measurements during and after pregnancy: A randomized placebo controlled trial. *Clin. Nutr.* **2011**, *30*, 156–164. [CrossRef] [PubMed]
88. Asemi, Z.; Jazayeri, S.; Najafi, M.; Samimi, M.; Mofid, V.; Shidfar, F. Effects of daily consumption of probiotic yoghurt on inflammatory factors in pregnant women: A randomized controlled trial. *Pak. J. Biol. Sci.* **2011**, *14*, 476–482. [CrossRef]
89. Asemi, Z.; Jazayeri, S.; Najafi, M.; Samimi, M.; Mofid, V.; Shidfar, F.; Shakeri, H.; Esmaillzadeh, A. Effect of daily consumption of probiotic yoghurt on oxidative stress in pregnant women: A randomized controlled clinical trial. *Ann. Nutr. Metab.* **2012**, *60*, 62–68. [CrossRef]
90. Lindsay, K.L.; Walsh, C.A.; Brennan, L.; McAuliffe, F.M. Probiotics in pregnancy and maternal outcomes: A systematic review. *J. Matern.-Fetal Neonatal Med.* **2013**, *26*, 772–778. [CrossRef]
91. Łagowska, K.; Malinowska, A.M.; Zawieja, B.; Zawieja, E. Improvement of glucose metabolism in pregnant women through probiotic supplementation depends on gestational diabetes status: Meta-analysis. *Sci. Rep.* **2020**, *10*, 17796. [CrossRef]
92. Zheng, J.; Feng, Q.; Zheng, S.; Xiao, X. The effects of probiotics supplementation on metabolic health in pregnant women: An evidence based meta-analysis. *PLoS ONE* **2018**, *13*, e0197771. [CrossRef]
93. Lan, X.; Li, B.; Zhao, J.; Stanton, C.; Ross, R.P.; Chen, W.; Yang, B. Probiotic intervention improves metabolic outcomes in gestational diabetes mellitus: A meta-analysis of randomized controlled trials. *Clin. Nutr.* **2024**, *43*, 1683–1695. [CrossRef]
94. Okesene-Gafa, K.A.; Moore, A.E.; Jordan, V.; McCowan, L.; Crowther, C.A. Probiotic treatment for women with gestational diabetes to improve maternal and infant health and well-being. *Cochrane Database Syst. Rev.* **2020**, *6*, CD012970. [CrossRef]
95. Zhang, J.; Ma, S.; Wu, S.; Guo, C.; Long, S.; Tan, H. Effects of probiotic supplement in pregnant women with gestational diabetes mellitus: A systematic review and meta-analysis of randomized controlled trials. *J. Diabetes Res.* **2019**, *2019*, 5364730. [CrossRef] [PubMed]
96. Mu, J.; Guo, X.; Zhou, Y.; Cao, G. The effects of probiotics/synbiotics on glucose and lipid metabolism in women with gestational diabetes mellitus: A meta-analysis of randomized controlled trials. *Nutrients* **2023**, *15*, 1375. [CrossRef] [PubMed]
97. Taylor, B.L.; Woodfall, G.E.; Sheedy, K.E.; O'Riley, M.L.; Rainbow, K.A.; Bramwell, E.L.; Kellow, N.J. Effect of probiotics on metabolic outcomes in pregnant women with gestational diabetes: A systematic review and meta-analysis of randomized controlled trials. *Nutrients* **2017**, *9*, 461. [CrossRef] [PubMed]
98. Yefet, E.; Bar, L.; Izhaki, I.; Iskander, R.; Massalha, M.; Younis, J.S.; Nachum, Z. Effects of probiotics on glycemic control and metabolic parameters in gestational diabetes mellitus: Systematic review and meta-analysis. *Nutrients* **2023**, *15*, 1633. [CrossRef]
99. Wang, C.C.; Tung, Y.T.; Chang, H.C.; Lin, C.H.; Chen, Y.C. Effect of probiotic supplementation on newborn birth weight for mother with gestational diabetes mellitus or overweight/obesity: A systematic review and meta-analysis. *Nutrients* **2020**, *12*, 3477. [CrossRef]
100. Davidson, S.J.; Barrett, H.L.; Price, S.A.; Callaway, L.K.; Dekker Nitert, M. Probiotics for preventing gestational diabetes. *Cochrane Database Syst. Rev.* **2021**, *2021*, CD009951. [CrossRef]
101. Callaway, L.K.; McIntyre, H.D.; Barrett, H.L.; Foxcroft, K.; Tremellen, A.; Lingwood, B.E.; Tobin, J.M.; Wilkinson, S.; Kothari, A.; Morrison, M.; et al. Probiotics for the prevention of gestational diabetes mellitus in overweight and obese women: Findings from the SPRING double-blind randomized controlled trial. *Diabetes Care* **2019**, *42*, 364–371. [CrossRef]
102. Lindsay, K.L.; Kennelly, M.; Culliton, M.; Smith, T.; Maguire, O.C.; Shanahan, F.; Brennan, L.; McAuliffe, F.M. Probiotics in obese pregnancy do not reduce maternal fasting glucose: A double-blind, placebo-controlled, randomized trial (Probiotics in Pregnancy Study). *Am. J. Clin. Nutr.* **2014**, *99*, 1432–1439. [CrossRef]
103. Jamilian, M.; Bahmani, F.; Vahedpoor, Z.; Salmani, A.; Tajabadi-Ebrahimi, M.; Jafari, P.; Hashemi Dizaji, S.; Asemi, Z. Effects of probiotic supplementation on metabolic status in pregnant women: A randomized, double-blind, placebo-controlled trial. *Arch. Iran Med.* **2016**, *19*, 687–682. [PubMed]
104. Wickens, K.L.; Barthow, C.A.; Murphy, R.; Abels, P.R.; Maude, R.M.; Stone, P.R.; Mitchell, E.A.; Stanley, T.V.; Purdie, G.L.; Kang, J.M.; et al. Early pregnancy probiotic supplementation with *Lactobacillus rhamnosus* HN001 may reduce the prevalence of gestational diabetes mellitus: A randomised controlled trial. *Br. J. Nutr.* **2017**, *117*, 804–813. [CrossRef]
105. Okesene-Gafa, K.A.M.; Li, M.; McKinlay, C.J.D.; Taylor, R.S.; Rush, E.C.; Wall, C.R.; Wilson, J.; Murphy, R.; Taylor, R.; Thompson, J.M.D.; et al. Effect of antenatal dietary interventions in maternal obesity on pregnancy weight-gain and birthweight: Healthy Mums and Babies (HUMBA) randomized trial. *Am. J. Obstet. Gynecol.* **2019**, *221*, 152.e1–152.e13. [CrossRef] [PubMed]
106. Pellonperä, O.; Mokkala, K.; Houttu, N.; Vahlberg, T.; Koivuniemi, E.; Tertti, K.; Rönnemaa, T.; Laitinen, K. Efficacy of fish oil and/or probiotic intervention on the incidence of gestational diabetes mellitus in an at-risk group of overweight and obese women: A randomized, placebo-controlled, double-blind clinical trial. *Diabetes Care* **2019**, *42*, 1009–1017. [CrossRef] [PubMed]

107. Aaltonen, J.; Ojala, T.; Laitinen, K.; Piirainen, T.J.; Poussa, T.A.; Isolauri, E. Evidence of infant blood pressure programming by maternal nutrition during pregnancy: A prospective randomized controlled intervention study. *J. Pediatr.* **2008**, *152*, 79–84.e2. [CrossRef] [PubMed]
108. Lindsay, K.L.; Brennan, L.; Kennelly, M.A.; Maguire, O.C.; Smith, T.; Curran, S.; Coffey, M.; Foley, M.E.; Hatunic, M.; Shanahan, F.; et al. Impact of probiotics in women with gestational diabetes mellitus on metabolic health: A randomized controlled trial. *Am. J. Obstet. Gynecol.* **2015**, *212*, 496.e1–496.e11. [CrossRef] [PubMed]
109. Movaghar, R.; Farshbaf-Khalili, A.; Hajizade, K.; MirzaRezaei, M.E.; Shahnazi, M. The effect of probiotics or synbiotics on the hypertensive disorders of pregnant women with gestational diabetes: A systematic review and meta-analysis. *J. Caring Sci.* **2022**, *11*, 94–104. [CrossRef]
110. Wu, R.; Luan, J.; Hu, J.; Li, Z. Effect of probiotics on pregnancy outcomes in gestational diabetes: Systematic review and meta-analysis. *Arch. Gynecol. Obstet.* **2024**, *310*, 769–781. [CrossRef]
111. Van den Nieuwboer, M.; Brummer, R.J.; Guarner, F.; Morelli, L.; Cabana, M.; Claasen, E. The administration of probiotics and synbiotics in immune compromised adults: Is it safe? *Benef. Microbes* **2015**, *6*, 3–17. [CrossRef]
112. Rautio, M.; Jousimies-Somer, H.; Kauma, H.; Pietarinen, I.; Saxelin, M.; Tynkkynen, S.; Koskela, M. Liver abscess due to a *Lactobacillus rhamnosus* strain indistinguishable from *L. rhamnosus* strain GG. *Clin. Infect. Dis.* **1999**, *28*, 1159–1160. [CrossRef]
113. Notario, R.; Leardini, N.; Borda, N.; Gambandé, T.; Cerutti, H. Absceso hepático y bacteriemia debida a *Lactobacillus rhamnosus* [Hepatic abscess and bacteremia due to *Lactobacillus rhamnosus*]. *Rev. Argent. Microbiol.* **2003**, *35*, 100–101.
114. Sherid, M.; Samo, S.; Sulaiman, S.; Husein, H.; Sifuentes, H.; Sridhar, S. Liver abscess and bacteremia caused by *lactobacillus*: Role of probiotics? Case report and review of the literature. *BMC Gastroenterol.* **2016**, *16*, 138. [CrossRef]
115. Pararajasingam, A.; Uwagwu, J. *Lactobacillus*: The not so friendly bacteria. *BMJ Case Rep.* **2017**, *2017*, bcr2016218423. [CrossRef] [PubMed]
116. Omar, A.M.; Ahmadi, N.; Ombada, M.; Fuscaldo, J.; Siddiqui, N.; Safo, M.; Nalamalapu, S. Breaking Bad: A case of *Lactobacillus* bacteremia and liver abscess. *J. Community Hosp. Intern. Med. Perspect.* **2019**, *9*, 235–239. [CrossRef] [PubMed]
117. Ramos-Coria, D.; Canto-Losa, J.; Carrillo-Vázquez, D.; Carbajal-Morelos, L.; Estrada-León, R.; Corona-Rodarte, E. *Lactobacillus gasseri* liver abscess and bacteremia: A case report. *BMC Infect. Dis.* **2021**, *21*, 518. [CrossRef] [PubMed]
118. Lnu, K.; Abdelsalam, A.; Bambach, W. *Lactobacillus* bacteremia: A tell-tale sign for diabetes? *J. Community Hosp. Intern. Med. Perspect.* **2022**, *12*, 86–88. [CrossRef]
119. Ioannou, P.; Ziogou, A.; Giannakodimos, I.; Giannakodimos, A.; Baliou, S.; Samonis, G. Infective endocarditis by *Lactobacillus* species-a narrative review. *Antibiotics* **2024**, *13*, 53. [CrossRef]
120. Salminen, M.K.; Rautelin, H.; Tynkkynen, S.; Poussa, T.; Saxelin, M.; Valtonen, V.; Järvinen, A. *Lactobacillus* bacteremia, clinical significance, and patient outcome, with special focus on probiotic *L. rhamnosus* GG. *Clin. Infect. Dis.* **2004**, *38*, 62–69. [CrossRef]
121. Mikucka, A.; Deptuła, A.; Bogiel, T.; Chmielarczyk, A.; Nurczyńska, E.; Gospodarek-Komkowska, E. Bacteraemia caused by probiotic strains of *Lacticaseibacillus rhamnosus*-case studies highlighting the need for careful thought before using microbes for health benefits. *Pathogens* **2022**, *11*, 977. [CrossRef]
122. Kullar, R.; Goldstein, E.J.C.; Johnson, S.; McFarland, L.V. *Lactobacillus* bacteremia and probiotics: A review. *Microorganisms* **2023**, *11*, 896. [CrossRef]
123. Rodríguez, J.M.; Fernández, L. Infectious mastitis during lactation: A mammary dysbiosis model. In *Prebiotics and Probiotics in Human Milk*; McGuire, M., McGuire, M., Bode, L., Eds.; Academic Press: London, UK, 2017; pp. 401–428.
124. Patel, S.H.; Vaidya, Y.H.; Patel, R.J.; Pandit, R.J.; Joshi, C.G.; Kunjadiya, A.P. Culture independent assessment of human milk microbial community in lactational mastitis. *Sci. Rep.* **2017**, *7*, 7804. [CrossRef]
125. Soto, A.; Martín, V.; Jiménez, E.; Mader, I.; Rodríguez, J.M.; Fernández, L. Lactobacilli and bifidobacteria in human breast milk: Influence of antibiotherapy and other host and clinical factors. *J. Pediatr. Gastroenterol. Nutr.* **2014**, *59*, 78–88 [CrossRef]
126. Jiménez, E.; Fernández, L.; Maldonado, A.; Martín, R.; Olivares, M.; Xaus, J.; Rodríguez, J.M. Oral administration of *Lactobacillus* strains isolated from breast milk as an alternative for the treatment of infectious mastitis during lactation. *Appl. Environ. Microbiol.* **2008**, *74*, 4650–4655. [CrossRef]
127. Arroyo, R.; Martin, V.; Maldonado, A.; Jiménez, E.; Fernández, L.; Rodríguez, J.M. Treatment of infectious mastitis during lactation: Antibiotics versus oral administration of lactobacilli isolated from breast milk. *Clin. Infect. Dis.* **2010**, *50*, 1551–1558. [CrossRef] [PubMed]
128. Maldonado-Lobón, J.A.; Díaz-López, M.A.; Carputo, R.; Duarte, P.; Díaz-Ropero, M.P.; Valero, A.D.; Sañudo, A.; Sempere, L.; Ruiz-López, M.D.; Bañuelos, Ó.; et al. *Lactobacillus fermentum* CECT 5716 reduces *Staphylococcus* load in the breastmilk of lactating mothers suffering breast pain: A randomized controlled trial. *Breastfeed. Med.* **2015**, *10*, 425–432. [CrossRef] [PubMed]
129. Espinosa-Martos, I.; Jiménez, E.; de Andrés, J.; Rodríguez-Alcala, L.M.; Tavarez, S.; Manzano, S.; Fernández, L.; Alonso, E.; Fontecha, J.; Rodríguez, J.M. Milk and blood biomarkers associated to the clinical efficacy of a probiotic for the treatment of infectious mastitis. *Benef. Microbes* **2016**, *7*, 305–318. [CrossRef] [PubMed]
130. Fernández, L.; Cardenas, N.; Arroyo, R.; Manzano, S.; Jiménez, E.; Martin, V.; Rodríguez, J.M. Prevention of infectious mastitis by oral administration of *Lactobacillus salivarius* PS2 during late pregnancy. *Clin. Infect. Dis.* **2016**, *62*, 568–573. [CrossRef]
131. Hurtado, J.A.; Maldonado-Lobón, J.A.; Díaz-Ropero, M.P.; Flores-Rojas, K.; Uberos, J.; Leante, J.L.; Affumicato, L.; Couce, M.L.; Garrido, J.M.; Olivares, M.; et al. Oral administration to nursing women of *Lactobacillus fermentum* CECT5716 prevents lactational mastitis development: A randomized controlled trial. *Breastfeed. Med.* **2017**, *12*, 202–209. [CrossRef]

132. Jiménez, E.; Manzano, S.; Schlembach, D.; Arciszewski, K.; Martin, R.; Ben Amor, K.; Roelofs, M.; Knol, J.; Rodríguez, J.M.; Abou-Dakn, M.; et al. *Ligilactobacillus salivarius* PS2 supplementation during pregnancy and lactation prevents mastitis: A randomised controlled trial. *Microorganisms* **2021**, *9*, 1933. [CrossRef]
133. Vazquez-Fresno, R.; Llorach, R.; Marinic, J.; Tulipani, S.; Garcia-Aloy, M.; Espinosa-Martos, I.; Jiménez, E.; Rodríguez, J.M.; Andres-Lacueva, C. Urinary metabolomic fingerprinting after consumption of a probiotic strain in women with mastitis. *Pharmacol. Res.* **2014**, *87*, 160–165. [CrossRef]
134. Yu, Q.; Xu, C.; Wang, M.; Zhu, J.; Yu, L.; Yang, Z.; Liu, S.; Gao, X. The preventive and therapeutic effects of probiotics on mastitis: A systematic review and meta-analysis. *PLoS ONE* **2022**, *17*, e0274467. [CrossRef]
135. Alemu, B.K.; Azeze, G.G.; Wu, L.; Lau, S.L.; Wang, C.C.; Wang, Y. Effects of maternal probiotic supplementation on breast milk microbiome and infant gut microbiome and health: A systematic review and meta-analysis of randomized controlled trials. *Am. J. Obstet. Gynecol. MFM* **2023**, *5*, 101148. [CrossRef]
136. Fernández, L.; Castro, I.; Arroyo, R.; Alba, C.; Beltrán, D.; Rodríguez, J.M. Application of *Ligilactobacillus salivarius* CECT5713 to achieve term pregnancies in women with repetitive abortion or infertility of unknown origin by microbiological and immunological modulation of the vaginal ecosystem. *Nutrients* **2021**, *13*, 162. [CrossRef] [PubMed]
137. Fernández, L.; Castro, I.; Arroyo, R.; Alba, C.; Beltrán, D.; Rodríguez, J.M. Immunomodulation of the vaginal ecosystem by *Ligilactobacillus salivarius* CECT 30632 improves pregnancy rates among women with infertility of unknown origin or habitual abortions. *Nutrients* **2023**, *15*, 362. [CrossRef] [PubMed]
138. Qian, D.; Lin, H.Y.; Wang, H.M.; Zhang, X.; Liu, D.L.; Li, Q.L.; Zhu, C. Involvement of ERK1/2 pathway in TGF-beta1-induced VEGF secretion in normal human cytotrophoblast cells. *Mol. Reprod. Dev.* **2004**, *68*, 198–204. [CrossRef] [PubMed]
139. Chung, I.B.; Yelian, F.D.; Zaher, F.M.; Gonik, B.; Evans, M.I.; Diamond, M.P.; Svinarich, D.M. Expression and regulation of vascular endothelial growth factor in a first trimester trophoblast cell line. *Placenta* **2000**, *21*, 320–324. [CrossRef]
140. Robertson, S.A.; Ingman, W.V.; O'Leary, S.; Sharkey, D.J.; Tremellen, K.P. Transforming growth factor β—A mediator of immune deviation in seminal plasma. *J. Reprod. Immunol.* **2002**, *57*, 109–128. [CrossRef]
141. Jepsen, I.E.; Saxtorph, M.H.; Englund, A.L.M.; Petersen, K.B.; Wissing, M.L.M.; Hviid, T.V.F.; Macklon, N. Probiotic treatment with specific lactobacilli does not improve an unfavorable vaginal microbiota prior to fertility treatment-A randomized, double-blinded, placebo-controlled trial. *Front. Endocrinol.* **2022**, *13*, 1057022. [CrossRef]
142. Thanaboonyawat, I.; Pothisan, S.; Petyim, S.; Laokirkkiat, P. Pregnancy outcomes after vaginal probiotic supplementation before frozen embryo transfer: A randomized controlled study. *Sci. Rep.* **2023**, *13*, 11892. [CrossRef]
143. Di Pierro, F.; Sinatra, F.; Cester, M.; Da Ros, L.; Pistolato, M.; Da Parè, V.; Fabbro, L.; Maccari, D.; Dotto, S.; Sossai, S.; et al. Effect of *L. crispatus* M247 Administration on pregnancy outcomes in women undergoing IVF: A controlled, retrospective, observational, and open-label study. *Microorganisms* **2023**, *11*, 2796. [CrossRef]
144. Tanha, F.D.; Rahmani, Z.; Rezaei, Z.; Asbagh, F.A.; Ebrahimi, M.; Quchani, S.H.; Feizabad, E.; Shahraki, Z. The effect of normalizing vaginal microbiome using Lactovag in improving pregnancy outcomes in frozen embryo transfer cycles: A randomized clinical trial. *Arch. Gynecol. Obstet.* **2023**, *308*, 1587–1592. [CrossRef]
145. Naghi Jafarabadi, M.; Hadavi, F.; Ahmadi, M.; Masoumi, M.; Zabihzadeh, S. Intravaginal probiotics before embryo transfer do not improve pregnancy rates in recurrent implantation failure cases: An RCT. *Int. J. Reprod. Biomed.* **2024**, *22*, 363–374. [CrossRef]
146. Cohen, P.A. Probiotic Safety-No Guarantees. *JAMA Intern. Med.* **2018**, *178*, 1577–1578. [CrossRef] [PubMed]
147. Freedman, S.B.; Schnadower, D.; Tarr, P.I. The probiotic conundrum: Regulatory confusion, conflicting studies, and safety concerns. *JAMA* **2020**, *323*, 823–824. [CrossRef] [PubMed]
148. Sanders, M.E.; Akkermans, L.M.; Haller, D.; Hammerman, C.; Heimbach, J.; Hörmannsperger, G.; Huys, G.; Levy, D.D.; Lutgendorff, F.; Mack, D.; et al. Safety assessment of probiotics for human use. *Gut Microbes* **2010**, *1*, 164–185. [CrossRef] [PubMed]
149. Wang, Y.; Liang, Q.; Lu, B.; Shen, H.; Liu, S.; Shi, Y.; Leptihn, S.; Li, H.; Wei, J.; Liu, C.; et al. Whole-genome analysis of probiotic product isolates reveals the presence of genes related to antimicrobial resistance, virulence factors, and toxic metabolites, posing potential health risks. *BMC Genom.* **2021**, *22*, 210. [CrossRef]
150. Aragón, A.; Jurado, R.; Jara, J.; Rodríguez, J.M.; Orgaz, B. Investigating the metabolism of estrogens in *Ligilactobacillus salivarius* strains isolated from human milk and vaginal microbiota. *Nutrients* **2024**, *16*, 861. [CrossRef]

foods

MDPI

Review

Leuconostoc citreum: A Promising Sourdough Fermenting Starter for Low-Sugar-Content Baked Goods

Maria-Florina Roșca, Adriana Păucean *, Simona Maria Man, Maria Simona Chiș, Carmen R. Pop, Anamaria Pop and Anca C. Fărcaș

Faculty of Food Science and Technology, University of Agricultural Sciences and Veterinary, 400372 Cluj-Napoca, Romania; maria-florina.rosca@student.usamvcluj.ro (M.-F.R.); simona.man@usamvcluj.ro (S.M.M.); simona.chis@usamvcluj.ro (M.S.C.); carmen-rodica.pop@usamvcluj.ro (C.R.P.); anamaria.pop@usamvcluj.ro (A.P.); anca.farcas@usamvcluj.ro (A.C.F.)
* Correspondence: adriana.paucean@usamvcluj.ro

Abstract: This review highlights *Leuconostoc citreum*'s promising possibilities as a proficient mannitol producer and its potential implications for sugar reduction, with a focus on its use in sourdough-based baked good products. Mannitol, a naturally occurring sugar alcohol, has gained popularity in food items due to its low calorie content and unique beneficial qualities. This study summarizes recent research findings and investigates the metabolic pathways and culture conditions that favor increased mannitol production by *Leuconostoc citreum*. Furthermore, it investigates the several applications of mannitol in baked goods, such as its function in increasing texture, flavor and shelf life while lowering the sugar content. Sourdough-based products provide an attractive niche for mannitol integration, as customer demand for healthier and reduced-sugar options increases.

Keywords: lactic acid bacteria; *Leuconostoc citreum*; sourdough; sugar reduction; baked goods

1. Introduction

1.1. *Leuconostoc*—History, Morphology and General Importance

The name *Leuconostoc* is a combination of the Greek word *leucus*, meaning clear, and the generic Latin noun *nostoc*, which refers to an alga [1] (*Leuconostoc*, colorless nostoc) [2].

Leuconostoc spp. are Gram-positive lactic acid bacteria (LAB) that are significant for the economy. Different environmental *Leuconostoc* strains naturally develop from greenery and roots, which act as their ecological habitat [3], and they spread to diverse niches such as raw milk or cold food goods [2]. Their population is often small, making up less than 1% of the overall microbial community [3].

Leuconostoc strains, which belong to the low G and C branch of the Gram-positive bacteria, are non-motile, non-spore-forming cocci that are typically seen in pairs or short chains. They lack catalase and arginine dehydrolase and are heterofermentative, creating CO_2 from the metabolism of glucose in addition to d-lactate, ethanol or acetate. They are mesophilic facultative anaerobic bacteria that grow at 10 °C and exhibit mesophilic features [3].

Farrow et al. (1989) [4] classified *Leuconostoc citreum* as a lactic acid bacteria (LAB) species. It was recognized as a novel species by DNA similarity with other *Leuconostoc* spp. strains but is not one of the original species of the genus *Leuconostoc*. *Leuconostoc citreum* has been isolated from a variety of plant-based food and beverage fermentations including kimchi [5], sourdough [6] and fermented corn beverages [7]. They accomplished a taxonomic distinction between *Leuconostoc citreum* and other *Leuconostoc* species by using primarily DNA-related evidence, which led to the reclassification of strains formerly categorized under *Leuconostoc mesenteroides* [4]. Spherical or lenticular cells (cca. 0.5 to 0.7 by 0.7 to 1.2 pm) are seen in pairs or short chains in *Leuconostoc citreum*. The majority of strains generate a lemon-yellow color and grow between 10 °C and 30 °C; certain strains grow at 37 °C, but when the temperature reaches 40 °C they stop growing. Because it cannot generate acid from ribose, *Leuconostoc citreum* is easily distinguished from *Leuconostoc mesenteroides* subsp. *mesenteroides* and *Leuconostoc pseudomesenteroides*. *Leuconostoc citreum* strains are also negative for α- and β-D-galactosidase and β-D-xylosidase, but *Leuconostoc mesenteroides* subsp. *mesenteroides* and *Leuconostoc pseudomesenteroides* are positive. Another distinctive feature is the capacity to use citrate. The ability of *Leuconostoc citreum* to metabolize citrate is reflected in its name ("*citreum*" refers to citrate) [4].

Leuconostoc spp. are of tremendous industrial importance since they serve as novel starter cultures for fermented foods (fermented vegetables, dairy products, meats and fresh alcoholic drinks) as well as host strains to develop value-added bioproducts in the field of biotechnology [8]. *Leuconostoc* can be detected in mixed-strain cultures at mesophilic temperatures but are notably absent in starter cultures that operate under thermophilic conditions. The use of starter cultures, including *Leuconostoc citreum*, ensures consistent and reliable results [9].

Leuconostoc citreum is a heterofermentative lactic acid bacterium (LAB) that creates lactic acid, acetic acid, alcohol, mannitol, CO_2 and aromatic compounds [10]. Additionally, genes for the production of dextran polysaccharide from sucrose were reported for *Leuconostoc citreum* EFEL2700 [11], while for *Leuconsotoc citreum* SK24.002 an exopolysaccharide with high resistance for dextranase was found [12].

Fermentation can enhance food's nutritional quality by improving digestibility and increasing the levels of vitamins and minerals compared to the initial content. This process also generates antimicrobial compounds like organic acids, hydrogen peroxide, diacetyls and bacteriocins, which both inhibit undesirable bacterial growth and extend the food's shelf life. Fermented food products with higher lactic acid content may improve the utilization of calcium, phosphorus and iron, while also increasing the absorption of iron and vitamin D. These foods contain a range of enzymes, each with its unique role in enhancing food quality, generally through better digestion or enhanced bioavailability. The acidification indirectly activates both the endogenous phytase and the microbial enzymes increasing the mineral bioavailability [9].

Leuconostoc species are especially considered as good hosts for heterologous protein expression in the food biotechnology industry because they have been identified as safe (GRAS), are not capable of generating endotoxin, grow well under aerobic and anaerobic conditions and secrete certain proteins such as dextransucrase despite being digested by proteolytic enzymes [3]. In this regard, certain leuconostocs have complete genes for the manufacture of certain vitamins and have the potential to improve the nutritional content of fermented foods. *Leuconostoc citreum* and *Leuconostoc mesenteroides* subsp. *mesenteroides* have the entire collection of genes for vitamin B_2 (riboflavin) production [13].

Furthermore, given their widespread dispersion in the environment and the small number of diseases they produce, it is considered that these bacteria represent very little harm to healthy humans. It is plausible to conclude that *Leuconostoc* is usually safe because of the extensive history of human exposure and ingestion [2]. Moreover, the Food and Drug Administration (FDA) has designated bacteria of the *Leuconostoc* genus as Generally Recognized as Safe (GRAS) due to a history of human exposure and ingestion, while the European Food Safety Authority (EFSA) has granted them the designation of Qualified Presumption of Safety (QPS) [14].

But undoubtedly, one of the most important properties of some *Leuconostoc* ssp. is the capacity to convert fructose to mannitol in sourdough. Therefore, this feature can be exploited as a technical solution to naturally manufacture low-calorie sweeteners and decrease overall sugar content in sweet baked items while maintaining sweetness and flavor. Compared to the other LAB species, *Leuconostoc citreum* and *Leuconostoc mesenteroides* are highly efficient mannitol producers, with yields ranging from 70 to 90% [15].

Excess sugar consumption is thought to be a major cause of many disorders, and sweet baked goods are one of the main sources of sugar in human nutrition. Several approaches could be developed to reduce sugar. Currently, there are two well-known strategies: using bulking agents and high-intensity sweeteners to substitute for sugar, or using sweet bulking substances like polyols. Polyols, such as mannitol, are the most commonly utilized sweet bulking agents in the field of sugar reduction. Polyols are sugar alcohols, which are either obtained by chemical or biochemical reduction of sugars or during fermentation by lactic acid bacteria or yeast. Nevertheless, polyols can be used only as a partial replacement for sugar, and the main obstacle in lowering sugar is ensuring product quality in terms of volume, texture, appearance, sweetness and shelf life. Sourdough technology should be regarded as an innovative technique to surmount the negative influence on the quality parameters of sugar-reduced bread goods. Both mannitol and exopolysaccharides production by LAB could contribute to quality improvement. Moreover, not only will consumers benefit from the utilization of sourdough in sugar-reduced products but also, from the point of view of the industry, this strategy provides clean-label products and brings economic benefits [16,17].

The aim of this review is to evaluate *Leuconostoc citreum*'s capacity for mannitol production and its potential application in reducing sugar levels, with a particular focus on sourdough in baked goods and bakery products.

1.2. Role in Fermented Plant-Origin Food Products

Several studies have reported on traditionally fermented vegetable-based foods, where *Leuconostoc* ssp., alone or in mixt cultures, are the main microbial agents of fermentation and provide specific sensorial and nutritional properties. *Leuconostoc* is the most common genus of LAB on plants, with *L. mesenteroides* as the most common isolate [18] from squash, peas, cucumbers, greens, cereals, etc. *Leuconostoc citreum* and *Leuconostoc mesenteroides* both contribute considerably to food fermentation. While they have some similarities, important features distinguish them.

Thereby, gundruk, sunki and khalpi are several examples of traditionally fermented foods in Nepal, Sikkim and Bhutan. Gundruk is an indigenous vegetable product from the Himalayas that is fermented, non-salted and has an acidic flavor. It can be enjoyed as both a pickle and a soup and shares similarities with other fermented acidic vegetable products like Korea's kimchi, Germany's sauerkraut and Japan's sunki [19]. Sunki, a traditional fermented food made from radish tap roots, is prepared through pit fermentation, a distinctive method of bio-preserving foods through lactic acid fermentation in the Sikkim Himalayas [9]. Khalpi is a fermented cucumber product made from *Cucumis sativus* L. It is a popular delicacy among the Brahmin Nepalis in Sikkim and stands as the sole known fermented cucumber product in the entire Himalayan region [19]. Kimchi is a traditional Korean fermented vegetable dish prepared from Chinese cabbage, radish, green onion, red pepper powder, garlic, ginger and fermented seafood. The fermentation of kimchi is influenced by temperature; it develops in one week at 15 °C and only takes three days to matures at 25 °C. However, a lower temperature is favorable during kimchi fermentation to avoid the excessive production of strong acid and overripening and to promote an extended period of optimal flavor [20]. The dominant microbial species from kimchi are *Leuconostoc mesenteroides* and *Lactobacillus plantarum*. Various findings suggest that LAB involved in kimchi fermentation encompass a range of species, including *Leuconostoc mesenteroides*, *Leuconostoc citreum*, *L. gasicomitatum*, *Lactobacillus brevis*, *L. curvatus*, *L. sakei*, *L. plantarum*, *L. lactis*, *P. pentosaceus*, *W. confusa* and *W. koreensis* [9]. However, Choi et al., 2003 [21] reported that during the early and mid-stages of kimchi fermentation at 15 °C, *Leuconostoc citreum* was the dominant species when 120 LAB were isolated at random from two different kimchi samples. Moreover, *Leuconostoc citreum* IH 22 played an important role in kimchi fermentation as starter culture, with good capacity to dominate the spontaneous microflora. When this strain was cultivated on 5% sucrose medium, it demonstrated a significantly better tolerance to acid, which resulted in the synthesis of insoluble dextran. The psychrotroph *Leuconostoc citreum* HS-P4 strain was also isolated from kimchi, with outstanding characteristics in terms of growth rate and dextransucrase activity [5].

Typically found in the microbial community of plants, *Leuconostoc* spp. have been isolated from sourdough environments on many occasions [22]. Sourdough is another representative fermented product obtained through LAB metabolism, and some strains of *Leuconostoc citreum* can be involved in sourdough fermentation inducing specific transformations. Sourdough is obtained by fermenting a mixture of flour and water, and this process is facilitated by the naturally present or inoculated LAB and/or yeasts, which play a role in its acidification and the leavening characteristics of the dough. LAB enhance the quality of sourdough baked goods by producing metabolites during the fermentation process, which positively influence the products′ texture, flavor and shelf life [23].

Notably, *Leuconostoc citreum* produces mannitol, a low-calorie sugar alternative that is suitable for diabetic and low-sugar meals. Its specialized strains, such as *L. citreum* TR116, have an outstanding ability to reduce sugar content in baked goods such as bread, cookies, buns and cakes [5,6,24–26]. Various *Leuconostoc citreum* strains derived from sourdough synthesized dextrans with varying amounts of α-(1-6), α-(1-3) and α-(1-2) linkages, along with mutan and fructans (Amari et al., 2015). These exopolysaccharides (EPSs) increase the viscosity of liquid systems, while in sourdough baked goods, they improve the rheological properties and the crumb structure [24]. These modifications could be explained by some EPS properties such as the water binding capacity and the capacity to combine with proteins in doughs [22].

Cells of *Leuconostoc* ssp. can survive in an unfavorable medium for long periods of time. However, the cells could be exposed to stress conditions like pH and temperature variations and oxidative or osmotic changes, and the intensity of these stress factors could lead to cell death or to cell adaptation through suitable chemical reactions in an effort to counteract the adverse effects and recover the capacity for growth or survival. For instance, unfavorable environmental circumstances promote interactions such as the production of glycocalyx in the presence of saccharose and trace minerals, culminating in the formation of a biofilm that defends cells from harmful agents. Evidence has indicated that cell growth ended when internal pH levels of 5.4 to 5.7 were attained, regardless of culture medium composition. The limiting external cellular pH, on the other hand, was heavily impacted by the growth media, namely the kind and concentration of organic acids. On the other hand, low cell lysis was reported at pHs above 6.5 in the presence of Ca^{2+} and Na^{+}. Notably, *Leuconostoc* could metabolize citrate when the pH ranged from 6.3 to 4.5, and diacetyl and acetoin were only produced at acidic pHs [3,27].

Stress proteins were overexpressed as a reaction of *Leuconostoc mesenteroides* cells to heat shock, while optimal conditions (30 °C) allowed lower pH values, higher citrate utilization and higher ethanol and acetic acid production. However, limiting or excessive contents of nitrogenous compounds could negatively affect *L. mesenteroides*′ metabolism.

2. Metabolism Characteristics

2.1. Metabolism of Sugars in *Leuconostoc ssp.* and *Leuconostoc citreum*

Cogan and Jordan (1994) claimed that because leuconostocs, like other LAB, lack a tricarboxylic acid cycle and a cytochrome system, they cannot receive energy from oxidative phosphorylation. They gain energy instead by substrate-level phosphorylation, which occurs during the fermentation of carbohydrates to lactic acid, ethanol or acetate and CO_2 [10].

In the early 1950s, the researchers described the process of glucose fermentation in leuconostocs. The removal of CO_2 from glucose is followed by the split of the resultant pentose into two-carbon and three-carbon segments, which is facilitated by phosphoketolase. This enzymatic activity produces glyceraldehyde-3-phosphate and acetylphosphate. This mechanism, known as the phosphoketolase route, is seen in heterofermentative lactobacilli as well (Figure 1). The three-carbon section eventually changed to lactate, whereas the two-carbon part converted to ethanol. These end products are critical in replenishing NAD^+ and sustaining continued fermentation [7]. The overall stoichiometry is as follows:

$$\text{Glucose} \rightarrow 1\ \text{Lactate} + 1\ \text{Ethanol} + 1\ CO_2 + 1\ \text{ATP} \quad (1)$$

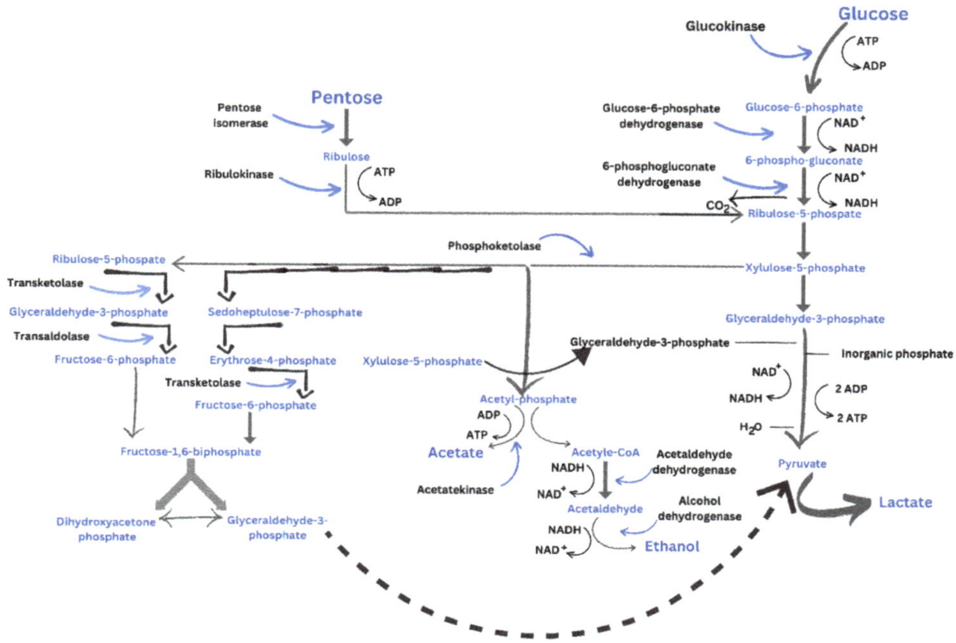

Figure 1. Glucose fermentation pathway. Heterofermentative metabolism (phosphoketolase pathway)—adapted after Teleky et al. 2020 [28].

This process is known as the heterofermentative path because it produces compounds other than lactate, as opposed to the homofermentative or glycolytic pathway seen in lactococci, enterococci and several lactobacilli, which only produce lactate:

$$1\ \text{Glucose} = 2\ \text{Lactate} + 2\ \text{ATP} \quad (2)$$

In the phosphoketase route, one ATP molecule is created for every mole of glucose metabolized, whereas the homofermentative pathway produces two ATP molecules. As a result, in terms of glucose fermentation, the phosphoketolase route is only half as effective as the homofermentative process. Although only the early and terminal enzymes of this route have been thoroughly studied and purified, their fundamental characteristics are well documented [10].

In contrast to glucose fermentation, fructose fermentation produces mannitol and acetate in addition to CO_2, lactate and ethanol. The fermentation balance is close to the following:

$$3\ \text{Fructose} \rightarrow 1\ \text{Mannitol} + 2\ \text{Lactate} + 0.5\ \text{Acetate} + 1.5\ \text{Ethanol} + 2\ CO_2 \quad (3)$$

Mannitol and NAD^+ are produced in a process catalyzed by mannitol dehydrogenase from fructose and NADH. This process produces more ATP because less acetylphosphate is converted to ethanol to renew the NAD^+ needed to continue the fermentation. For every mole of acetylphosphate that is not converted to ethanol, two moles of mannitol are generated. Fructokinase and phosphoglucoisomerase are found in leuconostocs fermenting fructose, showing that it is phosphorylated to fructose-6-phosphate before being isomerized to glucose-6-phosphate [10]. When mannitol synthesis is activated, *Leuconostoc citreum* produces acetic acid in order to restore ATP. Acetate is formed from acetylphosphate by an acetate kinase [5,29].

Regarding mannitol, it is a low-calorie sweetener and has the potential to replace sucrose, lactose, glucose or fructose in a wide range of food products. It can be metabolized independently of insulin, making it a viable choice for diabetic-friendly foods. *Leuconostoc pseudomesenteroides*, *Leuconostoc mesenteroides* and *Leuconostoc citreum* are notable for their ability to produce mannitol during the fermentation of fructose [3,6].

Leuconostoc ssp. has the capacity to produce exopolysaccharides (EPSs) as the result of environmental stress. EPSs are part of the biofilm layer surrounding the bacterial cells, together with proteins, nucleic acids and lipopolysaccharides. EPSs are categorized as homo- or hetero-polysaccharides, depending on whether they are made up of a single type of monosaccharide building block or several different types. Of the homo-polysaccharides, glucans and fructans are the most often discovered, and both are used as ingredients in the food industry. In the presence of sucrose, *Leuconostoc* ssp. produce α-glucans and β-fructans. The step-by-step addition of glucose moieties to an expanding α-glucan chain is the main mechanism. The glycosidic link in sucrose is first broken during the reaction, creating a covalent glucosyl-enzyme intermediate where the glucosyl is bonded in the donor subsite. The way the acceptor glucan targets the covalent glucosyl-enzyme intermediate dictates the kind of glycosidic bond that is formed [30]. For instance, the primary components of a new water-soluble dextran produced by *Leuconostoc citreum* SK24.002 are α-1,3 and α-1,6 linked d-glucopyranose units at a ratio of 4:5 [31]. In the case of *Leuconostoc citreum* E497 isolated from cereals, the dextran with 11% α-(1→2)-linked branches and 3.5% α-(1→3)-linked branches were produced. Additionally, *Leuconostoc citreum* NM105 may be an excellent prebiotic producer. To further support human health, food formulations may incorporate the natural form of this high-(1–2) branch dextran or the glucooligosaccharides form generated through partial acid hydrolysis. Furthermore, by use of an acceptor reaction with maltose and sucrose, the dextransucrase from *Leuconostoc citreum* NM105 has the potential to be employed in the synthesis of prebiotic oligosaccharides [31]. Moreover, in experimental media with added sucrose and maltose, *Leuconostoc citreum* HJ-P4, HJ-S13 strains were able to produce panose, while dextran synthesis was inhibited. In the meantime, after the isomaltooligosaccharide synthesis occurrence, sucrose releases an equivalent quantity of free fructose, imparting a sweet flavor [5]. It was reported that *Leuconostoc citreum* EFEL2700 demonstrated a typical hetero-type lactic acid fermentation and had the ability to metabolize a wide range of carbohydrates. The genome revealed genes involved in the metabolism of glucose, fructose, sucrose and mannose. Genes for the production of polysaccharides (dextran) from sucrose were also discovered [11].

In wheat flour sourdough supplemented with sucrose, *Leuconostoc citreum* FDR 421 was effective in sucrose conversion into glucose and fructose, while the last one was transformed to mannitol, resulting in the formation of acetic acid. The authors reported a smaller molar ratio of mannitol to acetate (close to 1 compared to 2:1), probably due to the strain's property of using oxygen as an electron acceptor during the specific conditions of the sourdough propagation, as well as due to the metabolic features of the strain. Fructose was not completely metabolized but the mannitol concentration was positively correlated with the fructose content. As was expected, maltose was also partially metabolized during sourdough fermentation. Sucrose addition facilitate dextran formation. Due to the presence of sucrose and maltose, which dextransucrase uses as an acceptor molecule to create maltooligosaccharides, a low dextran yield was recorded. The formation of maltoologisaccharides can complete dextran production in wheat substrates [22]. *Leuconostoc citreum* TR116 revealed similar behavior regarding glucose, fructose and maltose consumption during wheat sourdough fermentation and a high ability to produce mannitol [29]. *Leuconostoc citreum* DCM65 was also effective in mannitol and acetic acid production in wheat sourdough by fructose consumption [32].

When fava beans were inoculated with *Leuconostoc citreum* TR116 to obtain sourdough, glucose, fructose and sucrose were metabolized. Significant amounts of mannitol yielded. Using fructose as a substitute electron acceptor, mannitol dehydrogenase reduced NADH to mannitol. On the other hand, verbascose and stachyose/raffinose in fava bean sourdough showed a general declining tendency during the fermentation process, whereas galactose accumulated. These galactooligosaccharides are known as antinutritional compounds; thus, a little decrease is beneficial. It is also possible that this trend regarding galactooligosaccharides was due to the α-galactosidase activity of fava beans. Concerning EPS production, *Leuconostoc citreum* DSM 5577 revealed glucan synthesis in sucrose-supplemented fava bean sourdough [33].

2.2. Citrate Metabolism in Leuconostoc ssp. and Leuconostoc citreum

Leuconostocs like *lactococci* that use citrate as their primary energy source require a fermentable sugar as well. Citrate is metabolized by *Leuconostoc* via the same metabolic mechanism as citrate-positive strains of *L. lactis* [13].

The citrate metabolism pathway (Figure 2) includes the following stages: (a) citrate uptake facilitated by citrate permease; (b) citrate splitting into acetate and oxaloacetate via citrate lyase; (c) pyruvate formation by decarboxylating oxaloacetate; (d) pyruvate reduction to lactate. When citrate and sugar are co-metabolized, it supports growth by producing more energy (adenosine triphosphate (ATP)), through the acetate kinase pathway [10,13].

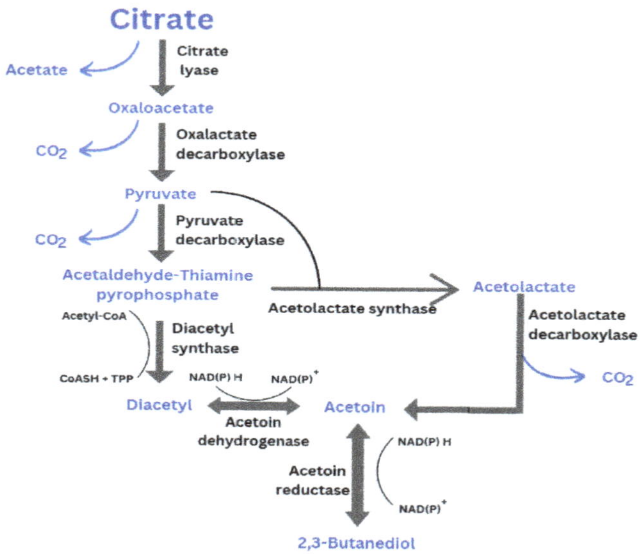

Figure 2. The citrate metabolism pathway in lactic acid bacteria. CoA—coenzyme A; CoASH—free coenzyme A; TPP—thiamine pyrophosphate. Adapted after Cogan and Jordan (1994) [10] and Endo et al. (2021) [13].

Leuconostoc has the ability to convert diacetyl into acetoin and 2,3-butanediol. Notably, citrate could have an important role in sourdough fermentation, providing volatile organic compounds with aromatic properties. Citrate is also metabolized by *Leuconostoc citreum*, which produces volatile chemicals, including acetoin, dyacetil and C4 compounds that contribute to the flavor of fermented products [6,11]. With a distinguished buttery, creamy odor, acetoin is produced by converting citrate to pyruvate, while its production is necessary to equilibrate the redox balance of cell metabolism. Since acetoin is an important aroma that is produced during sourdough fermentation and has a favorable link with wheat bread, larger concentrations of it are more likely to be well liked by customers.

It was reported that *Leuconostoc citreum* TR116 formed diacetyl in barley and oat malt wort during fermentation via acetolactate. But notably, the quantity of asparagine in fermented malt worts was correlated with the diacetyl concentration. Thus, diacetyl could come from oxaloacetate and fumarate, which are both precursors for pyruvate. 2-methylbutanol and 3-methylbutanol were also found in cereal wort fermented by *Leuconostoc citreum* TR116. An increase was reported in these aroma compounds when leucine and isoleucine were present in higher amounts in the wort. Ethylester, an ester with fruity, sweet sensory perceptions was also detected. It is formed when acetate reacts with the end products of ethanol fermentation [26].

3. Leuconostoc Citreum and Its Influence in Sourdough

Sourdough fermentation, a traditional biotechnological process, is a venerable practice profoundly rooted in the food business, acting as an early precursor to the revolutionary arrival of yeast fermentation in bread leavening [28,34]. Sourdough with wild lactic acid bacteria and naturally occurring yeasts has long been used to improve the texture, taste and shelf life of bread. This historic method is gaining popularity since it not only imparts specific organoleptic traits but also coincides with contemporary consumer expectations for cleaner, additive-free food items [10].

The influence of sourdough fermentation on wheat flour-based products is complex, with the acidic environment changing the structure of important dough elements such as arabinoxylans, starch and gluten [28,35]. The nutritional quality is influenced by sourdough fermentation, and this may result in changes to the levels of compounds, as well as the enhancement or inhibition of nutrient bioavailability (Figure 3) [36]. Lactic acid bacteria offers a wide variety of bread quality benefits, ranging from increased shelf life and texture to increased nutritional content and flavor [28,35]. Carbohydrates are absorbed during fermentation, which is vital for the growth of LAB and starch metabolization leading to a reduction in pH, which enhances organic acid production [37].

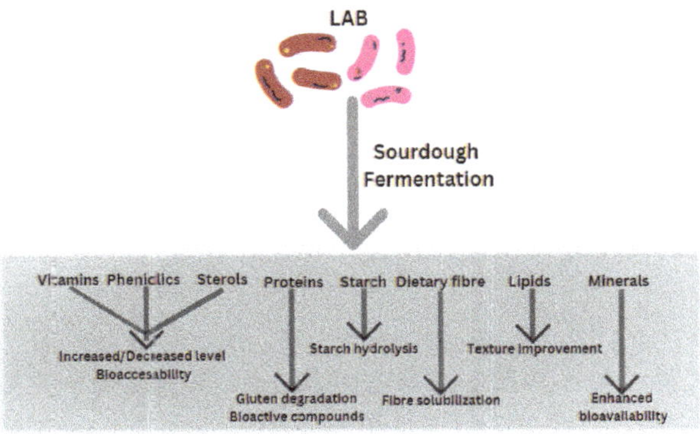

Figure 3. The impact of LAB during sourdough fermentation on nutrient effects. Adapted after Teleky et al., 2020) [38].

There is a rising tendency in the food industry to use *Leuconostoc citreum* strains as starter cultures, a phenomenon that has seen a growing presence in documented publications. This increased attention is due to their beneficial metabolic properties, including their capacity to create mannitol, exopolysaccharides (EPS) and other antimicrobial compounds. These advantageous properties have opened the door for the incorporation of *Leuconostoc citreum* strains into a wide range of food items. However, it is worth mentioning that one specific strain, *L. citreum* TR116, has received a lot of attention recently because of its extraordinary potential for functionalizing food and improving its quality and characteristics [7].

Leuconostoc citreum TR116 is famous for its distinct capacity to express the mannitol-2-dehydrogenase enzyme, which efficiently catalyzes the transformation of fructose into mannitol, as highlighted by Gänzle in 2015. What truly distinguishes *Leuconostoc citreum* TR116 from its competitors is its exceptional proficiency in consistently achieving a fructose-to-mannitol conversion rate that surpasses 80% across a wide spectrum of food matrices [15]. This exceptional attribute positions TR116 as a suitable choice for the creation of low-sugar food products. Sugar-reduced items are in significant demand due to a rising trend of non-communicable illnesses which include cardiovascular disease, obesity and type 2 diabetes [39]. However, lowering the amount of sugar in food items results in a loss of quality since, in addition to sweetness, sugar also affects texture, structure, microbiological shelf life and flavor [40]. In specific contexts, such as the production of quinoa-based milk alternatives, the utilization of TR116 in a single-strain fermentation process has yielded products enriched with improved nutritional value [41]. These products are characterized by a noteworthy reduction in sugar content, up to 40%, alongside a substantial reduction in glycemic response, a reduction of 35%, as evidenced by the research conducted by Jeske et al. in 2018.

In a study conducted by Rice et al. in 2020, the impact of *Leuconostoc citreum* TR116 was rigorously examined across various worts, including those derived from barley, oats and wheat. The study found that *Leuconostoc citreum* TR116 not only reduced sugar content but also produced fruity and brandy-like aroma compounds, enhancing the sensory experience. Additionally, *Leuconostoc citreum* TR116 was used in other studies to improve the quality of sugar-reduced sweet baked goods, highlighting its versatility in sugar reduction and flavor enhancement in various food applications [26]. Among the specific aroma compounds formed by this strain are ethylacetate, diacetyl, acetaldehyde, 2–3 butanediol.

In the case of burger buns, a substantial 50% reduction in sugar content was achieved, while retaining characteristics (reduced volume, denser crumb, prolonged shelf life, a brown crust and sweetness) comparable to those of full-sugar burger buns, as demonstrated in the study by Sahin et al. in 2019. The authors discovered that incorporating fermented sourdough with *Leuconostoc citreum* TR116, known for mannitol production, showed promising results in preserving these quality characteristics. Moreover, they suggested that increasing mannitol production could be achieved through continuous-flow sourdough fermentation by introducing more fructose into the system over time [6]. Similarly, Sahin et al. (2019) [42] demonstrated that using sourdough fermented with *Leuconostoc citreum* TR116 in cake formulations resulted in no discernible losses in sensory qualities or techno-functional features despite a 50% drop in sugar content. Furthermore, while focused on low-sugar sourdough cookies with sugar levels as low as 5%, Sahin et al., in 2019 [25], discovered that these goods retained rich flavor, probably due to the incorporation of acids and free amino acids, which supported the Maillard reaction and the subsequent flavor compounds. Beyond its role in sugar reduction, *Leuconostoc citreum* TR116 has been used to improve bread quality, notably through the addition of fermented high-protein fava bean flour, as Hoehnel et al. investigated in 2020 [24]. Moreover, in this specific fermented matrix, the authors report the presence of 15 phenolic compounds with antifungal activity that may contribute to the extension of the shelf life.

Other strains of *Leuconostoc citreum* were also tested and promising results were obtained. For instance, inoculation of *Leuconostoc citreum* DCM65 in wheat flour resulted in high mannitol production, and a positive correlation between mannitol and acetic acid amounts was also found. Moreover, this strain was able to counterbalance the negative effects of sugar reduction in soft buns due to its capacity to produce both mannitol and EPS [32].

Table 1 summarizes the more effective strains of *Leuconostoc citreum* in different sourdough matrices used for sugar-reduced bakery products and baked goods. The main products of fermentation and the effects on the products' quality characteristics are presented also.

Table 1. Influence of sourdough fermented with different *Leuconostoc citreum* strains on product quality.

Strain	Type of Product, Ingredients	Sourdough Fermentation Parameters	Products of Fermentation	Effects on the Product Quality
L. citreum DCM65 [32]	**Soft buns**: flour, tap water, sugar (0, 3, 6 or 9% addition), canola oil, salt, yeast, wheat gluten and sourdough (made of flour, tap water, sucrose, fructose and DCM65)	30 °C/24 h pH = 4	Mannitol—9.8 mg/g EPS—dextran Acetic acid—2.5 mg/g Lactic acid—3.1 mg/g	- Increased shelf life (visible molds only after 12–16 days) - Smaller slice area and finer pore structure than the reference buns - Higher pore cells and positive correlation with mannitol and acetic acid - Smaller volume of *L. citreum* DCM65 buns compared to reference buns, in inverse relationship with mannitol and acetic acid contents - EPS decreased the crumb firmness
L. citreum TR116 [6]	**Burger buns**: wheat flour, yeast, sucrose, salt, wheat gluten, wheat starch, ascorbic acid, sunflower oil, sodium stearoyl lactate, tap water and sourdough (made of: SD1: 50% flour, 50% tap water; SD2: 40% flour, 50% tap water, 10% fructose)	30 °C/48 h pH = 4.2	Maltose ~ * 4.95 g/100 g DM (SD1) ~2.62 g/100 g DM (SD2) Mannitol Acetic acid ~2.62 g/100 g DM (SD1) ~0.52 g/100 g DM (SD2) Lactic acid ~2.87 g/100 g DM (SD1) ~1.70 g/100 g DM (SD2)	- Specific volume, crumb hardness and chewiness revealed no significant differences between sugar-reduced sourdough burger buns and the control - No significant differences in sweetness and sourness between sugar-reduced buns and control
L. citreum TR116 [25]	**Biscuits**: flour with protein content of 8% and a moisture of 14%, sucrose, shortening, sodium stearoyl lactylate, salt, baking powder and sterile tap water (SD: 50% flour, 50% tap water. SD_{FRU}: 40% flour, 10% fructose, 50% tap water; SD_{FS}: 40% flour, 5% fructose, 5% sucrose, 50% tap water)	30 °C/30 h pH 5.41–5.47	Mannitol ~0.04 g/100 g (SD) ~0.48 g/100 g (SD_{FRU}) ~0.41 g/100 g (SD_{FS}) EPS ~0.71 g/kg (SD_{FS}) Maltose Acetic acid ~0.204 g/100 g (SD_{FS}) ~0.195 g/100 g (SD_{FRU}) Lactic acid ~0.112 g/100 g (SD) ~0.076 g/100 g (SD_{FRU}, SD_{FS})	- Improvement of the the viscoelastic properties, dough stickiness, hardness biscuit firmness improvement - Increased sweetness and flavor intensity - Improved biscuit color
L. citreum TR116 [42]	**Cakes**: flour, sterile tap water, sugar, canola oil, salt, yeast, wheat gluten, sourdough (SD: 50% sterile tap water, 50% flour;SD_{FRU}: 50% sterile tap water, 40% flour, 10% fructose)	30 °C/30 h	Mannitol ~0.70 g/100 g DM Acetic acid Lactic acid	- Improved the pasting properties - Similar specific volume to the full-sugar control - Significantly softer crumb - Increased sweetness perception (+93%), aroma (+30%) and flavor (+25.5%)
L. citreum C2.27 [43]	**Bread**: wheat flour, sterile tap water and liquid sourdough (made of wheat flour, sterile tap water and bacterial suspension)	30 °C/16 h pH 4.48	Acetic acid ~4.30 mmol/kg Lactic acid ~15.52 mmol/kg	- Increased specific volume - Decreased resilience and fracturability - Lower hardness of bread - Lighter crust
L. citreum FDR241 [22]	**Bread sourdough**: wheat flour, water, sucrose	30 °C/24 h pH 4	Mannitol ~2.15 mmol/kg Maltose ~3.70% Fructose ~1.73% Glucose ~0.18% Acetic acid 1.65% Lactic acid 3.38% EPS—dextran ~0.71%	- EPS improved structure and volume - Increased softness - Delay in staling
L. citreum TR116 [24]	**Bread**: Wheat flour, fava bean ingredients (dehulled flour, DF, and high-protein flour, PR), baker's yeast, NaCl, oil, water	30 °C/24 h pH 5.47	Mannitol ~0.63–0.77% DM Raffinose ~0.84–0.93% DM Verbascose ~0.49–0.54% DM Galactose ~0.75–0.99 % DM Acetic acid ~0.84–1.14 % DM Lactic acid ~1.23–1.33% DM	- Improved gluten quality - Increased volume of bread - Decreased crumb hardness - Crumb structure with a slight tendency towards bigger cells - Lower crust lightness

*—Approximately; DM—dry matter; SD1—first type of sourdough; SD2—second type of sourdough; SD—sourdough; SD_{FRU}—sourdough with fructose; SD_{FS}—sourdough with fructose and sucrose.

Leuconostoc citreum C2.27 isolated from durum wheat semolina was used to obtain liquid sourdoughs and a typical Apulian bread (puccia). Results showed an increased volume and an optimal fermentation quotient compared with *Weissella confusa* C5.7, which was also studied during the experiment. In addition, better textural properties and a softer and more elastic crumb in bread were recorded in the case of *Leuconostoc citreum* C2.27. Notably, when salt was omitted from the bread formulation, any distinct changes in the structural characteristics between the puccia bread made with sourdough and with baker's yeast were not identified. The bread had a strong flavor and taste even without salt. These results recommend this strain for even yeast-free bread production [43].

When *Leuconostoc citreum* FDR241 isolated from rye bran was used for wheat sourdough fermentation, the sourdoughs attained pH values of 4.0 and moderate titratable acidity, exhibiting consistent lactic acid bacterial cell density and acidification characteristics. Throughout the propagation process, the amount of carbohydrates consumed was constant, resulting in the synthesis of mannitol and almost equimolar amounts of lactic and acetic acid. The mannitol amount was correlated with the estimated fructose resulting from back-slopping and sucrose addition. The addition of sucrose facilitated the production of oligosaccharides and dextran, which increased the sourdough's viscosity by 2 to 2.6 times [22].

Mannitol as Sugar Substitute

In bakery products, naturally produced polyols can contribute to sweetness and flavor, while EPS acts as a bulking agent that enhances dough stability and the texture of the final product [42].

An innovative technical strategy for overcoming quality losses in low-sugar baked goods might be a controlled sourdough fermentation with selected LAB and yeast strains [42]. In a sourdough system, the in situ generation of polyols and exopolysaccharides is required to convert monosaccharides into low-calorie sweetening agents (polyols) and to lower the quantities of monosaccharides by polymerization into long-chain carbohydrates (EPS) [6].

Polyols are also known as "sugar alcohols", "sugar substitutes" and "hydrogenated carbohydrates". The name "polyol" comes from the word "polyalcohol" or "polyhydric alcohol". The suggested names are "polyols" or "hydrogenated carbohydrates", which specifically emphasize that these compounds are carbohydrates [44]. They are sugar-free sweeteners because they include carbs but no sugars [45]. Polyols are sorbitol, maltitol, lactitol, xylitol, erythritol, mannitol and isomalt. D-mannitol is the most prevalent polyol in nature, due to its production by a variety of species including bacteria, yeasts, fungi, algae, lichens and various plants. It is thought that a French chemist discovered this polyol in the early 19th century [46].

$$\mathrm{D-Glucose} \xrightarrow{\text{glucose isomerase}} \mathrm{D-Fructose} \xrightarrow[\text{MDH}]{\text{NADH} \curvearrowright \text{NAD+}} \mathrm{D-Mannitol} \tag{4}$$

Production of mannitol from glucose in two steps using enzymes originating from hyperthermophilic bacteria (glucose isomerase and mannitol dehydrogenase (MDH)) (image created using BioRender application https://app.biorender.com, accessed on 30 October 2023, redone after Martínez-Miranda et al., 2022) [46].

Polyols, commonly referred to as sugar alcohols, are a family of chemical compounds formed by a multidimensional process that involves either the chemical or biological reduction of sugars or, alternatively, fermentation facilitated by the presence of LAB or yeast [42].

In a 2018 study, Sahin et al. discovered that substituting sugar with mannitol in biscuits increased dough hardness when compared to the control. This might cause difficulties in rolling and shaping the dough, influencing the baking process, and perhaps resulting in restricted expansion. Mannitol used as sugar alternative helps to improve dough stability. The solubility of the component influences dough stability. The greater the solubility, the more solvent is present in the solution, resulting in a drop in dough viscosity and poorer stability after heating. Mannitol has a smaller influence on dough viscosity than sucrose or xylitol since its solubility is substantially lower. This increases its stability. As more gas cells remain in the dough, the specific volume of the final product increases. Mannitol, on the other hand, diminishes the flexibility, cohesiveness and chewiness of the cake crumb and adds weak sensory characteristics such as dryness [29].

Because of their bulking qualities and functions, several polyols such as erythritol, xylitol, sorbitol and mannitol, as well as other oligosaccharides, have been developed as alternative sweeteners for food products. Mannitol, in particular, has a wide range of applications outside of the food industry, including medicines, where it is used in the manufacture of chewable tablets and granulated powders. Mannitol is commercially manufactured by reducing fructose or sucrose under high pressure and temperature conditions, with sorbitol created as a by-product [47].

The utilization of mannitol, a sugar substitute, can help address the need for functional ingredients that mimic the rheological characteristics normally offered by sucrose in wheat dough systems. Natural mannitol production by LAB in sourdough is a critical component in the reduction in sugar content for baked goods, allowing for the incorporation of beneficial aspects associated with this sugar substitute. Research on sourdough has indicated an enhancement in bread quality, as discussed by Axel et al. (2015) and Arend et al. (2007) [48,49], and an amplification of flavor and aroma [50]. *Leuconostoc* ssp., among all mannitol-producing LAB, can reduce fructose to mannitol by up to 95% [51].

In general, mannitol can specifically be synthesized from glucose/fructose combinations without producing sorbitol as a by-product, eliminating the need for refined substrates or difficult product purification steps [52]. Under proper oxygen availability circumstances, glucose may be utilized as an energy and carbon source and fructose as an electron acceptor since it can be converted to mannitol by a particular mannitol dehydrogenase. If sugar absorption occurs simultaneously, a bacterium can create up to 2 mol mannitol from 1 mol glucose (and 2 mol fructose). Other metabolic products will include lactic acid, acetic acid and/or ethanol and carbon dioxide [52].

4. Conclusions

Leuconostoc citreum is a lactic acid bacterium that may perform lactic fermentation and operate in sourdough to generate mannitol, a polyol that helps in reducing sugar in baked goods.

This review has highlighted the promising potential of *Leuconostoc citreum* in mannitol production and its significant role in reducing sugar levels, while improving quality, nutritional profiles and sensory qualities, particularly when applied in sourdough for baked goods. Its efficiency in mannitol production, with yields of 70–90%, very good capacity to produce antifungal compounds (acids, ethanol, phenolic compounds, etc.) that could extend shelf life, exopolysaccharide formation with a beneficial role in product texture and specific aroma compounds formation with a positive role for product flavor and taste are among the important features that recommend *Leuconostoc citreum*'s strains as sourdough fermenting agents in some bakery products.

This novel technique fulfills the growing demand for low-sugar baked goods, appealing to health-conscious customers as well as those wanting great taste and quality.

Author Contributions: Conceptualization, M.-F.R. and A.P. (Adriana Păucean); methodology, C.R.P. and S.M.M.; validation, M.S.C., A.P. (Anamaria Pop) and A.C.F.; resources, M.-F.R. and M.S.C.; writing—original draft preparation, M.-F.R. and A.P. (Adriana Păucean); writing—review and editing, A.P. (Adriana Păucean) and S.M.M.; visualization, A.C.F.; supervision, C.R.P. and A.P. (Anamaria Pop). All authors have read and agreed to the published version of the manuscript.

Funding: This research was funded by National Research Development Projects to finance excellence (PFE)-14/2022-2024 granted by the Romanian Ministry of Research and Innovation.

Institutional Review Board Statement: Not applicable.

Informed Consent Statement: Not applicable.

Data Availability Statement: No new data were created or analyzed in this study.

Conflicts of Interest: The authors declare no conflicts of interest.

References

1. Thunell, R.K. Taxonomy of the Leuconostocs. *J. Dairy Sci.* **1995**, *78*, 2514–2522. [CrossRef]
2. Ogier, J.C.; Casalta, E.; Farrokh, C.; Saïhi, A. Safety assessment of dairy microorganisms: The Leuconostoc genus. *Int. J. Food Microbiol.* **2008**, *126*, 286–290. [CrossRef] [PubMed]
3. Hemme, D.; Foucaud-Scheunemann, C. Leuconostoc, characteristics, use in dairy technology and prospects in functional foods. *Int. Dairy J.* **2004**, *14*, 467–494. [CrossRef]
4. Eom, H.J.; Seo, D.M.; Han, N.S. Selection of psychrotrophic *Leuconostoc* spp. producing highly active dextransucrase from lactate fermented vegetables. *Int. J. Food Microbiol.* **2007**, *117*, 61–67. [CrossRef] [PubMed]
5. Sahin, A.W.; Rice, T.; Zannini, E.; Axel, C.; Coffey, A.; Lynch, K.M.; Arendt, E.K. Leuconostoc citreum TR116: In-situ production of mannitol in sourdough and its application to reduce sugar in burger buns. *Int. J. Food Microbiol.* **2019**, *302*, 80–89. [CrossRef] [PubMed]
6. Sahin, A.W.; Rice, T.; Coffey, A. Genomic analysis of Leuconostoc citreum TR116 with metabolic reconstruction and the effects of fructose on gene expression for mannitol production. *Int. J. Food Microbiol.* **2021**, *354*, 109327. [CrossRef] [PubMed]
7. Farrow, J.A.E.; Facklam, R.R.; Collins, M.D. Nucleic acid homologies of some vancomycin-resistant leuconostocs and description of Leuconostoc citreum sp. nov. and Leuconostoc pseudomesenteroides sp. nov. *Int. J. Syst. Bacteriol.* **1989**, *39*, 279–283. [CrossRef]
8. Yeon, S.; Kim, S.; Jang, Y.; Seo, S.; Soo, N. Screening of endogenous strong promoters of Leuconostoc citreum EFEL2700 based on transcriptome analysis and its application for food-grade production of β-galactosidase. *J. Biotechnol.* **2021**, *325*, 1–6. [CrossRef]
9. Swain, M.R.; Anandharaj, M.; Ray, R.C.; Rani, R.P. Fermented Fruits and Vegetables of Asia: A Potential Source of Probiotics. *Biotechnol. Res. Int.* **2014**, *2014*, 250424. [CrossRef]
10. Cogan, T.M.; Jordan, K.N. Metabolism of Leuconostoc Bacteria. *J. Dairy Sci.* **1994**, *77*, 2704–2717. [CrossRef]
11. Kim, S.A.; Jang, Y.J.; Heo, J.E.; Li, L.; Moon, J.S.; Han, N.S. Complete genome sequence of Leuconostoc citreum EFEL2700, a host strain for transformation of pCB vectors. *J. Biotechnol.* **2018**, *287*, 52–58. [CrossRef] [PubMed]
12. Miao, M.; Bai, A.; Jiang, B.; Song, Y.; Cui, S.W.; Zhang, T. Characterisation of a novel water-soluble polysaccharide from Leuconostoc citreum SK24.002. *Food Hydrocoll.* **2014**, *36*, 265–272. [CrossRef]
13. Endo, A.; Maeno, S.; Liu, S.Q. *Lactic Acid Bacteria: Leuconostoc spp.*; Elsevier: Amsterdam, The Netherlands, 2021; Volume 4.
14. Sese, S. Assessment of safety characteristics, postbiotic potential, and technological stress response of Leuconostoc strains from different origins for their use in the production of functional dairy foods. *LWT* **2022**, *165*, 113722. [CrossRef]
15. Gänzle, M.G. Lactic metabolism revisited: Metabolism of lactic acid bacteria in food fermentations and food spoilage. *Curr. Opin. Food Sci.* **2015**, *2*, 106–117. [CrossRef]
16. Sahin, A.W.; Zannini, E.; Coffey, A.; Arendt, E.K. Sugar reduction in bakery products: Current strategies and sourdough technology as a potential novel approach. *Food Res. Int.* **2019**, *126*, 108583. [CrossRef] [PubMed]
17. Luo, X.; Arcot, J.; Gill, T.; Louie, J.C.Y.; Rangan, A. A review of food reformulation of baked products to reduce added sugar intake. *Trends Food Sci. Technol.* **2019**, *86*, 412–425. [CrossRef]
18. Stiles, M.E.; Holzapfel, W.H. Lactic acid bacteria of foods and their current taxonomy. *Int. J. Food Microbiol.* **1997**, *36*, 1–29. [CrossRef]
19. Yan, P. Effect of inoculating lactic acid bacteria starter cultures on the nitrite concentration of fermenting Chinese paocai. *Food Control* **2008**, *19*, 50–55. [CrossRef]
20. Nam, Y.; Chang, H.; Kim, K.; Woon, S.; Bae, J. International Journal of Food Microbiology Metatranscriptome analysis of lactic acid bacteria during kimchi fermentation with genome-probing microarrays. *Int. J. Food Microbiol.* **2009**, *130*, 140–146. [CrossRef]
21. Choi, I.K.; Jung, S.H.; Kim, B.J.; Park, S.Y.; Kim, J.; Han, H.U. Novel Leuconostoc citreum starter culture system for the fermentation of kimchi, a fermented cabbage product. *Antonie Van Leeuwenhoek Int. J. Gen. Mol. Microbiol.* **2003**, *84*, 247–253. [CrossRef]
22. Coda, R.; Xu, Y.; Moreno, D.S.; Mojzita, D.; Nionelli, L.; Rizzello, C.G.; Katina, K. Performance of Leuconostoc citreum FDR241 during wheat flour sourdough type I propagation and transcriptional analysis of exopolysaccharides biosynthesis genes. *Food Microbiol.* **2018**, *76*, 164–172. [CrossRef] [PubMed]
23. Oshiro, M.; Zendo, T.; Nakayama, J. Diversity and dynamics of sourdough lactic acid bacteriota created by a slow food fermentation system. *J. Biosci. Bioeng.* **2021**, *131*, 333–340. [CrossRef] [PubMed]
24. Hoehnel, A.; Bez, J.; Sahin, A.W.; Coffey, A.; Arendt, E.K.; Zannini, E. Leuconostoc citreum TR116 as a Microbial Cell Factory to Functionalise High-Protein Faba Bean Ingredients for Bakery Applications. *Foods* **2020**, *9*, 1706. [CrossRef] [PubMed]
25. Sahin, A.W.; Rice, T.; Zannini, E.; Lynch, K.M.; Coffey, A.; Arendt, E.K. The incorporation of sourdough in sugar-reduced biscuits: A promising strategy to improve techno-functional and sensory properties. *Eur. Food Res. Technol.* **2019**, *245*, 1841–1854. [CrossRef]

26. Rice, T.; Sahin, A.W.; Heitmann, M.; Lynch, K.M.; Jacob, F.; Arendt, E.K.; Coffey, A. Application of mannitol producing Leuconostoc citreum TR116 to reduce sugar content of barley, oat and wheat malt-based worts. *Food Microbiol.* **2020**, *90*, 103464. [CrossRef] [PubMed]
27. Cicotello, J.; Wolf, I.V.; D'Angelo, L.; Guglielmotti, D.M.; Quiberoni, A.; Suárez, V.B. Response of Leuconostoc strains against technological stress factors: Growth performance and volatile profiles. *Food Microbiol.* **2018**, *73*, 362–370. [CrossRef] [PubMed]
28. Teleky, B.E.; Martău, G.A.; Vodnar, D.C. Physicochemical effects of lactobacillus plantarum and lactobacillus casei cocultures on soy–wheat flour dough fermentation. *Foods* **2020**, *9*, 1894. [CrossRef]
29. Sahin, A.W.; Axel, C.; Zannini, E.; Arendt, E.K. Xylitol, mannitol and maltitol as potential sucrose replacers in burger buns. *Food Funct.* **2018**, *9*, 2201–2212. [CrossRef]
30. Leemhuis, H.; Pijning, T.; Dobruchowska, J.M.; van Leeuwen, S.S.; Kralj, S.; Dijkstra, B.W.; Dijkhuizen, L. Glucansucrases: Three-dimensional structures, reactions, mechanism, α-glucan analysis and their implications in biotechnology and food applications. *J. Biotechnol.* **2013**, *163*, 250–272. [CrossRef]
31. Yang, Y.; Peng, Q.; Guo, Y.; Han, Y.; Xiao, H.; Zhou, Z. Isolation and characterization of dextran produced by Leuconostoc citreum NM105 from manchurian sauerkraut. *Carbohydr. Polym.* **2015**, *133*, 365–372. [CrossRef]
32. Müller, D.C.; Schipali, S.; Näf, P.; Kinner, M.; Schwenninger, S.M.; Schönlechner, R. Potential of a Techno-Functional Sourdough and Its Application in Sugar-Reduced Soft Buns. *Fermentation* **2022**, *8*, 42. [CrossRef]
33. Wang, Y.; Sorvali, P.; Laitila, A.; Maina, N.H.; Coda, R.; Katina, K. Dextran produced in situ as a tool to improve the quality of wheat-faba bean composite bread. *Food Hydrocoll.* **2018**, *84*, 396–405. [CrossRef]
34. Angioloni, A.; Romani, S.; Pinnavaia, G.G.; Rosa, M.D. Characteristics of bread making doughs: Influence of sourdough fermentation on the fundamental rheological properties. *Eur. Food Res. Technol.* **2006**, *222*, 54–57. [CrossRef]
35. Zhang, Y.; Hong, T.; Yu, W.; Yang, N.; Jin, Z.; Xu, X. Structural, thermal and rheological properties of gluten dough: Comparative changes by dextran, weak acidification and their combination. *Food Chem.* **2020**, *330*, 127154. [CrossRef] [PubMed]
36. Poutanen, K.; Flander, L.; Katina, K. Sourdough and cereal fermentation in a nutritional perspective. *Food Microbiol.* **2009**, *26*, 693–699. [CrossRef] [PubMed]
37. Chiş, M.S.; Păucean, A.; Man, S.M.; Vodnar, D.C.; Teleky, B.E.; Pop, C.R.; Stan, L.; Borsai, O.; Kadar, C.B.; Urcan, A.C.; et al. Quinoa sourdough fermented with lactobacillus plantarum ATCC 8014 designed for gluten-free muffins—A powerful tool to enhance bioactive compounds. *Appl. Sci.* **2020**, *10*, 7140. [CrossRef]
38. Teleky, B.E.; Martău, A.G.; Ranga, F.; Cheţan, F.; Vodnar, D.C. Exploitation of lactic acid bacteria and Baker's yeast as single or multiple starter cultures of wheat flour dough enriched with soy flour. *Biomolecules* **2020**, *10*, 778. [CrossRef]
39. Clemens, R.A.; Jones, J.M.; Kern, M.; Lee, S.Y.; Mayhew, E.J.; Slavin, J.L.; Zivanovic, S. Functionality of Sugars in Foods and Health. *Compr. Rev. Food Sci Food Saf.* **2016**, *15*, 433–470. [CrossRef]
40. Sahin, A.W.; Axel, C.; Arendt, E.K. Understanding the function of sugar in burger buns: A fundamental study. *Eur. Food Res. Technol.* **2017**, *243*, 1905–1915. [CrossRef]
41. Jeske, S.; Zannini, E.; Lynch, K.M.; Coffey, A.; Arendt, E.K. Polyol-producing lactic acid bacteria isolated from sourdough and their application to reduce sugar in a quinoa-based milk substitute. *Int. J. Food Microbiol.* **2018**, *286*, 31–36. [CrossRef]
42. Sahin, A.W.; Rice, T.; Zannini, E.; Lynch, K.M.; Coffey, A.; Arendt, E.K. Sourdough technology as a novel approach to overcome quality losses in sugar-reduced cakes. *Food Funct.* **2019**, *10*, 4985–4997. [CrossRef] [PubMed]
43. De Bellis, P.; Rizzello, C.G.; Sisto, A.; Valerio, F.; Lonigro, S.L.; Conte, A.; Lorusso, V.; Lavermicocca, P. Use of a selected leuconostoc citreum strain as a starter for making a "yeast-free" bread. *Foods* **2019**, *8*, 70. [CrossRef] [PubMed]
44. Livesey, G. Health potential of polyols as sugar replacers, with emphasis on low glycaemic properties. *Nutr. Res. Rev.* **2003**, *16*, 163–191. [CrossRef] [PubMed]
45. Bhise, S. Polyols to improve quality and shelf life of baked products: A review. *Int. J. Adv. Sci. Tech. Res. Issue* **2013**, *3*, 262–272.
46. Martínez-Miranda, J.G.; Chairez, I.; Durán-Páramo, E. Mannitol Production by Heterofermentative Lactic Acid Bacteria: A Review. *Appl. Biochem. Biotechnol.* **2022**, *194*, 2762–2795. [CrossRef] [PubMed]
47. Yun, J.W.; Kim, D.H. A comparative study of mannitol production by two lactic acid bacteria. *J. Ferment. Bioeng.* **1998**, *85*, 203–208. [CrossRef]
48. Axel, C.; Röcker, B.; Brosnan, B.; Zannini, E.; Furey, A.; Coffey, A.; Arendt, E.K. Application of Lactobacillus amylovorus DSM19280 in gluten-free sourdough bread to improve the microbial shelf life. *Food Microbiol.* **2015**, *47*, 36–44. [CrossRef]
49. Arendt, E.K.; Ryan, L.A.M.; Dal Bello, F. Impact of sourdough on the texture of bread. *Food Microbiol.* **2007**, *24*, 165–174. [CrossRef]
50. Montanari, C.; Bargossi, E.; Lanciotti, R.; Chinnici, F.; Gardini, F.; Tabanelli, G. Effects of two different sourdoughs on the characteristics of Pandoro, a typical Italian sweet leavened baked good. *LWT* **2014**, *59*, 289–299. [CrossRef]
51. Otgonbayar, G.E.; Eom, H.J.; Kim, B.S.; Ko, J.H.; Han, N.S. Mannitol production by leuconostoc citreum kacc 91348p isolated from kimchi. *J. Microbiol. Biotechnol.* **2011**, *21*, 968–971. [CrossRef]
52. Carvalheiro, F.; Moniz, P.; Duarte, L.C.; Esteves, M.P.; Gírio, F.M. Mannitol production by lactic acid bacteria grown in supplemented carob syrup. *J. Ind. Microbiol. Biotechnol.* **2011**, *38*, 221–227. [CrossRef] [PubMed]

foods

MDPI

Review

Microbial Fermentation Processes of Lactic Acid: Challenges, Solutions, and Future Prospects

Yueying Huang [1], Yu Wang [1], Nan Shang [2] and Pinglan Li [1,*]

[1] Key Laboratory of Functional Dairy, College of Food Science and Nutritional Engineering, China Agricultural University, Beijing 100083, China
[2] College of Engineering, China Agricultural University, Beijing 100083, China
* Correspondence: lipinglan@cau.edu.cn

Abstract: The demand for lactic acid and lactic acid-derived products in the food, pharmaceutical, and cosmetic industries is increasing year by year. In recent decades, the synthesis of lactic acid by microbials has gained much attention from scientists due to the superior optical purity of the product, its low production costs, and its higher production efficiency compared to chemical synthesis. Microbial fermentation involves the selection of feedstock, strains, and fermentation modes. Each step can potentially affect the yield and purity of the final product. Therefore, there are still many critical challenges in lactic acid production. The costs of feedstocks and energy; the inhibition of substrates and end-product; the sensitivity to the inhibitory compounds released during pretreatment; and the lower optical purity are the main obstacles hindering the fermentation of lactic acid. This review highlights the limitations and challenges of applying microbial fermentation in lactic acid production. In addition, corresponding solutions to these difficulties are summarized in order to provide some guidance for the industrial production of lactic acid.

Keywords: lactic acid production; microbial fermentation; production costs; difficulties in production process

1. Introduction

Lactic acid (LA) is an essential organic acid with several applications in food, pharmaceuticals, and cosmetics [1,2]. Furthermore, LA is the monomer for synthesizing polylactic acid (PLA) plastic, a bioabsorbable and compostable plastic. PLA plastic has been considered one of the solutions which can be used to alleviate the "white pollution" caused by the massive use of petroleum-based plastics. It is an excellent substitute for conventional plastics in various applications, such as food packaging, field vegetable mulch, and surgical sutures. The widespread use of LA has led to a surge in its demand (Figure 1). The total demand for LA is expected to reach 19.601 million tons by 2025, which is estimated to be worth around USD 9.8 billion in the international market [1,3].

LA exists in two optical configurations: L-LA and D-LA. D-LA cannot be metabolized by humans and may even cause poisoning. DL-LA is a mixture of D-LA and L-LA. It is not only unavailable for food and pharmaceuticals, but also does not meet the requirements for the production of PLA plastics. Therefore, the pure isomer (L-LA or D-LA) is more valuable for utilization than the DL-LA [4,5]. LA can be synthesized by biological or chemical methods. The most common method of chemical synthesis is the hydrolysis of lactonitrile with strong acids, while biosynthesis uses microbial fermentation techniques. Compared with chemical synthesis, the synthesis of LA via microbial fermentation is gaining increasing attention from researchers due to advantages such as the low cost, high efficiency, environmental friendliness, and high optical purity of the product [6]. At present, the common microorganisms used to ferment LA mainly include *Lactobacillus*, *Bacillus*, *Escherichia coli*, *Rhizopus oryzae*, *Aspergillus niger*, *Saccharomyces cerevisiae*, *Lactobacillus cruzi*, etc. [7]. Although LA produced by microbial fermentation is superior to that produced through chemical synthesis, several challenges still remain (Figure 2). In terms of cost, the feedstocks initially used for LA fermentation were refined sugars or starches, which are more expensive. After that, to promote sustainability, agricultural by-products became the new substrate choices for LA fermentation. Agricultural by-products, such as straw, rice husk, bran, etc., are rich in cellulose and can be hydrolyzed into microbially available sugar to further produce high value-added products, such as LA. However, it was found that the production of inhibitory compounds after the pretreatment of these materials and the production of mixed sugars that the strains could not fully utilize caused a decrease in LA production [8,9]. There are also a number of hindering factors in the fermentation process, such as high initial sugar concentration, which inhibits the growth of the strain [10]. The accumulation of LA in the late stage of fermentation has a toxic effect on microbial cells [11]. By-products are produced and the optical purity of LA is reduced due to changes in the cultural environment and culture conditions [12]. Therefore, to meet the growing demand for LA, there is a requirement to overcome the hindrances to LA fermentation and develop lower-cost and more efficient technologies to produce LA with high optical purity. The main objective of this review is to discuss the current challenges faced by the microbial production of LA and to introduce possible solutions to overcome these challenges, with the aim of providing some guidance for the industrial production of LA.

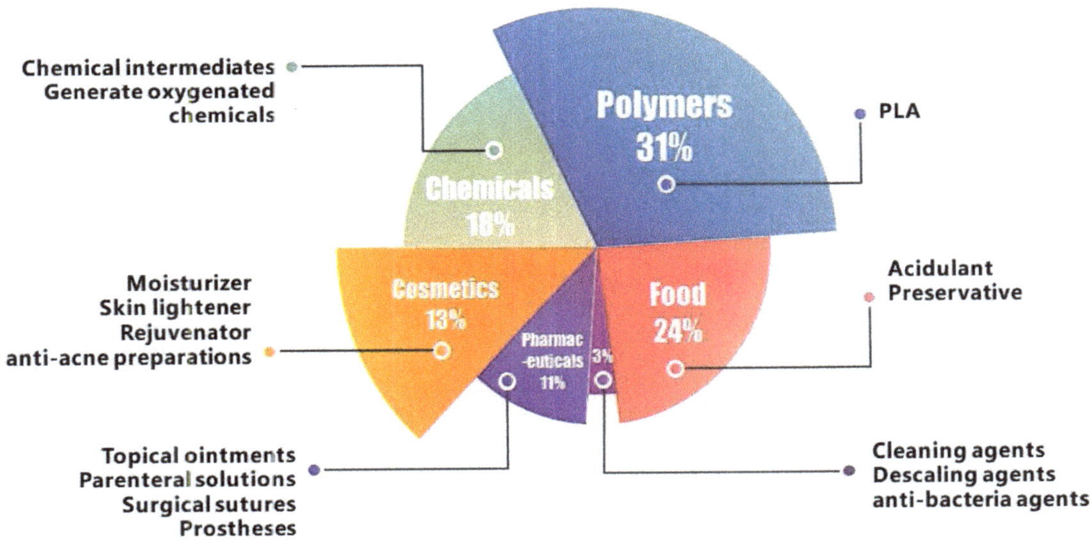

Figure 1. The applications of and demand for lactic acid in different industries.

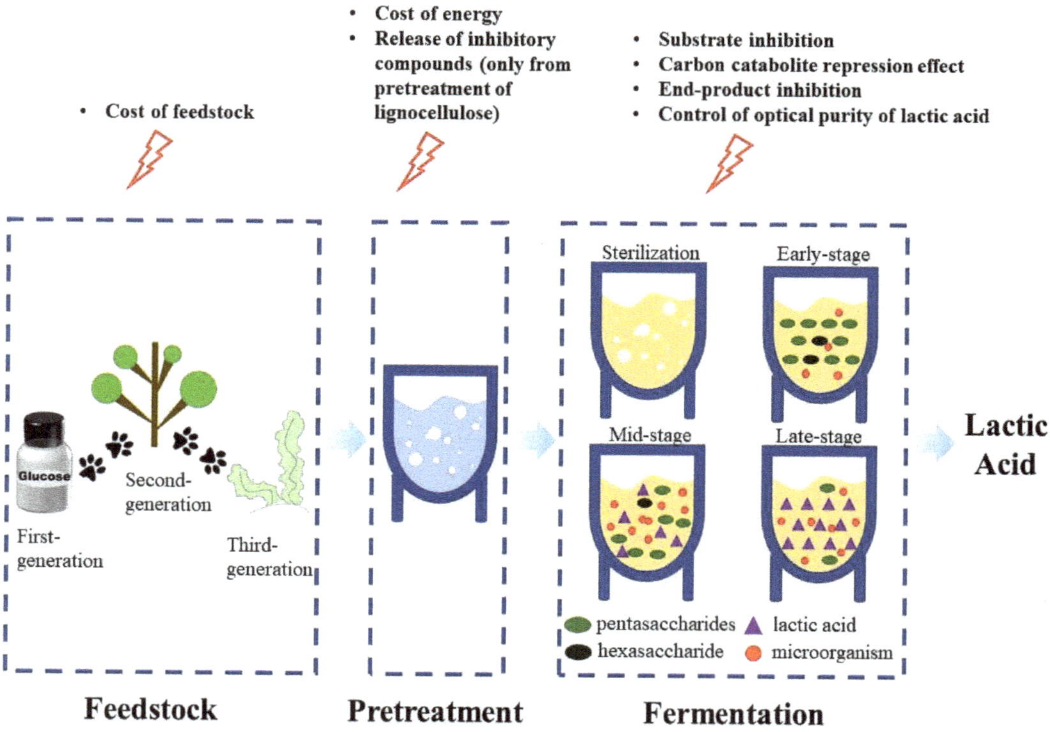

Figure 2. Current challenges in microbial production of lactic acid.

2. Ways to Reduce the Costs of LA Fermentation

2.1. Inexpensive Feedstock Application for Fermentation

In LA production, feedstock costs usually account for 40–70% of the total costs [13]. To minimize the costs, carbon feedstock that can be fermented to yield LA has been investigated in various studies. The development of feedstock has progressed through three stages. The first generation was feedstock made with refined sugar or starch crops; second-generation feedstock consisted of lignocellulose biomass or other waste resources derived from food production and daily life; and third-generation feedstock comprises microalgae or macroalgae taken from the ocean, which is easily cultured in captivity. First-generation feedstocks are unsuitable for mass production of LA due to their competitive relationship with human food and their high costs [4]. The second-generation feedstocks are the most widely researched; these include by-products of agricultural products such as straw, bagasse, corn cobs; waste from food processing such as molasses and cheese whey; and food waste from life, such as expired bread and fruit peels. These substances can usually be hydrolyzed to produce large amounts of sugar for microbial fermentation into LA. For instance, sugarcane bagasse contains 35–40% cellulose and 25–30% hemicellulose, which can be hydrolyzed to hexoses and pentoses. *Bacillus coagulans* DSM ID 14-300 fermented acid-hydrolyzed bagasse at 52 °C to obtain LA at a concentration of 55.99 g/L with a productivity of 1.7 g/(L·h) and an optical purity of 99.4% [14]. In another study, corn cobs were used for LA production. The quantity of LA synthesized by *B. coagulans* H-1 was 0.85 g/g after 36 h of batch fermentation [15]. Molasses, the waste stream from the sugar process, is also a suitable substrate for LA fermentation. The cost of fermentation using molasses and corn pulp powder was estimated to be only 37.7% of the cost of production using an MRS medium. Microbial consortium CEE-DL15 utilized sugarcane molasses to produce

LA. After 25 h of batch fermentation, 112.34 g/L LA was obtained from 350 g/L sugarcane molasses, with a maximum productivity of 4.49 g/(L·h) [16]. Whey, a waste product of the cheese production process, also contains about 49% lactose. Juodeikiene, et al. [17] used β-galactosidase for the enzymatic digestion of lactose in cheese whey. A maximum acid production of 47.0–51.2 g/L was achieved after whey fermentation by *Pediococcus acidilactici* KTU05-7 and *Pediococcus pentosaceus* KTU05-9 at pH 4.4–4.5 with the addition of 2% (*w*/*v*) neutralizer $CaCO_3$. In addition, solid waste disposal has also been a key research issue in recent years. Food waste is rich in nutrients, including 30–60% starch, 5–10% protein and 10–40% lipids, and is a good substrate for LA production [18]. Yousuf et al. [19] obtained LA from food waste without an extra inoculum or pH control, with a yield of 0.46 g/g- total solids at 37 °C. Similarly, the highest output of 30.25 g/L LA was achieved from household food waste with *Lactobacillus rhamnosus* ATCC 7469 [20]. A variety of lignocellulose-based biomasses have been successfully applied in LA fermentation. However, these agricultural by-products are affected by a series of factors, including season, climate, and region. They may also produce some inhibitory substances during the pretreatment process, which may affect the industrial production of LA. The third-generation feedstock solves these problems to some extent. Microalgae and macroalgae have short growth cycles and are easy to cultivate. They are rich in carbohydrates and do not contain lignin, which is unavailable to microorganisms [21]. The overall quantity of LA synthesized from *Ulva fasciata*, *Gracilaria corticata*, and *Kappaphycus alvarezii* by *Lactobacillus plantarum* MTCC 1407 were, respectively, 0.4 g/g, 0.66 g/g, and 0.66 g/g DW of seaweed [22]. Nagarajan et al. [23] compared the ability of *Lactobacillus plantarum* to ferment glucose, sugar cane bagasse, cheese whey, and microalgae for LA production. Microalgal hydrolysate attained the highest productivity (12.56 g/(L·h) in continuous fermentation), along with great LA concentrations (37.9 g/L) and yields (0.91 g/g). However, in most studies on the fermentation of LA from algae, the yield is generally found to be lower than that of lignocellulosic biomass. How to increase the efficiency of LA production from algae will be the focus of future research. In general, a variety of inexpensive feedstocks are available for microbial fermentation to produce LA with high yields.

Table 1 summarizes the collection of LA concentrations, yields, and productivity levels by applying different lactic acid-producing bacterial strains and fermentation strategies using various feedstocks.

Table 1. Recent studies on lactic acid production from low-cost carbon sources.

Feedstock	Strain (s)	Strategy	Lactic Acid			References
			Concentrations (g·L)	Yield (g·g^{-1})	Productivity (g·(L·h) $^{-1}$)	
Brewer's spent grain, malt rootlets	*L. rhamnosus* ATCC 7469	Batch fermentation	25.73	-	0.95	[24]
		Fed-batch fermentation	58.01	-	1.19	
Beechwood	*Lactobacillus delbrueckii* subsp. *bulgaricus* (ATCC® 11842)	Simultaneous saccharification and fermentation (batch)	62.00	0.69	0.86	[25]
Food waste	*E.coli* JU15	Batch fermentation	22.66	0.99	0.73	[26]
Food waste	*Lacobacillus manihotivorans* DSM 13343	Batch fermentation	18.69	0.73	-	[27]
	L. plantarum DSM 20174		17.03	0.69	-	

Table 1. *Cont.*

Feedstock	Strain (s)	Strategy	Lactic Acid			References
			Concentrations (g·L)	Yield (g·g^{-1})	Productivity (g·(L·h)$^{-1}$)	
Orange peel waste	*Lactobacillus delbrueckii* ssp. *bulgaricus* CECT 5037	Batch fermentation	39.00	-	-	[28]
Coffee ground waste	*L. rhamnosus* ATCC 10863	Batch fermentation	24.95	-	0.59	[29]
Sugarcane bagasse	*Lactobacillus pentosus* ATCC 8041	Batch fermentation	65.00	0.93	1.01	[30]
Sugarcane molasses	microbial consortium CEE-DL15	Batch fermentation	112.34	0.81	4.94	[16]
Bakery waste hydrolysates and lucerne green juice	*B. coagulans*	Batch fermentation Continuous fermentation	62.20 55.00	0.57 0.48	2.59 11.28	[31]
Cheese whey powder	*Lactobacillus bulgaricus* CGMCC 1.6970	Batch fermentation Continuous fermentation	70.70 113.18	- -	1.47 2.36	[32]
Sargassum cristaefolium	*L. plantarum* 23	Batch fermentation	26.00	0.81	7.53	[33]
Ulva sp.	*L. rhamnosus*	Batch fermentation	30.93	0.85	7.53	[33]
Gracilaria sp.	*Weissella paramesenteroides* 24	Batch fermentation	29.00	0.94	3.58	[33]
Eucheuma denticulatum	*Bacillus coagulans* ATCC 7050	Prehydrolysis simultaneous saccharification and fermentation (batch)	14.00	-	-	[34]
Hydrodictyon reticulum	*Lactobacillus paracasei* LA104	Simultaneous saccharification and cofermentation	37.11	0.46	1.03	[35]

2.2. Change the Production Process to Save Energy Consumption

Traditional microbial fermentation generally requires sterilization and cooling of the medium, both of which consume a significant amount of energy and increase the cost of production. Therefore, researchers have attempted non-sterilized fermentation for LA production. Since most LA producers are used at medium temperatures, non-sterilization fermentation can easily lead to contamination. Anaerobic fermentation is one of the ways to avoid the growth of miscellaneous bacteria. For example, a high D-LA concentration (113.18 g/L) with an average productivity of 2.36 g/(L·h) was obtained in fed-batch fermentation from cheese whey (without sterilization) at 42 °C in an anaerobic environment [32]. In addition, the simultaneous enzymatic saccharification and fermentation of biomass can reduce energy costs. With simultaneous saccharification and fermentation (SSF), not only can energy costs be reduced by avoiding the significantly energy-consuming biomass saccharification, but also by providing sufficient sugar release to achieve higher productivity than single-step conversion. However, SSF is less efficient due to the low activity of saccharifying enzymes at mesophilic temperatures. Therefore, microorganisms that can ferment at thermophilic temperatures are widely used in this fermentation process. They can function at higher temperatures than those used in normal microbial fermentation, and have a good ability to produce LA. One study reported that a strain of *Enterococcus faecium*

QU 50 isolated from Egyptian fertile soil was effective in fermenting xylose to produce LA in wide pH (6.0–10.0) and temperature (30–52 °C) ranges. In this study, the maximum yield of LA (1 g/g) with an optical purity ≥99.2% was attained at 50 °C [36]. Similarly, *B. coagulans* LA204 achieved an LA yield of 0.54 g/(g corn cob), with an average output of 0.57 g/(L·h), using corncob and corn steep liquor powder as substrates with fed-batch simultaneous saccharification and fermentation [37]. Another study conducted 5 L and 50 L bioreactor scale-up trials and found that the SSF of corn stover by the *P. acidilactici* DQ2 at 48 °C still achieved a high LA concentration (~101 g/L) and high productivity (~1.06 g/(L·h)), demonstrating its strong potential for industrial applications [38]. Furthermore, a recent study achieved simultaneous liquefaction, saccharification, and fermentation. *Enterococcus faecalis* DUT1805 had superior thermotolerance and was active at higher temperatures (50 °C or higher). In addition, the highest output—73.75 g/L LA with a yield of 87% of initial starch—was attained by aging paddy rice with hull [39]. If this process can be successfully applied in industry, it will save much time and energy during production. Moreover, searching for strains that can produce LA directly from starchy material can also save energy consumed during the saccharification step [40]. Overall, the researchers intend to minimize energy costs by simplifying and optimizing the fermentation process.

3. Problems in the Fermentation Process and Their Solutions

3.1. Select Appropriate Strains and Fermentation Processes to Accommodate High Concentrations of Substrates

Traditional split-wholesale fermentation is performed by optimizing the substrate in the medium to the optimum concentration of the strain so that it ferments as much product as possible in the experimental environment, but this is often time-consuming and energy-intensive. In order to increase the production efficiency, it is desired that the strains produce large amounts of LA in a medium with high substrate concentration. However, high substrate concentrations may suppress the strain, as evidenced by extended lag periods, increased osmotic stress, etc. [10]. Therefore, several studies have addressed the inhibition of strains at high substrate concentrations by screening and mutating the strains, as well as changing the fermentation process. The mutant strain *Lactobacillus paracasei* NCBIO01-M2 was obtained by atmospheric pressure room-temperature plasma treatment and high-throughput screening technology. It can flexibly regulate the ratio of unsaturated fatty acids and intracellular compatible solute pools, and, therefore, has a high tolerance to osmotic stress. The 248 g/L glucose in the medium was wholly consumed during the batch fermentation, producing 223.7 g/L LA with a yield of 5.53 g/(L·h) [41]. As for the production process, fed-batch fermentation and SSF are two processes with mature applications at present. Fed-batch fermentation leads to a higher utilization of the substrate compared to continuous fermentation. It provides a suitable growth environment for the bacteria by controlling the concentration of nutrients during fermentation so that the production status of the bacteria can be maintained at optimal levels. Pejin et al. [24] reported the production of L-LA by *L. rhamnosus* ATCC 7469 from brewer's spent grain hydrolysate and malt rootlet extract. The highest LA concentration in the batch fermentation was 25.73 g/L, with a yield of 0.95 g/(L·h) and a conversion of 86.31%; on the other hand, an increase in LA concentration to 58.01 g/L, an increase in yield to 1.19 g/(L·h), and a conversion rate of 88.54% was achieved with fed-batch fermentation. SSF enabled the timely utilization of sugars from starch hydrolysis, and the hydrolysis reaction proceeded positively. Using this fermentation mode, the initial sugar concentration in the medium was low. Additionally, the degree of hydrolysis of polysaccharides was increased, which facilitated the growth and fermentation of microorganisms. The shortcoming of this process is that saccharification requires a high temperature. Therefore, high-temperature-tolerant strains need to be selected to meet this process. The strain *B. coagulans* WCP10-4, isolated from soil, could be fermented by SSF at 50 °C, consuming 200 g/L of corn starch to produce 202.0 g/L of L-LA [42]. In conclusion, strains with high glucose tolerance, as well as SSF and fed-batch fermentation, have great advantages in terms of LA production.

3.2. Methods to Control the Carbon Catabolite Repression Effect

The substances obtained by hydrolysis of raw materials, such as lignocellulose, are usually a mixture of hexoses and pentoses. Most strains have a carbon catabolite repression (CCR) effect on the fermentation of mixed sugars, which affects the utilization of sugars and reduces the final production of LA [43]. CCR can be mitigated either by the selection of production strains or by adjusting the composition of the medium. First, CCR-negative strains can be selected as inoculants. The metabolism of *Lactobacillus lactis* 2369 (CCR negative) was shown to follow a mixed sugar fermentation pathway during the fermentation of rice straw. Glucose was its preferred substrate, fully utilized within 48 h, followed by xylose utilization of more than 40% at the end of fermentation [44]. Secondly, mixed bacterial fermentation using microbial consortia can exploit the abilities of different strains to consume different sugars and alleviate the CCR effect. The microbial consortia isolated from cattle stomach were domesticated to obtain the microbial consortia DUT47, which has robust tolerance to inhibitors. In the experiment, the microbial consortia DUT47 simultaneously utilized glucose and xylose. LA was produced from corn stover under non-detoxified, non-sterile conditions, with a yield of 0.50 g/(g corn stover) and a productivity of 0.32 g/(L·h) [45]. In general, most studies have focused on rescuing the CCR by screening or modifying strains to adapt them to the composition of the medium. Recently, several studies have proposed another feasible option to eliminate the CCR of microorganisms by adjusting the composition and the ratio of the substrate. *E. faecium* QU 50 exhibited a significant CCR in the fermentation of glucose/xylose mixtures, with a much lower xylose consumption ratio than glucose. However, CCR was avoided in the fermentation of cellobiose/xylose mixtures. *E. faecium* QU 50 can simultaneously utilize various sugars in a mixed sugar medium consisting of 5 g/L glucose, 40 g/L cellobiose, and 20 g/L xylose. In high-density open continuous fermentation with cell cycling technology, the productivity of LA was 6.49 g/(L·h). No by-products were detected up to 250 h of fermentation, and the strain was stable for up to 500 h of fermentation [46]. The researchers reached a slightly different conclusion in another study than that mentioned above. The study tested the ability of the strain to ferment cellobiose/glucose mixtures. The results showed that cellobiose promoted the consumption of glucose, but glucose inhibited the consumption of cellobiose, and there was a significant CCR. Therefore, a medium with glucose and xylose as the main sugars along with a small amount of cellobiose is suitable for LA fermentation of *B. coagulans* Azu-10 [47]. In short, to effectively mitigate the CCR, the formulation of the medium can be designed in a targeted manner according to the sugar utilization characteristics of the producing strains.

3.3. Release of Inhibitory Compounds during Feedstock Pretreatment and the Detoxification Methods

The pretreatment of lignocellulose generates some inhibitory compounds, including furan derivatives, weak acids, and phenolic compounds [48]. Furan derivatives include 5-hydroxymethylfurfural (5-HMF) and furfural, which negatively impact cell growth and sugar uptake [49]. Phenolic compounds such as vanillin, syringaldehyde, and ferulic and coumaric acid have been shown to have inhibitory effects on cellulolytic enzymes, DNA transcription, and protein synthesis [1,50]. Weak acids such as acetic and formic acid can hinder the growth of microbial cells [51]. The traditional way to remove these compounds is to wash them with several times their volume in water, but this is not easily implemented in industry. In recent years, various physical, chemical, and biological methods have been applied to remove inhibitory substances, showing promising results [37,52,53]. Although the removal of inhibitory substances from the medium allows the production strain to ferment properly, this step also increases the cost of production. The detoxification step can be omitted by developing strains capable of tolerating a wide range of concentrations of the inhibitory compounds. Adaptive evolution of the strain is one way to enhance its tolerance. *B. coagulans* was inoculated with a 20% volume fraction of wheat straw hydrolysate for a passaging culture. After 63 generations, the strain was stable in

45% (*v*/*v*) hydrolysate, while the parental strain failed to produce LA at concentrations above 25% (*v*/*v*) hydrolysate [54]. Similar domestication was equally effective for the microbial consortia [45]. In addition, finding production strains that are both able to produce LA and degrade inhibitors can also allow their effect on LA production to be avoided. The first use of *B. coagulans* for LA production and bio-detoxification was tested by Alves et al. [14]. The LA yield of sugarcane bagasse hemicellulose hydrolysate, fermented by *B. coagulans* DSM ID 14-300, was 0.87 g/g, while the removal of 5-hydroxymethylfurfural and furfural from the hydrolysate reached 99.6% and 100%, respectively. Moreover, the anaerobic environment may also enhance the tolerance of the strain to inhibitory compounds [55].

3.4. Suitable Strains and Fermentation Processes for Solving Product Inhibition Problems

The accumulation of the end product, lactic acid, during fermentation affects the metabolism of microorganisms, and cytotoxicity occurs around the pKa value (3.86) [56]. Reducing the effect of end-product inhibition can be achieved by increasing the acid tolerance of the strain and optimizing the fermentation process. The end-product LA causes the pH of the medium to drop significantly. Therefore, the industry maintains the pH at 5.5–6.5 by adding neutralizers such as $Ca(OH)_2$ and $CaCO_3$. However, this increases the difficulty and cost of subsequent LA isolation and purification [57]. Improving the acid tolerance of strains can be achieved by adaptive evolution or genetic modification. Adaptive evolution of *S. cerevisiae* SR8LDH was carried out in a medium (pH 3.1) containing 20 g/L glucose and 8% LA. The evolved strain *S. cerevisiae* BK01 was fermented at a high sugar concentration (200 g/L), and the LA yield was increased by 17% [58]. After the adaptive domestication of *P. acidilactici*, the amount of bacteria doubled when the pH of the fermentation broth was reduced to 4.0 [59]. Gene knockdown by RNA interference allowed for sequence-specific down-regulation on a genome-wide basis, according to the phenotype. It achieved slight down-regulation of multi-copy genes without considering the genome's copy number. The identified acid-tolerant gene target was transferred to the lactic acid-overproducing strain *S. cerevisiae* CEN.PK in the form of a knockout, resulting in a 33% increase in cell growth, a 58% increase in glucose consumption, and a 60% increase in L-LA production [60]. Strains tolerant to acidic environments due to their structures are also promising for LA production. *S. cerevisiae* is tolerant to acid. The D-LA-producing strain *S. cerevisiae* JHY5210 was obtained by expressing the D-lactate dehydrogenase gene (*Lm. ldhA*). Genes associated with ethanol production (*ADH1* and *PDC1*), glycerol production (*GPD1* and *GPD2*), and D-lactate degradation (*JEN1* and *DLD1*) were also deleted from the strains to reduce by-product production [61]. However, 0.6 g/g of ethanol was still produced during the fermentation of *S. cerevisiae* JHY5210. In the following study, *S. cerevisiae* JHY605, which was missing five *ADH* genes, *GPD1*, and *GPD2*, was used as the starting strain. To further improve the traits of the strain, *DLD1* was deleted and *TDH3* was integrated into its locus as a promoter to control *Lm. ldhA*. The final strain, JHY5610, was obtained with a 44% higher level of D-LA production. By adaptive evolution of the JHY5610 strain, the D-LA yield of the obtained strain, JHY5710, was again enhanced by 30% [62]. In summary, the combination of adaptive evolution and genetic modification can substantially increase the acid tolerance of the strain, resulting in efficient conversion of the carbon source to LA.

In terms of process, fermentation processes such as continuous fermentation, fed-batch fermentation, cell immobilization fermentation, and cell-recycled batch fermentation have been developed and applied to LA fermentation. These strategies reduce the effect of end-product accumulation in the fermentation system on the strain. Continuous fermentation is the continuous addition of fresh medium to the bioreactor at a specific rate while allowing the fermentation broth containing metabolites to flow out so that the yield of the fermentation system is maintained at a high level. The maximum LA productivity of 3.69 g/(L·h) was reached at a pH of 6.0, a temperature of 50 °C, and a dilution rate of 0.167 h^{-1} [63]. However, continuous fermentation may also have the disadvantage of inadequate utilization of nutrients. Fed-batch fermentation leads to a higher utilization of the substrate compared to continuous fermentation. It provides a suitable growth environment for the bacteria by controlling the concentration of nutrients during fermentation so that the production status of the bacteria can be maintained at optimal levels. Pejin et al. [24] reported the production of L-LA by *L. rhamnosus* ATCC 7469 from brewer's spent grain hydrolysate and malt rootlet extract. The highest LA concentration in the batch fermentation was 25.73 g/L, with a yield of 0.95 g/(L·h) and a conversion of 86.31%, while an LA concentration increased to 58.01 g/L, a yield increased to 1.19 g/(L·h), and a conversion of 88.54% were achieved with fed-batch fermentation. Cell immobilization technology was utilized to immobilize microbial cells on water-insoluble carriers, allowing them to carry out their life activities in a specific spatial range. It has the advantage of being less susceptible to contamination, and has a high fermentation rate. It has been shown that cell immobilization technology in batch fermentation and continuous fermentation results in relatively high LA yields [64]. However, it has also been pointed out that cell immobilization bioreactors may decrease productivity after long-term use due to limited mass transfer and accumulation of dead cells [65]. Furthermore, the cell-recycled batch fermentation process is also a way to reduce product inhibition. After a period of fermentation, the product is promptly removed, and a new batch of fermentation broth is added to reuse the bacteriophage cells. This fermentation mode omits the inoculant preparation step and eliminates the lag period of cells, thus reducing the fermentation time. Using membrane-integrated repeated batch fermentation, *B. coagulans* IPE22 was fermented with wheat straw hydrolysate as the carbon source. The LA productivity increased from 1.01 g/(L·h) to 2.35 g/(L·h) after six fermentation batches. It was speculated that the synergistic effect of cell reuse and microbial domestication may have improved the quality of the strain and facilitated its fermentation of the sugar mixture [66]. Furthermore, the insertion of dynamic membranes into the fermentation reactor allowed the microbial cells to be retained and reused, and the permeability of the dynamic membranes can be maintained by regular backwashing [67]. Because of the high cost of processing equipment for cell immobilization technology and cell cycling technology, the application of these method is still in the laboratory stage.

3.5. Methods to Control of Optical Purity of Lactic Acid

The stereospecificity and optical purity of LA produced by microbial fermentation depend mainly on the chosen strain and its lactate dehydrogenase (LDH) specificity. It has been found that both L-LDH and D-LDH enzymes can be active in some LA fermentation strains, producing LA with low optical purity [68]. Improving the optical purity of LA in fermentation products can be achieved by genetic modifications and fermentation conditions of the strain. Temperature is one of the key factors affecting the fermentation of strains for the production of L-LA and D-LA. One study tested the optical purity of LA produced by fermenting carbohydrates using activated sludge (a by-product of municipal wastewater treatment plants) as an inoculant at different temperatures. It was found that both the production and consumption of L-LA and D-LA are optimal at moderate temperatures, while L-LA consumption and D-LA production are suppressed at high temperatures. Therefore, the authors developed a two-stage, temperature-controlled method, i.e., high-temperature preheating at 50 °C for 4 h, followed by medium-temperature fermentation at 37 °C for 34 h, resulting in L-LA with an optical purity of 98.3% [69]. Another study also showed that temperature affects the strains' selectivity in the production of L-LA and D-LA. *Lactobacillus coryniformis*, which is high in D-LA production, shows a sharp increase in L-LA concentration at high temperatures [70]. Modifying the strain genes can directly control the production of either pure L-LA or pure D-LA. Kuo et al. [12] blocked the metabolic pathway of D-LA by knocking out the D-lactate dehydrogenase from *L. paracasei*. The obtained *ldhD*-deficient strain 7BL was able to produce optically pure L-LA, and no D-LA production was detected during its fermentation of lignocellulosic hydrolysate.

4. Selecting Suitable Fermentation Processes for LA Production

The conventional two-step batch fermentation method has the disadvantages of substrate and/or product inhibition and high energy consumption. In order to solve these problems and improve production efficiency, some fermentation processes have been developed and applied in practical production (these are described in Sections 3.1 and 3.4). This includes fed-batch fermentation, continuous fermentation, cell immobilization fermentation, and cell-recycled batch fermentation. These can be used to address some of the challenges encountered in LA production. Since each of these processes has advantages and disadvantages, they are applicable in different situations. All substrates are converted into a fermentable form before fermentation, so the type of substrate has little influence on the choice of fermentation process. It is worth noting that inhibitory compounds released during the pretreatment of lignocellulose may not be completely removed. Therefore, this may affect the strain growth, and cell immobilization fermentation and cell-recycled batch fermentation are not recommended. According to the characteristics of these processes, the different ranges of applications are listed in Table 2.

Table 2. Fermentation processes for various pretreated substrates and their scopes of application.

Fermentation Process	Feedstock	Pretreatment	Applicable Scope
Fed-batch fermentation	First-generation feedstocks	No need	• High concentrations of substrates • High acid tolerance of the strain
	Second-generation feedstocks	Solids (e.g., food waste, corn stover): mill (except powder)/dissolve/mix (with other base feedstocks)/detoxification (lignocellulose as substrate)/sterilization/saccharification (or SSF) Liquid (e.g., whey, molasses): mix (with other base materials)/sterilization/saccharification (or SSF)	
	Third-generation feedstocks	Mill/dissolve/mix (with other base materials)/detoxification/sterilization/saccharification (or SSF)	
Continuous fermentation	First-generation feedstocks Second-generation feedstocks Third-generation feedstocks	Same as above	• High-cell-density fermentation • Low concentrations of substrates
Cell immobilization fermentation	First-generation feedstocks	Same as above	• Suitable for batch fermentation or continuous fermentation • Fermentation environment and fermentor are strictly sterilized • Pursuit of minimal fermentation time
	Second-generation feedstocks	Same as above (pretreatment of feedstocks (e.g., lignocellulose) that may produce inhibitory substances is not recommended)	
	Third-generation feedstocks	Same as above	
Cell-recycled batch fermentation	First-generation feedstocks	Same as above	• High-cell-density fermentation • Fermentation environment and fermentor are strictly sterilized • Pursuit of minimal fermentation time
	Second-generation feedstocks	Same as above (pretreatment of feedstocks (e.g., lignocellulose) that may produce inhibitory substances is not recommended)	
	Third-generation feedstocks	Same as above	

5. Conclusions

The main challenge of LA production is achieving high LA concentrations, yields, productivity, and optical purity using economical renewable resources. The cost of LA production can be significantly reduced by selecting suitable raw materials and fermentation strategies. While facing a fermentation process that is hindered by substrate concentration, product concentration, and the presence of inhibitory compounds, strains must be screened for better tolerance or domesticated to adapt to the fermentation environment. In addition, the continuous development of new processes and the creation of a suitable production environment for the production strains are also critical to guarantee LA yields and production rates. Therefore, exploring excellent production strains and designing efficient processes will be the focus of future research on microbial fermentation for LA production.

Author Contributions: Conceptualization, writing—original draft preparation, Y.H.; writing—review and editing, supervision, Y.W., N.S. and P.L. All authors have read and agreed to the published version of the manuscript.

Funding: This study was supported by the National Natural Science Foundation of China (grant number 32172172, 32201994).

Data Availability Statement: Data are contained within the article.

Conflicts of Interest: The authors have declared no conflict of interest.

References

1. Nwamba, M.C.; Sun, F.; Mukasekuru, M.R.; Song, G.; Harindintwali, J.D.; Boyi, S.A.; Sun, H. Trends and hassles in the microbial production of lactic acid from lignocellulosic biomass. *Environ. Technol. Innov.* **2021**, *21*, 101337. [CrossRef]
2. Djukic-Vukovic, A.; Mladenovic, D.; Ivanovic, J.; Pejin, J.; Mojovic, L. Towards sustainability of lactic acid and poly-lactic acid polymers production. *Renew. Sustain. Energy Rev.* **2019**, *108*, 238–252. [CrossRef]
3. Rombouts, J.L.; Kranendonk, E.; Regueira, A.; Weissbrodt, D.G.; Kleerebezem, R.; Loosdrecht, M. Selecting for lactic acid producing and utilising bacteria in anaerobic enrichment cultures. *Biotechnol. Bioeng.* **2020**, *117*, 1281–1293. [CrossRef] [PubMed]
4. Ahmad, A.; Banat, F.; Taher, H. A review on the lactic acid fermentation from low-cost renewable materials: Recent developments and challenges. *Environ. Technol. Innov.* **2020**, *20*, 101138. [CrossRef]
5. Acid, M.L. Microbial Lactic Acid, Its Polymer Poly(lactic acid), and Their Industrial Applications. *Microbiol. Monogr.* **2010**, *14*, 323–346.
6. Feng, L.Y.; Yuan, F.Y.; Liu, F.; Wang, T.T.; Chen, Y.G. Research Progress on Lactic Acid Production from Food Waste by Fermentation. *J. Tongji Univ. Nat. Sci.* **2021**, *49*, 1688–1700.
7. Abedi, E.; Hashemi, S.M.B. Lactic acid production—Producing microorganisms and substrates sources-state of art. *Heliyon* **2020**, *6*, e04974. [CrossRef]
8. Szliszka, E.; Czuba, Z.P.; Domino, M.; Mazur, B.; Zydowicz, G.; Krol, W. Ethanolic extract of propolis (EEP) enhances the apoptosis- inducing potential of TRAIL in cancer cells. *Molecules* **2009**, *14*, 738–754. [CrossRef]
9. Guo, W.; He, R.; Ma, L.; Jia, W.; Li, D.; Chen, S. Construction of a constitutively expressed homo-fermentative pathway in *Lactobacillus brevis*. *Appl. Microbiol. Biotechnol.* **2014**, *98*, 6641–6650. [CrossRef]
10. Abdel-Rahman, M.A.; Tashiro, Y.; Sonomoto, K. Recent advances in lactic acid production by microbial fermentation processes. *Biotechnol. Adv.* **2013**, *31*, 877–902. [CrossRef]
11. Grabar, T.B.; Zhou, S.; Shanmugam, K.T.; Yomano, L.P.; Ingram, L.O.N. Methylglyoxal Bypass Identified as Source of Chiral Contamination in l(+) and d(−)-lactate Fermentations by Recombinant *Escherichia coli*. *Biotechnol. Lett.* **2006**, *28*, 1527–1535. [CrossRef]
12. Kuo, Y.C.; Yuan, S.F.; Wang, C.A.; Huang, Y.J.; Guo, G.L.; Hwang, W.S. Production of optically pure L-lactic acid from lignocellulosic hydrolysate by using a newly isolated and D-lactate dehydrogenase gene-deficient *Lactobacillus paracasei* strain. *Bioresour. Technol.* **2015**, *198*, 651–657. [CrossRef]
13. Abdel-Rahman, M.A.; Sonomoto, K. Opportunities to overcome the current limitations and challenges for efficient microbial production of optically pure lactic acid. *J. Biotechnol.* **2016**, *236*, 176–192. [CrossRef]
14. Alves de Oliveira, R.; Schneider, R.; Vaz Rossell, C.E.; Maciel Filho, R.; Venus, J. Polymer grade l-lactic acid production from sugarcane bagasse hemicellulosic hydrolysate using *Bacillus coagulans*. *Bioresour. Technol. Rep.* **2019**, *6*, 26–31. [CrossRef]
15. Jiang, S.; Xu, P.; Tao, F. l-Lactic acid production by *Bacillus coagulans* through simultaneous saccharification and fermentation of lignocellulosic corncob residue. *Bioresour. Technol. Rep.* **2019**, *6*, 131–137. [CrossRef]
16. Sun, Y.; Xu, Z.; Zheng, Y.; Zhou, J.; Xiu, Z. Efficient production of lactic acid from sugarcane molasses by a newly microbial consortium CEE-DL15. *Process Biochem.* **2019**, *81*, 132–138. [CrossRef]

17. Juodeikiene, G.; Zadeike, D.; Bartkiene, E.; Klupsaite, D. Application of acid tolerant *Pedioccocus* strains for increasing the sustainability of lactic acid production from cheese whey. *LWT Food Sci. Technol.* **2016**, *72*, 399–406. [CrossRef]
18. Pleissner, D.; Lin, C.S.K. Valorisation of food waste in biotechnological processes. *Sustain. Chem. Process* **2013**, *1*, 21. [CrossRef]
19. Yousuf, A.; Bastidas-Oyanedel, J.R.; Schmidt, J.E. Effect of total solid content and pretreatment on the production of lactic acid from mixed culture dark fermentation of food waste. *Waste Manag.* **2018**, *77*, 516–521. [CrossRef]
20. Song, L.; Liu, S.; Liu, R.; Yang, D.; Dai, X. Direct lactic acid production from household food waste by lactic acid bacteria. *Sci. Total Environ.* **2022**, *840*, 156479. [CrossRef]
21. Lin, H.-T.V.; Huang, M.-Y.; Kao, T.-Y.; Lu, W.-J.; Lin, H.-J.; Pan, C.-L. Production of Lactic Acid from Seaweed Hydrolysates via Lactic Acid Bacteria Fermentation. *Fermentation* **2020**, *6*, 37. [CrossRef]
22. Sudhakar, M.P.; Dharani, G. Evaluation of seaweed for the production of lactic acid by fermentation using *Lactobacillus plantarum*. *Bioresour. Technol. Rep.* **2022**, *17*, 100890. [CrossRef]
23. Nagarajan, D.; Nandini, A.; Dong, C.-D.; Lee, D.-J.; Chang, J.-S. Lactic Acid Production from Renewable Feedstocks Using Poly(vinyl alcohol)-Immobilized *Lactobacillus plantarum* 23. *Ind. Eng. Chem. Res.* **2020**, *59*, 17156–17164. [CrossRef]
24. Pejin, J.; Radosavljevic, M.; Pribic, M.; Kocic-Tanackov, S.; Mladenovic, D.; Djukic-Vukovic, A.; Mojovic, L. Possibility of L-(+)-lactic acid fermentation using malting, brewing, and oil production by-products. *Waste Manag.* **2018**, *79*, 153–163. [CrossRef]
25. Karnaouri, A.; Asimakopoulou, G.; Kalogiannis, K.G.; Lappas, A.; Topakas, E. Efficient d-lactic acid production by *Lactobacillus delbrueckii* subsp. bulgaricus through conversion of organosolv pretreated lignocellulosic biomass. *Biomass Bioenergy* **2020**, *140*, 105672. [CrossRef]
26. Santos-Corona, A.M.; Lázaro-Mixteco, P.E.; Vargas-Tah, A.A.; Castro-Montoya, A.J. Lactic acid production from food waste using the lactogenic *Escherichia coli* strain JU15: Optimization of reducing sugar recovery. *J. Chem. Technol. Biotechnol.* **2021**, *97*, 668–675. [CrossRef]
27. Rawoof, S.A.A.; Kumar, P.S.; Devaraj, K.; Devaraj, T.; Subramanian, S. Enhancement of lactic acid production from food waste through simultaneous saccharification and fermentation using selective microbial strains. *Biomass Convers. Biorefin.* **2020**, *12*, 5947–5958. [CrossRef]
28. Bustamante, D.; Tortajada, M.; Ramón, D.; Rojas, A. Production of D-Lactic Acid by the Fermentation of Orange Peel Waste Hydrolysate by Lactic Acid Bacteria. *Fermentation* **2019**, *6*, 1. [CrossRef]
29. Koo, J.-R.; Min Park, H.; Kyung Kim, S.; Shik Yun, H. Lactic Acid Fermentation from Coffee Ground Waste Hydrolysate by *Lactobacillus Rhamnosus*. *J. Renew. Mater.* **2019**, *7*, 365–372. [CrossRef]
30. Wischral, D.; Arias, J.M.; Modesto, L.F.; de Franca Passos, D.; Pereira, N., Jr. Lactic acid production from sugarcane bagasse hydrolysates by *Lactobacillus pentosus*: Integrating xylose and glucose fermentation. *Biotechnol. Prog.* **2019**, *35*, e2718. [CrossRef]
31. Alexandri, M.; Blanco-Catala, J.; Schneider, R.; Turon, X.; Venus, J. High L(+)-lactic acid productivity in continuous fermentations using bakery waste and lucerne green juice as renewable substrates. *Bioresour. Technol.* **2020**, *316*, 123949. [CrossRef]
32. Liu, P.; Zheng, Z.; Xu, Q.; Qian, Z.; Liu, J.; Ouyang, J. Valorization of dairy waste for enhanced D-lactic acid production at low cost. *Process Biochem.* **2018**, *71*, 18–22. [CrossRef]
33. Nagarajan, D.; Oktarina, N.; Chen, P.T.; Chen, C.Y.; Lee, D.J.; Chang, J.S. Fermentative lactic acid production from seaweed hydrolysate using *Lactobacillus* sp. And *Weissella* sp. *Bioresour. Technol.* **2022**, *344*, 126166. [CrossRef]
34. Chai, C.Y.; Tan, I.S.; Foo, H.C.Y.; Lam, M.K.; Tong, K.T.X.; Lee, K.T. Sustainable and green pretreatment strategy of *Eucheuma denticulatum* residues for third-generation l-lactic acid production. *Bioresour. Technol.* **2021**, *330*, 124930. [CrossRef]
35. Nguyen, C.M.; Kim, J.S.; Hwang, H.J.; Park, M.S.; Choi, G.J.; Choi, Y.H.; Jang, K.S.; Kim, J.C. Production of l-lactic acid from a green microalga, *Hydrodictyon reticulum*, by *Lactobacillus paracasei* LA104 isolated from the traditional Korean food, makgeolli. *Bioresour. Technol.* **2012**, *110*, 552–559. [CrossRef]
36. Abdel-Rahman, M.A.; Tashiro, Y.; Zendo, T.; Sakai, K.; Sonomoto, K. *Enterococcus faecium* QU 50: A novel thermophilic lactic acid bacterium for high-yield l-lactic acid production from xylose. *FEMS Microbiol. Lett.* **2015**, *362*, 1–7. [CrossRef]
37. Liu, J.; Cai, Y.; Liu, L.; Zhan, R.; Zhao, S.; Liang, Y.; Peng, N.; Zhu, J.; Li, H.; Xiao, N.; et al. Enhanced lactic acid production by *Bacillus coagulans* through simultaneous saccharification, biodetoxification, and fermentation. *Biofuels Bioprod. Bioref.* **2020**, *14*, 533–543. [CrossRef]
38. Zhao, K.; Qiao, Q.; Chu, D.; Gu, H.; Dao, T.H.; Zhang, J.; Bao, J. Simultaneous saccharification and high titer lactic acid fermentation of corn stover using a newly isolated lactic acid bacterium *Pediococcus acidilactici* DQ2. *Bioresour. Technol.* **2013**, *135*, 481–489. [CrossRef]
39. Sun, Y.; Yang, Y.; Liu, H.; Wei, C.; Qi, W.; Xiu, Z. Simultaneous liquefaction, saccharification, and fermentation of L-lactic acid using aging paddy rice with hull by an isolated thermotolerant *Enterococcus faecalis* DUT1805. *Bioprocess Biosyst. Eng.* **2020**, *43*, 1717–1724. [CrossRef]
40. Shahri, S.Z.; Vahabzadeh, F.; Mogharei, A. Lactic acid production by loofah-immobilized *Rhizopus oryzae* through one-step fermentation process using starch substrate. *Bioprocess Biosyst. Eng.* **2020**, *43*, 333–345. [CrossRef]
41. Tian, X.; Wang, Y.; Chu, J.; Mohsin, A.; Zhuang, Y. Exploring cellular fatty acid composition and intracellular metabolites of osmotic-tolerant mutant *Lactobacillus paracasei* NCBIO-M2 for highly efficient lactic acid production with high initial glucose concentration. *J. Biotechnol.* **2018**, *286*, 27–35. [CrossRef] [PubMed]

42. Zhou, X.; Ye, L.; Wu, J.C. Efficient production of L-lactic acid by newly isolated thermophilic *Bacillus coagulans* WCP10-4 with high glucose tolerance. *Appl. Microbiol. Biotechnol.* **2013**, *97*, 4309–4314. [CrossRef] [PubMed]
43. Nagarajan, D.; Chen, C.-Y.; Ariyadasa, T.U.; Lee, D.-J.; Chang, J.-S. Macroalgal biomass as a potential resource for lactic acid fermentation. *Chemosphere* **2022**, *309*, 136694. [CrossRef] [PubMed]
44. Sivagurunathan, P.; Raj, T.; Chauhan, P.S.; Kumari, P.; Satlewal, A.; Gupta, R.P.; Kumar, R. High-titer lactic acid production from pilot-scale pretreated non-detoxified rice straw hydrolysate at high-solid loading. *Biochem. Eng. J.* **2022**, *187*, 108668. [CrossRef]
45. Sun, Y.; Li, X.; Wei, C.; Qi, W.; Xiu, Z. An aptly industrialized bioprocess for lactic acid production from corn stover using thermotolerant microbial consortia. *Bioprocess Biosyst. Eng.* **2021**, *44*, 2445–2454. [CrossRef]
46. Abdel-Rahman, M.A.; Tan, J.; Tashiro, Y.; Zendo, T.; Sakai, K.; Sonomoto, K. Non-carbon loss long-term continuous lactic acid production from mixed sugars using thermophilic *Enterococcus faecium* QU 50. *Biotechnol. Bioeng.* **2020**, *117*, 1673–1683. [CrossRef]
47. Abdel-Rahman, M.A.; Hassan, S.E.-D.; Alrefaey, H.M.A.; Elsakhawy, T. Efficient Co-Utilization of Biomass-Derived Mixed Sugars for Lactic Acid Production by *Bacillus coagulans* Azu-10. *Fermentation* **2021**, *7*, 28. [CrossRef]
48. Moreno, A.D.; Tomás-Pejó, E.; Ballesteros, M.; Negro, M.J. Pretreatment Technologies for Lignocellulosic Biomass Deconstruction Within a Biorefinery Perspective. In *Biofuels: Alternative Feedstocks and Conversion Processes for the Production of Liquid and Gaseous Biofuels*; Biomass, Biofuels, Biochemicals; Elsevier: Amsterdam, The Netherlands, 2019; pp. 379–399.
49. Yi, X.; Zhang, P.; Sun, J.; Tu, Y.; Gao, Q.; Zhang, J.; Bao, J. Engineering wild-type robust *Pediococcus acidilactici* strain for high titer L- and D-lactic acid production from corn stover feedstock. *J. Biotechnol.* **2016**, *217*, 112–121. [CrossRef]
50. Zhai, R.; Hu, J.; Saddler, J.N. Extent of Enzyme Inhibition by Phenolics Derived from Pretreated Biomass Is Significantly Influenced by the Size and Carbonyl Group Content of the Phenolics. *ACS Sustain. Chem. Eng.* **2018**, *6*, 3823–3829. [CrossRef]
51. Kim, D. Physico-Chemical Conversion of Lignocellulose: Inhibitor Effects and Detoxification Strategies: A Mini Review. *Molecules* **2018**, *23*, 309. [CrossRef]
52. Roque, L.R.; Morgado, G.P.; Nascimento, V.M.; Ienczak, J.L.; Rabelo, S.C. Liquid-liquid extraction: A promising alternative for inhibitors removing of pentoses fermentation. *Fuel* **2019**, *242*, 775–787. [CrossRef]
53. Biswas, R.; Uellendahl, H.; Ahring, B.K. Wet Explosion: A Universal and Efficient Pretreatment Process for Lignocellulosic Biorefineries. *BioEnergy Res.* **2015**, *8*, 1101–1116. [CrossRef]
54. Ouyang, S.; Zou, L.; Qiao, H.; Shi, J.; Zheng, Z.; Ouyang, J. One-pot process for lactic acid production from wheat straw by an adapted *Bacillus coagulans* and identification of genes related to hydrolysate-tolerance. *Bioresour. Technol.* **2020**, *315*, 123855. [CrossRef]
55. Cubas-Cano, E.; Gonzalez-Fernandez, C.; Ballesteros, M.; Tomas-Pejo, E. *Lactobacillus pentosus* CECT 4023 T co-utilizes glucose and xylose to produce lactic acid from wheat straw hydrolysate: Anaerobiosis as a key factor. *Biotechnol. Prog.* **2019**, *35*, e2739. [CrossRef]
56. Park, H.J.; Bae, J.H.; Ko, H.J.; Lee, S.H.; Sohn, J.H. Low-pH production of D-lactic acid using newly isolated acid tolerant yeast *Pichia kudriavzevii* NG7: PARK et al. *Biotechnol. Bioeng.* **2018**, *115*, 2232–2242. [CrossRef]
57. Juturu, V.; Wu, J.C. Microbial production of lactic acid: The latest development. *Crit. Rev. Biotechnol.* **2016**, *36*, 967–977. [CrossRef]
58. Jang, B.K.; Ju, Y.; Jeong, D.; Jung, S.K.; Kim, C.K.; Chung, Y.S.; Kim, S.R. l-Lactic Acid Production Using Engineered *Saccharomyces cerevisiae* with Improved Organic Acid Tolerance. *J. Fungi* **2021**, *7*, 928. [CrossRef]
59. Yan, Z.; Chen, M.; Jia, J.; Bao, J. Increasing Acid Tolerance of an Engineered Lactic Acid Bacterium *Pediococcus acidilactici* for L-Lactic Acid Production. *Fermentation* **2022**, *8*, 96. [CrossRef]
60. Lee, J.J.; Crook, N.; Sun, J.; Alper, H.S. Improvement of lactic acid production in *Saccharomyces cerevisiae* by a deletion of ssb1. *J. Ind. Microbiol. Biotechnol.* **2016**, *43*, 87–96. [CrossRef]
61. Baek, S.H.; Kwon, E.Y.; Yong, H.K.; Hahn, J.S. Metabolic engineering and adaptive evolution for efficient production of D-lactic acid in *Saccharomyces cerevisiae*. *Appl. Microbiol. Biotechnol.* **2016**, *100*, 2737–2748. [CrossRef]
62. Baek, S.H.; Kwon, E.Y.; Bae, S.J.; Cho, B.R.; Kim, S.Y.; Hahn, J.S. Improvement of d-Lactic Acid Production in *Saccharomyces cerevisiae* Under Acidic Conditions by Evolutionary and Rational Metabolic Engineering. *Biotechnol. J.* **2017**, *12*, 1700015. [CrossRef] [PubMed]
63. Ahring, B.K.; Traverso, J.J.; Murali, N.; Srinivas, K. Continuous fermentation of clarified corn stover hydrolysate for the production of lactic acid at high yield and productivity. *Biochem. Eng. J.* **2016**, *109*, 162–169. [CrossRef]
64. Pimtong, V.; Ounaeb, S.; Thitiprasert, S.; Tolieng, V.; Sooksai, S.; Boonsombat, R.; Tanasupawat, S.; Assabumrungrat, S.; Thongchul, N. Enhanced effectiveness of *Rhizopus oryzae* by immobilization in a static bed fermentor for L-lactic acid production. *Process Biochem.* **2017**, *52*, 44–52. [CrossRef]
65. Shi, Z.; Wei, P.; Zhu, X.; Cai, J.; Huang, L.; Xu, Z. Efficient production of l-lactic acid from hydrolysate of Jerusalem artichoke with immobilized cells of *Lactococcus lactis* in fibrous bed bioreactors. *Enzym. Microb. Technol.* **2012**, *51*, 263–268. [CrossRef]
66. Zhang, Y.; Chen, X.; Qi, B.; Luo, J.; Shen, F.; Su, Y.; Khan, R.; Wan, Y. Improving lactic acid productivity from wheat straw hydrolysates by membrane integrated repeated batch fermentation under non-sterilized conditions. *Bioresour. Technol.* **2014**, *163*, 160–166. [CrossRef]
67. Tang, J.; Wang, X.C.; Hu, Y.; Ngo, H.H.; Li, Y. Dynamic membrane-assisted fermentation of food wastes for enhancing lactic acid production. *Bioresour. Technol.* **2017**, *234*, 40–47. [CrossRef]
68. Viana, R.; Yebra, M.J.; Galán, J.; Monedero, V.; Pérez-Martínez, G. Pleiotropic effects of lactate dehydrogenase inactivation in *Lactobacillus casei*. *Res. Microbiol.* **2005**, *156*, 641–649. [CrossRef]

69. Jian, Q.; Li, X.; Chen, Y.; Liu, Y.; Pan, Y. Production of high optical purity l-lactic acid from waste activated sludge by supplementing carbohydrate: Effect of temperature and pretreatment time. *Environ. Technol.* **2016**, *37*, 2457–2466. [CrossRef]
70. Gu, S.A.; Jun, C.; Joo, J.C.; Kim, S.; Lee, S.H.; Kim, Y.H. Higher thermostability of l-lactate dehydrogenases is a key factor in decreasing the optical purity of d-lactic acid produced from *Lactobacillus coryniformis*. *Enzym. Microb. Technol.* **2014**, *58–59*, 29–35. [CrossRef]

MDPI AG
Grosspeteranlage 5
4052 Basel
Switzerland
Tel.: +41 61 683 77 34

Foods Editorial Office
E-mail: foods@mdpi.com
www.mdpi.com/journal/foods